MODERN ASPECTS OF ELECTROCHEMISTRY
No. 56

Series Editors:
Ralph E. White
Department of Chemical Engineering
University of South Carolina
Columbia, SC 29208

Constantinos G. Vayenas
Department of Chemical Engineering
University of Patras
Patras 265 00
Greece

For further volumes:
http://www.springer.com/series/6251

Previously from Modern Aspects of Electrochemistry

Modern Aspects of Electrochemistry No. 54

Electrochemical Production of Metal Powders

Edited by Stojan S. Djokić, Professor of Chemical & Materials Engineering at the University of Alberta

Topics in Number 54 include:

- General theory of disperse metal electrodeposition
- Electrodeposition of pure metal powders e.g., Cu, Ag, Pb, Co, Fe and Ni
- Formation of porous Cu electrodes by constant and periodically changing regimes of electrolysis
- Morphology, phase and chemical composition of electrodeposited Co-Ni, Fe-Ni and Mo-Ni-O powders
- Electrodeposition of dispersed nanoparticles
- Electroless deposition of metal powders from aqueous solution

Modern Aspects of Electrochemistry No. 55

Biomedical Applications

Edited by Stojan S. Djokić, Professor of Chemical & Materials Engineering at the University of Alberta

Topics in Number 55 include:

- CoCrMo alloy for biomedical applications
- Electroless synthesis of metallic nanostructures for biomedical technologies
- Biodegradable Mg alloys: Corrosion, surface modification and biocompatibility
- Microcantilever sensors: Electrochemical aspects and biomedical applications
- Surface treatments with silver and its compounds for biomedical applications

Mordechay Schlesinger
Editor

Applications of Electrochemistry in Medicine

 Springer

Editor
Mordechay Schlesinger
Department of Physics
University of Windsor
Windsor, ON, Canada

ISSN 0076-9924
ISBN 978-1-4614-6147-0   ISBN 978-1-4614-6148-7 (eBook)
DOI 10.1007/978-1-4614-6148-7
Springer New York Heidelberg Dordrecht London

Library of Congress Control Number: 2013930593

© Springer Science+Business Media New York 2013
This work is subject to copyright. All rights are reserved by the Publisher, whether the whole or part of the material is concerned, specifically the rights of translation, reprinting, reuse of illustrations, recitation, broadcasting, reproduction on microfilms or in any other physical way, and transmission or information storage and retrieval, electronic adaptation, computer software, or by similar or dissimilar methodology now known or hereafter developed. Exempted from this legal reservation are brief excerpts in connection with reviews or scholarly analysis or material supplied specifically for the purpose of being entered and executed on a computer system, for exclusive use by the purchaser of the work. Duplication of this publication or parts thereof is permitted only under the provisions of the Copyright Law of the Publisher's location, in its current version, and permission for use must always be obtained from Springer. Permissions for use may be obtained through RightsLink at the Copyright Clearance Center. Violations are liable to prosecution under the respective Copyright Law.
The use of general descriptive names, registered names, trademarks, service marks, etc. in this publication does not imply, even in the absence of a specific statement, that such names are exempt from the relevant protective laws and regulations and therefore free for general use.
While the advice and information in this book are believed to be true and accurate at the date of publication, neither the authors nor the editors nor the publisher can accept any legal responsibility for any errors or omissions that may be made. The publisher makes no warranty, express or implied, with respect to the material contained herein.

Printed on acid-free paper

Springer is part of Springer Science+Business Media (www.springer.com)

Preface

This volume in the series of "Modern Aspects of Electrochemistry" is devoted to electrochemistry in the service of medicine and the relation of electrochemistry to medicine. It is indeed timely for such a topic to be discussed in this series. Medicine is the second oldest profession and as such deserves our closer attention. Further, in the ongoing process of "globalization" we are witnessing not only the tendency of commercial unification of the globe but the rapidly emerging interdependency of different scientific and technical disciplines as well. Indeed the much bandied-about "n" word—nanotechnology—may well evolve as the underlying thread in this technological melding process.

With the advent of relativity and quantum mechanics in the early part of the twentieth century and the development of molecular biology in the second half, it is now accepted that mathematics, physics, chemistry, and biology constitute but different parts of the same broader scientific discipline. Such interdependency and subsequent confluence of different disciplines may become one of the hallmarks of the twenty-first century.

Despite the theoretical scientific unity mentioned above, it is often the case that the impact a given branch of science has upon another requires a rather special vehicle to become common knowledge. Given that, it is hardly surprising that the impact of modern electrochemistry in medicine is not yet properly recognized by all. This volume and its featured articles are designed to focus on electrochemistry as it relates to medicine. It is hoped that this special volume will give the readership a broader view of the important role electrochemistry plays in medicine, as well as a glimpse into the future possibilities as both of these now related disciplines develop.

The volume contains 11 chapters. The first and last of these are written by the editor with his graduate students and they introduce and describe the relationship between the two disciplines as above. In the second chapter, Asaf and Blum present applications of electrochemistry in the design and development of medical technologies and devices. In the third Boffa and Koschinski discuss intracoronary stents: medical devices at the interface of biology and electrochemistry. The fourth by Metters and Banks describes screen printed electrodes which open new vistas in sensing, an important advance with implications for medical application to medical diagnosis. In the fifth Heller and Feldman present Electrochemical Glucose Sensors and Their Application in Diabetes Management. The sixth features the work of Hernandez, Lendeckel, and Scholz and is entitled Electrochemistry of Adhesion and Spreading of Lipid Vesicles on Electrodes. In the seventh Alvarez, Wishwanathan, Burkholz, Khairan, and Jacob discuss Bio-Electrochemistry and Chalcogens. The eighth is by Tsakova dealing with Conductive Polymer-Based Materials for Medical Electroanalytic Applications. The ninth's theoretical part is the work of Travo, Huang, Cheng, and Rangan while the experimental one is by Ertorer and Mittler and its subject matter is Experimental and Theoretical Issues of Nanoplasmonics in Medicine. The tenth chapter is authored by Kim, Lines, Duffy, Alber, and Crawford and it introduces Modeling and Measuring Extravascular Hemoglobin: Aging Contusions.

The reader should keep in mind that each chapter is self-contained and independent of the other chapters, which means that the chapters do not have to be read in consecutive order or as a continuum. Readers who are familiar with the material in certain chapters may skip those chapters and still derive maximum benefit from the chapters they read.

Finally, thanks are due to each of the 31 authors who helped make the volume possible.

Windsor, ON, Canada   Mordechay Schlesinger

Contents

1 **Applications of Electrochemistry in Medicine** 1
Robert Petro and Mordechay Schlesinger

2 **Applications of Electrochemistry in the Design and Development of Medical Technologies and Devices** 35
Roy Asaf and Shany Blum

3 **Intracoronary Stents: Medical Devices at the Interface of Biology and Electrochemistry** 55
Michael B. Boffa and Marlys L. Koschinsky

4 **Screen Printed Electrodes Open New Vistas in Sensing: Application to Medical Diagnosis** 83
Jonathan P. Metters and Craig E. Banks

5 **Electrochemical Glucose Sensors and Their Application in Diabetes Management** 121
Adam Heller and Ben Feldman

6 **Electrochemistry of Adhesion and Spreading of Lipid Vesicles on Electrodes** 189
Victor Agmo Hernández, Uwe Lendeckel, and Fritz Scholz

7 **Bio-Electrochemistry and Chalcogens** 249
Enrique Domínguez Álvarez, Uma M. Viswanathan, Torsten Burkholz, Khairan Khairan, and Claus Jacob

8 Conductive Polymer-Based Materials for Medical Electroanalytic Applications ... 283
Vessela Tsakova

9 Experimental and Theoretical Issues of Nanoplasmonics in Medicine ... 343
Daniel A. Travo, Ruby Huang, Taiwang Cheng, Chitra Rangan, Erden Ertorer, and Silvia Mittler

10 Modeling and Measuring Extravascular Hemoglobin: Aging Contusions ... 381
Oleg Kim, Collin Lines, Susan Duffy, Mark Alber, and Gregory Crawford

11 Model of Tumor Growth and Response to Radiation ... 403
L.J. Liu, S.L. Brown, and M. Schlesinger

Index ... 443

Contributors

Mark Alber Department of Applied and Computational Mathematics and Statistics, University of Notre Dame, Notre Dame, IN, USA
Department of Medicine, Indiana University School of Medicine, Indianapolis, IN, USA

Enrique Domínguez Álvarez Universidad de Navarra, Faculty of Pharmacy, Pamplona, Spain

Roy Asaf The Edmond and Lily Safra Children's Hospital, Sheba Medical Center at Tel Hashomer, Ramat-Gan, Israel

Craig E. Banks Division of Chemistry and Environmental Science, School of Chemistry and the Environment, Manchester Metropolitan University, Manchester, Lancashire, UK

Shany Blum Laboratory of Vascular Medicine, Faculty of Medicine, Technion—Israel Institute of Technology, Haifa, Israel

Michael B. Boffa Department of Chemistry and Biochemistry, University of Windsor, Windsor, ON, Canada

S.L. Brown Henry Ford Hospital, Detroit, MI, USA

Torsten Burkholz Division of Bioorganic Chemistry, School of Pharmacy, Saarland University, Saarbruecken, Germany

Taiwang Cheng Department of Physics, University of Windsor, Windsor, ON, Canada

Gregory Crawford Department of Physics, University of Notre Dame, Notre Dame, IN, USA

College of Science, University of Notre Dame, Notre Dame, IN, USA

Susan Duffy Department of Emergency Medicine and Pediatrics, Alpert Medical School of Brown University, Providence, RI, USA

Erden Ertorer Biomedical Engineering Program, Western University, London, ON, Canada

Department of Physics and Astronomy, Western University, London, ON, UK

Ben Feldman Abbott Diabetes Care, Alameda, CA, USA

Adam Heller Department of Chemical Engineering, University of Texas at Austin, Austin, TX, USA

Victor Agmo Hernández Department of Chemistry—BMC, Uppsala University, Uppsala, Sweden

Ruby Huang Department of Physics, University of Windsor, Windsor, ON, Canada

Claus Jacob Division of Bioorganic Chemistry, School of Pharmacy, Saarland University, Saarbruecken, Germany

Khairan Khairan Division of Bioorganic Chemistry, School of Pharmacy, Saarland University, Saarbruecken, Germany

Oleg Kim Department of Applied and Computational Mathematics and Statistics, University of Notre Dame, Notre Dame, IN, USA

Marlys L. Koschinsky Department of Chemistry and Biochemistry, University of Windsor, Windsor, ON, Canada

Uwe Lendeckel Institut für Medizinische Biochemie und Molekularbiologie, Universitätsmedizin Greifswald, Greifswald, Germany

Collin Lines Department of Physics, University of Notre Dame, Notre Dame, IN, USA

L.J. Liu Department of Physics, University of Windsor, Windsor, ON, Canada

Jonathan P. Metters Division of Chemistry and Environmental Science, School of Chemistry and the Environment, Manchester Metropolitan University, Manchester, Lancashire, UK

Silvia Mittler Department of Physics and Astronomy, Western University, London, ON, Canada

Robert Petro Department of Physics, University of Windsor, Windsor, ON, Canada

Chitra Rangan Department of Physics, University of Windsor, Windsor, ON, Canada

Mordechay Schlesinger Department of Physics, University of Windsor, Windsor, ON, Canada

Fritz Scholz Institut für Biochemie, Universität Greifswald, Greifswald, Germany

Daniel A. Travo Department of Physics, University of Windsor, Windsor, ON, Canada

Vessela Tsakova Institute of Physical Chemistry, Bulgarian Academy of Sciences, Sofia, Bulgaria

Uma M. Viswanathan Division of Bioorganic Chemistry, School of Pharmacy, Saarland University, Saarbruecken, Germany

Chapter 1
Applications of Electrochemistry in Medicine

Robert Petro and Mordechay Schlesinger

1.1 Electrochemistry and Medicine

Medicine is an applied field of science that has always strived to make use of new, and existing, technologies and techniques. Advances such as the synthesis of medications, X-ray and magnetic resonance (MR) imaging, mechanical implants, along with a host of others, improve health and increase the life span of individuals. Advancements in medicine have largely occurred in concert with technological advances resulting in a sophisticated interplay in the advancement of both. The ongoing goal of medicine to better treat patients requires a great deal of research from a wide variety of fields including electrochemistry.

As medicine as well as medical treatments and technologies continue to evolve, electrochemistry is playing an increasingly vital role in the creation and development of both medical treatments and devices. Electrochemical study of the human system has been of interest for several decades with research conducted on many fronts such as metallic implants, analysis of bodily fluids, and function of the body such as the immune system to name a few. Electrochemical

---

R. Petro (✉) • M. Schlesinger
Department of Physics, University of Windsor,
Windsor, ON N9B 3P4, Canada
e-mail: petror@uwindsor.ca; msch@uwindsor.ca

Table 1.1 Oxidation and reduction half-reactions for iron [Fe] and titanium [Ti] in oxygen

Oxidation of metals	Reduction of oxygen	Resulting reaction
$Fe \rightarrow Fe^{3+} + 3e^-$	$O_2 + 4e^- \rightarrow 2O^{2-}$	$4Fe + 3O_2 \rightarrow 2Fe_2O_3$
$Ti \rightarrow Ti^{4+} + 4e^-$	$O_2 + 4e^- \rightarrow 2O^{2-}$	$Ti + O_2 \rightarrow TiO_2$

considerations must be made whenever any surface or material is in contact with the internal systems of the human body. Blood, saliva, urine, and other extracellular fluid, which includes all body fluid outside the cells and contains ions among other hormones, proteins, etc., act as electrolytes, fluids that contain ions, and provide a conductive environment that promotes the electrochemical exchange of electrons. This exchange of electrons results in the oxidation, donation of electrons, of materials/metals within the human body and an associated reduction of the oxidizing agent, Table 1.1.

For certain materials/metals, such as iron [Fe], this electrochemical exchange results in the formation of a layer that is prone to dissolution and ultimately leads to the destruction of the metal in a process known as corrosion. Corrosion occurs as a result of a material reacting and reorganizing itself into a lower, more natural energy state. Most metals, with the exception of gold [Au], are not naturally found in their pure, refined, state and are generally found in the form of some sort of oxide. The process of corrosion may be accelerated if, in addition to the oxidation of the surface, the electrolyte provides electrical coupling between electrically dissimilar implanted metals [1–3]. This coupling is referred to as the formation of a galvanic cell and resulting galvanic corrosion can ultimately end in the dissolution of one of the metal species, oxidation of the anode, with the liberated metal being reduced on the other, the cathode. Hence, the performance of any implant, medical or dental, within the human body is dictated by the biofunctionality as well as biocompatibility of the material within the patient.

Electrochemical considerations are also made in the case of medications such as vaccines or intravenous medications. Here the consideration is made for better matching of the medication with the extracellular fluid as the medication itself may not be easily administered and/or absorbed by the human body. Excipients, the term used for such additives, include sodium chloride, sugars, and aluminum

1 Applications of Electrochemistry in Medicine

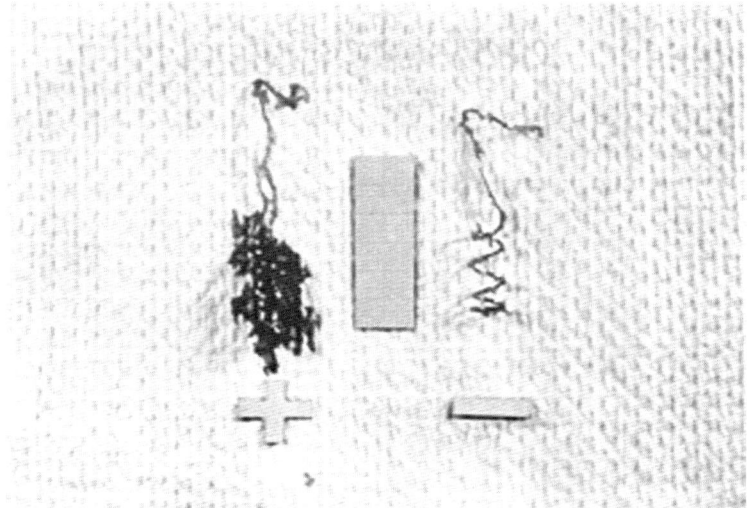

Fig. 1.1 Passage of a small amount of current between platinum helices submerged in citrated blood, with the positive electrode collecting the blood at a rate and weight depending on the faradays of current passed whereas the negative electrode not only resists thrombus deposition but produces some sort of protein variant in the blood which makes the blood permanently anti-coagulated over long-term experiments. Figure from Sawyer [6] reproduced by permission of ECS—The Electrochemical Society

salts, to name a few. Another example of electrochemical behavior in the body is the finding that the vascular interface, wall of a blood vessel, is negatively charged when looked at with a pair of probing electrodes attached to a high impedance potentiometer [4]. Interestingly, the reversal of this polarity usually relates to thrombosis (clotting). This discovery led to the development of direct current (DC) hemostasis, the stopping of bleeding, where the positive electrode, electrode providing the electrons, will generate thrombus and stop bleeding in tissues, while the negative electrode prevents thrombosis [4–6], Fig. 1.1. The thrombus deposited on an electrode or in tissues is directly related to the number of electrons which flowed between the electrode pair [6].

Pertaining to implant materials, electrochemical considerations are essential in the evaluation of both biofunctionality, which takes into account the longevity of the material and the ability of the material to perform a function, and biocompatibility, which is determined by the compatibility of the material with the human body. Though in most cases some overlap regarding functionality and compatibility exists, the strict biofunctional aspect of a material is often more easily fulfilled than the biocompatibility; that is not to say that currently available materials are ideally suited for biofunctionality purposes. Nevertheless, issues such as electrochemical stability are essential to biocompatibility as the presence of dissolution products in the surrounding tissue contribute to the formation of fibrous tissue around the implanted materials. Hence, a central concern regarding biocompatibility is corrosion.

Beyond the material interface, electrochemical considerations and techniques, such as electro-polishing [7], are essential in the development of materials and technologies for the human body. Technologies such as implants that dissolve and break down in the body after they have fulfilled their purpose [8–11], biosensors [12–14], and piezo-electric *in vivo* energy generation for implants such as pacemakers [15–19] are just some of the current fields of research that are part of the nexus between electrochemistry and medicine. This chapter will introduce the reader to electrochemistry with respect to *in vivo* considerations of biocompatibility and biofunctionality for material implantation within the human body as well as introduce some of the frontiers of current electrochemical research in medicine.

1.2 Biocompatibility of Metals in Blood and Tissue

Fundamentally, the physiological environment within the human body is inhospitable for most foreign species. The inhospitable nature of this environment is largely due to the presence of chloride ions, around 0.11 mol/L or 0.11 N (normal) in interstitial fluids [20, 21], though biologic macromolecules and proteins in extracellular fluids are able to influence corrosion considerably. In light of the hostile nature of the physiological environment, care must be taken to ensure any implanted material remains viable and provides the desired service for the patient.

Though the development, manufacture, understanding, and innovation of biocompatible materials is of intense ongoing research [2, 22–24], a relatively wide range of biocompatible materials are and have been available since the mid-1990s to satisfy the basic requirements for implantable devices [21]. Commonly implanted metal alloys include 316L stainless steel (316L SS), cobalt-chromium [CoCr], tantalum [Ta], and titanium [Ti]-based alloys, which rely on passivation by a thin surface layer of oxide [21]. The presence of this oxide provides the corrosion resistance of the metal within the physiological environment. The measurement and understanding of the corrosion process is best suited to electrochemical research techniques and has long been a part of the development of biocompatible alloys [25].

The corrosion resistance of both stainless steel and Co-Cr alloys is due to the Cr content of at least 12%, which forms a passive Cr-oxide [Cr_2O_3] layer. Though this layer provides corrosion resistance, stainless steel is the least corrosion resistant of the more commonly used metals, and is used largely for temporary implants only. A minimal amount of carbon is present within these alloys as Cr-carbides hinder corrosion resistance and are likely to form during the alloying process between 450 and 900°C [21].

Like the stainless steel, Co-Cr alloys form a passive Cr_2O_3 layer, though contrary to stainless steel, this layer provides significantly better corrosion resistance, due to a generally higher Cr content, which can be around 30% [21]. Additionally, carbon-containing cast Co-Cr alloy is characterized by a superior wear resistance and is therefore the metal of choice for the articulating surfaces of joint replacements. Particularly, the MP35N alloy was developed for the manufacture of the highly stressed anchorage stems of total hip prostheses [21]. The excellent corrosion resistance of Ta is due to its highly stable surface oxide layer, which prevents electron exchange between the metal and the adsorbed biological species [26]. Metal cladding of 316L SS with Ta has been used to improve corrosion resistance and biocompatibility of stents, though it has seen only limited use as base stent material due to poor mechanical properties [26].

The widespread use of Ti-based alloys within dentistry and medicine, along with other body modification uses such as piercings, is the result of the formation of a passive Ti-oxide [TiO_2] layer. Commonly used alloys of Ti include the 6%wt. aluminum, 7%wt. niobium [Ti-6Al-7Nb] alloy, which has equal corrosion resistance to pure titanium and increasingly replaces the $\alpha\beta$ Ti-Al-vanadium [Ti-6Al-4V] alloy [27], and Nitinol, a nickel-titanium [Ni-Ti] shape memory alloy that displays pseudo-elastic behavior when heated [21, 26]. The main drawback of titanium and its alloys is a lesser wear resistance compared to Co-Cr alloys; however, the inclusion of plastics and

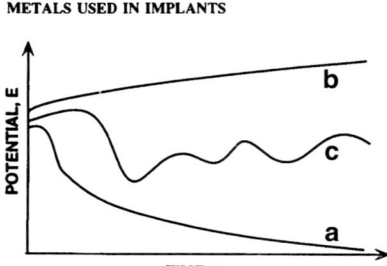

Fig. 1.2 Isolated electrode potentials as function of time for (**a**) a metal that exhibits film breakdown after immersion, (**b**) a metal with intact film, and (**c**) a metal that exhibits film breakdown after initial thickening. Figure reproduced from Gotman [21] by permission from the publisher for this copyrighted material Mary Ann Liebert, Inc. publishers

accompanying improvements in the design of joint replacements have allowed Ti alloys to find greater use within joint replacements.

The corrosion behavior of metal implants within the physiological environment can be characterized by following one of three curves of the rest potential of the metal plotted as a function of time, Fig. 1.2, when immersed in a simulated physiological solution, such as Hanks' or Ringer's solutions [21]. A rapid and maintained drop in the potential (curve a) is indicative of complete film breakdown after immersion and is a sign of general corrosion. Curves that rise and remain at a high level over a long period of time (curve b) indicate a passive metal or alloy, such as Ti, Ta, and Co-Cr alloys. Materials susceptible to pitting corrosion, where corrosion damage is localized to small areas and the remainder of the surface is largely unaffected, are characterized by a fluctuating curve (curve c) and is more typical of 316 SS due to the intermixed corrosion resistant Cr_2O_3 with less corrosion resistant ferric [Fe_2O_3] and Ni- [NiO] oxides.

Though is it often claimed that the passive film formed on Co-Cr and Ti alloys, including Nitinol, prevent corrosion in the body; in the strictest sense corrosion does occur as metal ions slowly diffuse through the protective oxide layer and accumulate in the surrounding tissue. This diffusion results in the implanted metal becoming surrounded by a layer of fibrous tissue of a thickness that is proportional to the amount and toxicity of the dissolution products and to the amount of motion between the implant and the adjacent tissues [21]. Pure titanium may elicit a minimal fibrous encapsulation under some conditions [28], whereas the proliferation of a fibrous layer as much as 2 mm thick is encountered with the use of Co-Cr and steel implants [21]. The formation of a fibrous layer prevents firm union between the implant and the surrounding tissue, which may be beneficial in cases where removal is desired and hence rigid fixation of the implant is not necessary. Other long known biomaterials that elicit a minimal fibrous response, under certain conditions, include high purity alumina and some special grades of polymers such as polytetrafluoroethylene, ultrahigh molecular weight high density polyethylene, and silicone rubber [29].

Aside from the fibrous layer, toxicity, more specifically genotoxicity, is of concern as genetic damage to cells around an implant pose a potential for significant risk in the case of long-term implantation. In an analysis using peripheral blood lymphocytes, mature type of white blood cell that circulate in the blood and are well known to directly respond to foreign bodies, as the testing medium to compare the genotoxicity of pure Ni, pure Ti, Nitinol, and 316L SS [30], it was found that pure Ni posed significantly greater immunological response than any other tested material, while pure Ti posed the least response. The release of Ni elements took the following decreasing distribution for the different powders extracted: pure Ni, 316L SS, Nitinol, and pure Ti, with 316L SS specimens proving more genotoxic than Nitinol and pure Ti in the tested *in vitro* conditions [30]. The biocompatibility seen in Nitinol alloys is attributed to the high Ti content, which is present at about an equal atomic ratio with Ni, where Ni also provides about 54–60 wt.% of the alloy.

The difficulty in achieving biocompatibility in tissue is not due to any one component of the physiological system but rather its dynamic nature. A complex biofilm of proteins and cells forms within seconds on materials in contact with blood [31] and fibrinogen adsorption, which adsorbs more on pure TiO_2 than 316L SS or Nitinol and is believed to be central to surface-induced thrombus formation and plays a very important role in subsequent acute and chronic inflammatory responses, is one of the initial events that occur when biomaterials come in contact with blood [32]. The presence of an implant during the dynamic process of wound healing will, in the best case scenario, only alter the morphology of the wound area due to the disruption of the collagen, capillaries, and blood vessels forming around the implant and resulting in differing patterns of vasculature and collagen fibers running parallel to the implant surface, Fig. 1.3 [28]. The scenario depicted in Fig. 1.3 is the ideal case using an

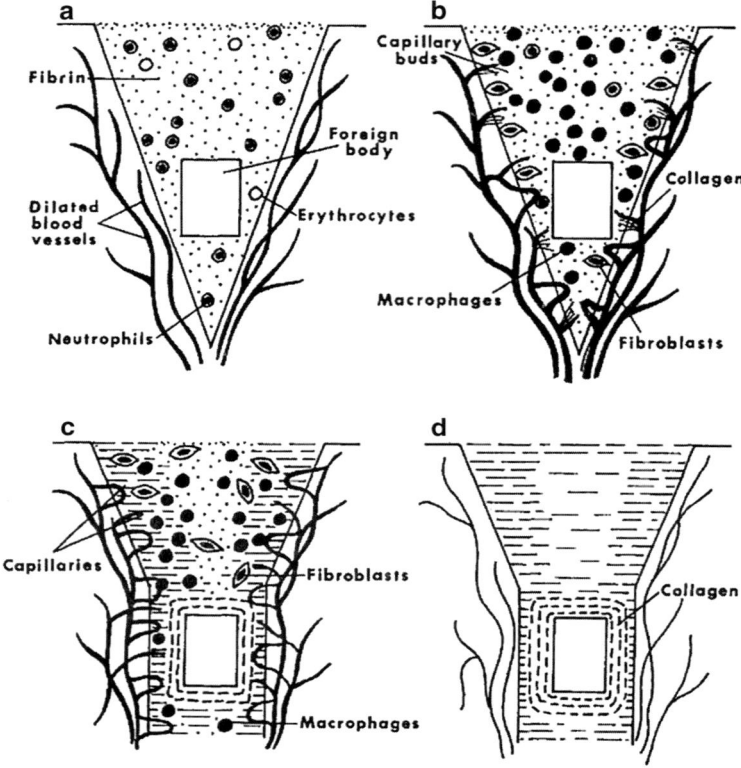

Fig. 1.3 (a–d) Wound healing in the presence of a foreign body, such as an implant. Figure from Williams [28] reproduced with kind permission from Springer Science and Business Media

entirely inert implant. In the more realistic scenario, Fig. 1.4, thicker fibrous encapsulation results; increasing healing time and taking longer to achieve stability. Corrosion of the implant during this time can lead to the implant providing a constant irritant resulting in a granuloma, Fig. 1.4c, and in sufficiently severe cases necrosis [28].

In addition to the dynamics of wound healing, the voltages of oxide-covered biomedical alloys are known to be naturally dynamic due to the electrochemical reorganization of the thin oxide films present on biomedical alloy surfaces during certain physiologically relevant situations [33]. In the case of stents, oxide disruption events may occur during initial immersion into solution, stent deployment,

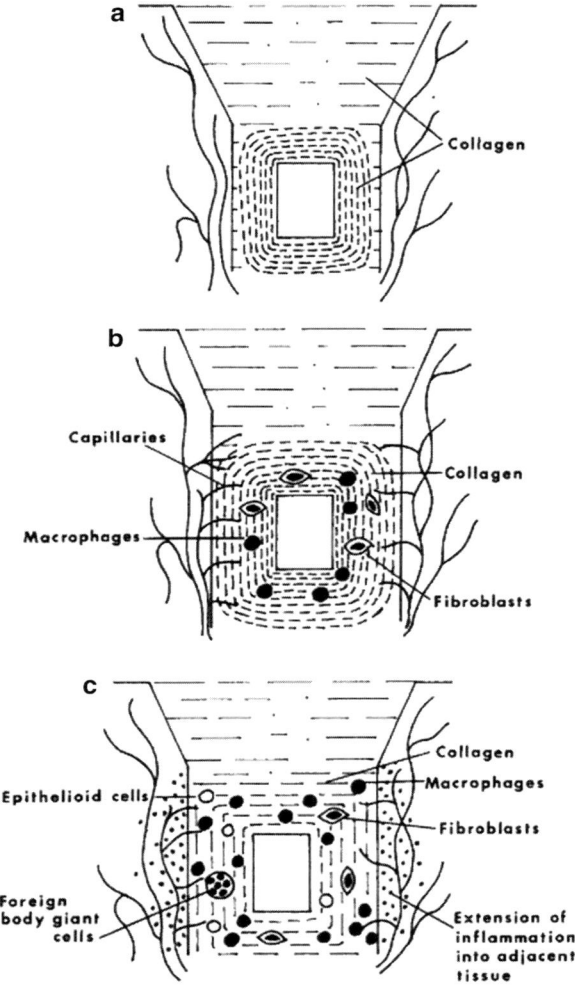

Fig. 1.4 More severe responses at implants (compared to Fig. 1.3d) (**a**) more extensive capsule; (**b**) more extensive and cellular capsule, with persistent macrophagic response alongside fibroblastic response; (**c**) extensive chronic inflammation (including foreign body giant cells and epithelioid cells) with granulation tissue. Figure from Williams [28] reproduced with kind permission from Springer Science and Business Media

and wear and/or corrosion damage at the asperities of contact surfaces, known as fretting. Anodic voltages may also arise at the bio-metal interface under highly oxidizing conditions such as those during wound healing when super oxide ions [O_2^-] and hydrogen peroxide [H_2O_2] are released from inflammatory cells [33].

Another important element of biocompatibility is thrombogenicity, the tendency of a material in contact with the blood to produce a thrombus (blood clot), or rather the resistance thereof. As previously alluded to with negative electrode preventing thrombosis [6], the thrombogenicity of metals is related to the standard electrode potential of the metal with those having higher electronegativity being more antithrombogenic [6, 21]. Given that corrosion resistance falls with greater electronegativity, compromise is necessary in cases of vascular implants. Nevertheless, among surgical metals both Co-Cr and Ti alloys display relatively good resistance to thrombosis formation [21]. To this end, the majority of cardiovascular stents currently on the market are made from 316L SS [26] with others being made from Cr alloys, such as Co-Cr and platinum [Pt]-Cr [34]. In addition to these, the Food and Drug Administration (FDA) in the United States has more recently approved Nitinol cardiovascular stents. The shape memory characteristic of Nitinol stents is due to the crystallographically reversible structural change known as martensitic transformation [21]. This change is brought about by a change in temperature where stents are fabricated at a high temperature, cooled to a lower temperature, and deformed to then achieve their original, functional, shape once heated in situ inside the body. Despite the increasingly large selection of implantable material and alloys, compromise is always necessary. More specifically, a stent is a compromise and there is no single stent that is ideal for all scenarios and a suitable stent must take into account the location to be stented and kind of lesion [35]. One method used to ease the compromise among bare metallic stents is through the application of coatings. Coatings are able to improve bio- and hemo-compatibilities, reduce surface energy, smooth surface texture, deliver medications, and enhance surface oxide layer; all of which can influence thrombosis and neointimal, arterial thickening, and can lead to restenosis, narrowing of a blood vessel [26, 36]. It should be noted that inorganic materials and endothelial, blood vessel lining, cells are exclusively used as coating materials, while the polymers are used as both as a coating material as well as the sole stent material [26]. A selection of stent coatings are outlined in Sect. 1.4.1.

Biocompatibility with tissue and blood is a complex, dynamic, and multifaceted issue for which no ideal material currently exists. Compromise is often required where a certain strength or corrosion resistance is necessary, such as in stents. For example, the shape memory alloy Nitinol includes Ni which provides an allergenic response in some individuals. Though alloying can help mitigate biological responses, cytotoxicity, genotoxicity, biocompatibility, corrosion, and other factors all contribute to the utility and behavior of any metal introduced into the body, Table 1.2.

Blood and soft tissues provide for many necessary considerations in the choice of metal used. These choices go beyond implants to the material from which medical tools are forged. In the case of bone further considerations must be made as mechanical strength, surface chemistry, and surface morphology play a greater role.

1.3 Osteo-Biocompatibility

In addition to biocompatibility with surrounding tissue, orthopedic implants require biocompatibility with bone as well mechanical strength. The biocompatibility of implanted materials with bone, or hard tissue, depends on many of the same factors as biocompatibility with soft tissues. Along with morphological and geometrical considerations, the mechanical stress system acting on the implant-bone system, material chemistry as well as the type of bone must be considered [28]. Electrochemical understanding of the formation of galvanic cells/couples between dissimilar metals within orthopedic implants is essential to prevent implant degradation [1–3]. For example, it is best if corrosion-sensitive elements, such as welds, are in the cathodic position so that it is not damaged by a galvanic couple [1]. Most importantly, as in the case of soft tissue, the application and role of the implant in contact with bone must be considered.

The healing process for bone begins with fluid from the circulatory system, exudate, forming within the defect, such as in the form of a blood clot. This material begins to slowly reorganize during osteogenesis, bone production, where new bone may grow, either directly from the existing bony walls, osteoconduction, or from isolated areas within the reorganizing tissue where appropriate bone cells, osteoblasts, and growth factors are present. The degree of reorganization and quantity/distribution of unmineralized soft connective tissue present will then depend on the presence, material, and morphology of an implant, resulting in either a layer/region of fibrous unmineralized

Table 1.2 Biological impact and associated degree of concern of various metals with respect to various physiological responses [22]

Periodic position	Element	Biocompatible	Carcinogenic	Genotoxic	Mutagenic	Cytotoxic	Allergenic	Prone to corrosion	Other
3d	Ti	Yes	No	No	No	Medium	No	No	No
	V	No	Yes	Yes	Yes	High	Disputed	No	No
	Cr	No	Disputed	Yes	Yes	High	Yes	No	No
	Mn	No	No	Yes	No	High	No	Yes	No
	Fe	No	No	Yes	Disputed	Medium	No	Yes	No
	Co	No	Yes	Yes	Yes	High	Yes	Yes	Yes
	Ni	No	Yes	Yes	Yes	High	Yes	Yes	Yes
	Cu	No	No	Yes	Yes	High	Yes	Yes	No
4d	Zr	Yes	No	No	No	Low	No	No	No
	Nb	Yes	No	No	No	Low	No	No	No
	Mo	No	Disputed	Yes	Yes	Low	Yes	Yes	Yes
	Tc	No				Radioactive			
	Ru	Yes	No	No	No	Medium	No	No	Yes
	Rh	No	Yes	Yes	Yes	High	Unknown	No	No
	Pd	No	Yes	No	Disputed	Medium	Yes	No	No
	Ag	No	No	No	No	High	Yes	No	Yes

5d	Hf	Unknown	Unknown	Unknown	Unknown	Medium	No	Unknown
	Ta	Yes	No	No	No	Low	No	No
	W	No	Yes	Yes	No	Medium	Yes	No
	Re	Unknown	Unknown	Unknown	Unknown	Unknown	No	Unknown
	Os	No	No	Yes	Yes	High	No	Yes
	Ir	No	No	No	Yes	High	No	No
	Pt	No	Yes	Yes	Yes	High	Yes	No
	Au	Yes	No	No	No	High	No	No
3p	Al	No	No	Yes	No	Low	No	Yes
4p	Zn	No	No	No	No	High	No	Yes
5p	Sn	Yes	No	No	No	Low	No	Yes

Note: "Other" refers to issues beyond those already listed such as haemolysis, neurological effects, etc

| Unknown/Disputed | Minimal/No Concern | Moderate Concern | Serious Concern |

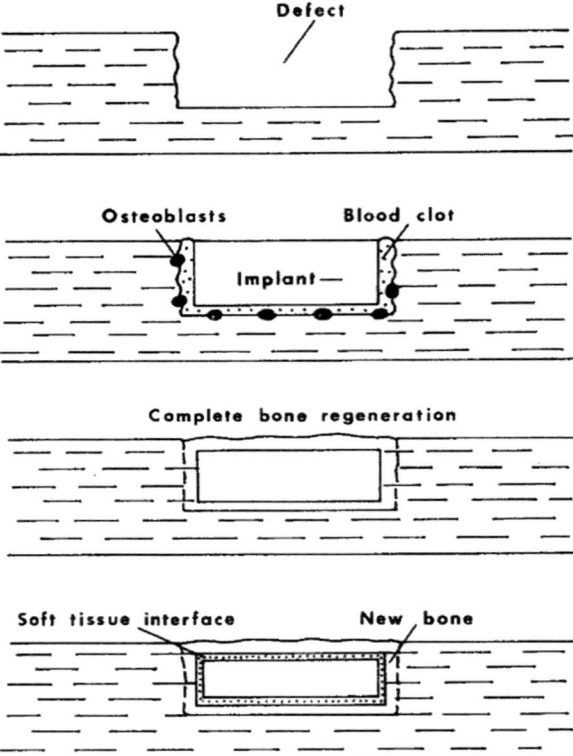

Fig. 1.5 Response in presence of monolithic implant, leading either to complete bone regeneration or a soft tissue interface. Figure from Williams [28] reproduced with kind permission from Springer Science and Business Media

tissue or good bone-implant contact, Fig. 1.5 [28]. It is important to note that an upper limit does exist on the defect size that can be bridged; above that limit the defect will be filled with unmineralized tissue.

The degree and connection of an implanted biomaterial with bone is dependant on the purpose of the procedure and implant. Implants may be used to hold bone fragments together while bone healing takes place, to facilitate the healing of bone within a defect, or inserted into a bone for the purpose of reconstruction in adjacent tissues, such as in the cases of joint replacements or dental implants [28, 37]. In the most common orthopedic applications of joint replacement and skeletal reconstruction, the strength of the material and its resulting biofunctionality become more prominent. Additionally, the behavior of implants with respect to bone can be categorized either as being relatively inert, or encouraging the formation of new bone, and/or sustaining growth from the surrounding bone up to the implant surface resulting in direct bone-material contact, osseointegration [28, 38], which is essential for reconstructive long-term implantation purposes.

In the case of reconstructive implants, Ti alloys are most widely used due to the metal's ability to become surrounded by bone tissue without interposed fibrous structure and without chronic inflammatory reactions [28], hence allowing for better bone inclusion. This does not mean that any alloy of Ti is suitable for any implantation and osseointegration. The risk of surface damage to the oxide coating of some reconstructive implants results in a risk of dissolution products entering the surrounding tissue. In bone, aluminum [Al] exerts the pathological effects of the inhibition of hydroxyapatite formation, growth and bone cell proliferation, as well as suppresses bone cell activity [39], while vanadium [V] has strong cytotoxicity [40], limiting the use of Ti alloys such as Ti-6Al-4V [41]. For this reason, other Ti alloys such as those containing zirconium [Zr], niobium [Nb], and Tantalum [Ta], such as the more commonly used Ti-6Al-7Nb, along with Ti-5Al-13Ta, Ti-13Zr-13Nb, have been explored for possible implantation. More fundamentally, the individual pure metals of Ti, hafnium [Hf], Nb, Ta, and rhenium [Re] have also been tested. Although few results exist for the behavior of individual metals, the limited numbers of available studies appear to show good biocompatibility and osteogenisis among these metals, with all metals showing similar fibrosis around the implants and bone-implant contact being better for Ti and Ta after 4 weeks implantation in Wistar rats [40]. Aside from alloying, passive coatings are another means to increase the biocompatibility and osseointegration of implant alloys such as those made of Co-Cr and stainless steel. These coatings include metallic coatings such as Zr and/or Ti on Co-Cr screws, Ti anodization, and plasma sprayed ceramic hydroxyapatite coatings. Further details on coatings for better osseointegration can be found in Sect. 1.4.2.

Aside from joint implants, the facilitation of bone growth to assist skeletal reconstruction is a necessity when bone loss secondary to trauma, infection, or tumor removal occurs. Due to the health concerns associated with harvesting autogenous cancellous bone, bone from the patient, and allograft bone, bone from a donor, the development of synthetic bone graft substitutes have become essential. Currently used bone graft substitutes, mainly porous calcium hydroxyapatite, a synthetic analogue to the mineral phase of bone itself, and other Ca phosphate ceramics, although osteoconductive and able to eventually be incorporated in the surrounding bone, are not sufficiently strong to stand in as a complete replacement for load-bearing bones [38]. The use of porous Ti metal structures, to act as strong scaffolds for bone ingrowth, is of interest as Ti has a relatively low elastic modulus minimizing the stress-shielding phenomenon and allows direct apposition of the bone tissue [28, 38]. Given the passive nature of Ti, the induction of osteogenesis, formation/growth of bone, necessitates the inclusion of a bone growth factor to provide the implant osteoinductivity, the ability of the implant to induce stem cells or osteoprogenitor cells to differentiate into osteoblasts [38]. For clinical benefit, these growth factors must be delivered directly to the site of regeneration via a carrier matrix and administered via a controlled delivery system [38, 42]. These methods include co-precipitation into a three-dimensional inorganic latticework of biomimetic Ca-P [43] and the novel inclusion within sol–gel processed silica glass controlled release matrices [38].

In the latter case, the highly porous, nanostructured, resorbable, solid silica glass dissolves in the body via hydrolysis, (1.1), producing aqueous silicon hydroxide among other dissolution products of silica and bioglasses; products which are nontoxic at low concentrations [38]. The incorporation of bone growth factors to induce cell differentiation is an essential to produce strong osteoinductive bone graft substitutes for load-bearing locations within the body. Incorporation of such implants in the repair of tissue is most necessary in orthopedic implants where a degree or permanents exists and replacement at a later time is not ideal.

$$SiO_{2(s)} + 2H_2O \rightarrow Si(OH)_{4(aq)} \tag{1.1}$$

1.4 Biodegradable Implants

Within the context of biocompatible implants that are highly corrosion resistant and withstand the conditions of the harsh physiological environment exists a desire for implants that naturally breakdown and are absorbed, assimilated, or expelled by the body after they have fulfilled their purpose. The most commonly used and well-known bio-absorbable implants are medical sutures, or more commonly stitches, that are made from polyglycolic acid (PGA), polylactic acid (PLA), or polydioxanone [10, 26]. The biodegradation/absorption of these materials is part of biofunctionality in a similar way as material strength. Similar to the use of bio-absorbable sutures, the goal of biodegradable/absorbable implants is to obviate the need of a secondary surgery and improve patient results. It is important to note that although bio-absorbable and biodegradable speak about the same result, the process by which each occurs differs. Bio-absorbable materials, mostly polymers, are materials that are most often broken down, or degrade, via hydrolysis and become part of a biological mechanism producing energy, water [H_2O], and carbon dioxide [CO_2] via the citric acid cycle, also known as the TCA cycle, Krebs cycle, or Szent-Györgyi–Krebs cycle. Conversely, biodegradable materials often denote those materials which are broken down and expelled or assimilated into the body; this is most often the case with metals. Additionally, the utility of resorbable implant components as a release mechanism for medications has been shown with respect to bone growth factors as well as other medications [10, 38].

1.4.1 Bio-Absorbable Polymers

The use of bio-absorbable devices has increased since the 1960s when the concept was introduced by Kulkarni et al. [10]. Bio-absorbable implants are used in many orthopedic surgeries in the form of screws, pins, plugs, and plates for orthopedic, oral, and craniofacial surgery and are most widely used for the fixation of fractures and ligaments [10, 11, 44]. The factors upon which biodegradation depends include the identity and phase, crystalline or amorphous, of the material, size of the implant, implantation site, bone or tissue,

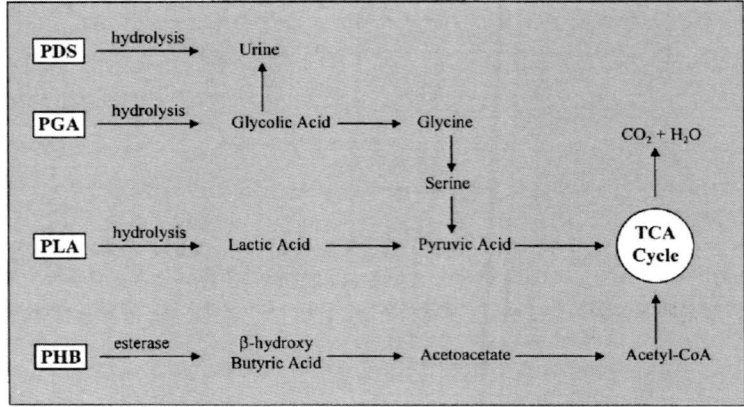

Fig. 1.6 Routes of elimination for commonly used polyhydroxyacids including poly(p-dioxanone) (PDS), polyglycolic acid (PGA), polylactic acid (PLA), and poly(b-hydroxybutyrate) (PHB). Enzymes, free radicals, and immune cells are also thought to play various roles in the degradation of bio-absorbable materials. Image reproduced from An et al. [10] with permission from permission from Elsevier

purpose of the implant, and mechanism of degradation, whether enzymatic cleavage or hydrolysis [10]. The mechanism of degradation/absorption of commonly used polyhydroxyacids, poly(p-dioxanone) (PDS), PGA, PLA, and poly(b-hydroxybutyrate) (PHB), Fig. 1.6, is terminated by the citric acid cycle, also known as the TCA cycle, Krebs cycle, or Szent-Györgyi–Krebs cycle, producing energy, water [H_2O], and carbon dioxide [CO_2] [10]. The decomposition of the polymers begin upon contact with tissues *in vivo* and degrade in as little as 3 weeks, self-reinforced polyglycolic acid (SR-PGA) exhibits complete strength loss in 1 month and fully degrades in about 3 months while the L-isomer of polylactic acid (PLLA), which is more crystalline than the amorphic D-isomer of polylactic acid (PDLA), may take years to break down [10]. Copolymers of PLA-PGA, or PDLA-PLLA, are less crystalline, less hydrophobic, and tend to have faster rates of hydrolysis [10].

The limitation of bio-absorbable polymers is the low tensile yield strength which is reported in the range of 11–72 MPa for PLLA and 57 MPa for PGA [10] compared to 290 MPa for stainless steel. Though PGA reinforced with PLA fibers has the ability to resist deformation under load, flexural strength, similar to stainless steel,

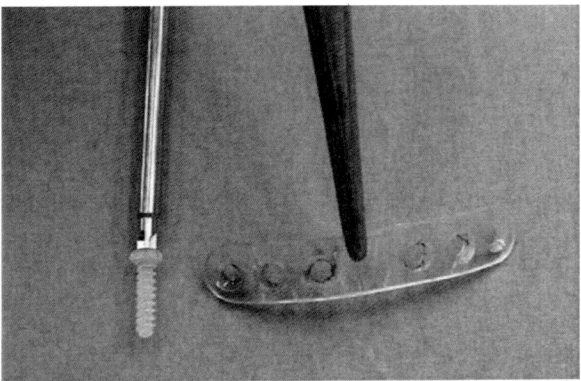

Fig. 1.7 Bioresorbable self-reinforced copolymer poly-L/D-lactide (SR-PLDLA) devices: plate and screw. Image reproduced from Ylikontiola et al. [11] with permission from permission from Elsevier

250 MPa for PGA and 280 MPa for stainless steel, the flexural modulus, or the ratio of stress to strain in flexural deformation or the tendency for a material to bend, of the material remains very low 1.6–27 GPa for various reinforced and non-reinforced polymers compared to 200 GPa for stainless steel [10]. Bio-absorbable implants, Fig. 1.7, are excellent in the healing of fractures [11, 44] and are compatible with bone fillers, such as hydroxyapatite, or as carriers for osteogenic substances or medications to aid in the repair of bones. Nevertheless, despite the flexibility and number of polymers available, the main drawback to these plastic materials is a lack of strength that is traditionally only available from metals.

1.4.2 Biodegradable Metals

Biodegradable materials, like all other implant materials, must possess mechanical properties, such as material strength, that are suitable for the proper biofunctionality of the implant devices. In applications where material strength is required, metals are used in the place of lower strength plastic alternatives. Two metals that are able to fulfill biofunctionality along with biodegradability requirements are iron [Fe] and magnesium [Mg], the latter having first been investigated in 1892 by Payr et al. [28]. Magnesium in particular is of interest within the context of biodegradable orthopedic implants, though both Fe and Mg have been investigated as stents, given that both are present naturally in the human body [26]. The biodegradation of Fe involves its oxidation into ferrous [Fe^{2+}] and ferric [Fe^{3+}] ions and subsequent dissolution into biological media whereas Mg corrodes into soluble Mg hydroxide [$Mg(OH)_2$], Mg chloride [$MgCl_2$], and hydrogen gas [H_2] [26].

In the case of stents, pure Fe (>99.5%) is the major component in biodegradable Fe stents whereas Mg is typically alloyed with other metals for use in stents, such as in the WE43 (4% yttrium, 3% rare earth, Balance Mg) and AZ31 (3% aluminum, 1% zinc, Balance Mg) Mg alloys [26, 45]. Though both balance materials are present within the human body, the impact from vascular applications of elevated local concentration of these elements is, at present, not fully known. A critical requirement for stent materials is a large elastic modulus, or tendency to be deformed elastically, so as to prevent recoil and closing of the stent once deployed [45]. Of the two biodegradable metals, Fe possesses superior radial strength because of its higher elastic modulus, whereas the low elastic modulus of Mg alloys results in poor radial strength requiring thicker struts and larger area of metal–artery interaction in order to provide proper vessel wall support [26]. Additionally Mg does not have very high ductility needed to withstand deformation during expansion and is radiolucent and cannot be imaged by X-rays [26, 45].

The limited use in vascular applications is offset by Mg and its alloys having excellent characteristics for use as orthopedic biomaterials. These characteristics include an elastic modulus, compressive yield strength, and density close to that of bone (density is approximately 16% more than bone); an ability to promote osteogenisis, fracture toughness greater than osteogenic ceramic materials such as synthetic hydroxyapatite, and most importantly good biocompatibility [9, 46–48]. Cytotoxicity tests have shown that Mg and most of its alloys do not typically reduce the viability of cells, including nerve and muscle cells [28], or their growth; only moderately inhibiting the growth of osteoblast cells at concentrations above 10 mM [49]. Specifically, cytotoxicity tests using 1% by weight alloying elements including aluminum [Al], zinc [Zn], tin [Sn], silicon [Si], zirconium [Zr], and yttrium [Y] on murine cells have shown that Mg–1Al, Mg–1Sn, and Mg–1Zn alloys do not significantly reduce fibroblast (L-929 & NIH3T3) viability; Mg–1Al, Mg–1Si, Mg–1Sn, Mg–1Y,

Mg–1Zn, and Mg–1Zr alloys are not significantly toxic to osteoblasts (MC3T3-E1); and that Mg–1Al and Mg–1Zn indicated no negative effect on the viability of blood vessel-related cells, (ECV304 & VSMC), though Si and Y negatively affected corrosion properties [50]. Other tests have confirmed the biocompatibility of pure Mg, AZ91D (9%Al, 1%Zn, Balance Mg), AZ31 (3%al, 1%Zn, Balance Mg), and Mg-manganese [Mg-Mn] among other alloys [9, 51, 52].

Unlike in long-term implants where corrosion resistance is an absolute requirement, degradation and corrosion is an absolute necessity in biodegradable implants; however, the corrosion rate cannot be so large as to cause premature failure of the implant or the excessive release of corrosion products. In particular, the release of hydrogen $[H_2]$ and nitrogen $[N_2]$ gases from the corrosion of Mg is of some concern. Though apparently inert, localized and gradually absorbed, if the corrosion occurs too quickly, the slowly absorbing N_2 can create gas pockets that require puncture or aspiration [28].

Understanding and better controlling the *in vivo* corrosion of Mg is currently one of the main focuses for the development of biodegradable Mg implants, Fig. 1.8. The corrosion rate of Mg alloys not only depends on the composition and microstructure of the alloy, with certain alloys being more prone to corrosion, but also on the composition of the electrolyte [9, 51].

One mitigation technique known as anodizing, an electrolytic passivation process used to increase the thickness of the natural oxide layer on a metal surface. In this process the metal to be anodized, anode, and an inert, often platinum [Pt] or stainless steel, cathode, electrode from which electrons flow, are placed in an acidic aqueous solution and a current is passed between. Within this process hydrogen gas $[H_2]$ evolution is seen on the cathode as a buildup of the oxide layer occurs on the anode. A buildup of such a layer on the AZ31 Mg alloy, in a trisodium phosphate $[Na_3PO_4]$ bath using a 100 Hz pulsed current for 5 min at room temperature, and subsequent corrosion testing in simulated body fluid found that coatings formed at 250 V were more corrosion resistant than those formed at higher voltages [52].

The determination of a perfect *in vivo* facsimile for the physiological environment has yet to be found with Mg alloys having persistently shown to have a lower corrosion rate *in vivo* than within *in vitro*-simulated corrosive environments [9, 51, 52]. This determination is made more difficult by the sensitivity of the corrosion to the identity of buffer solutions and the lack of consensus on whether the addition of proteins

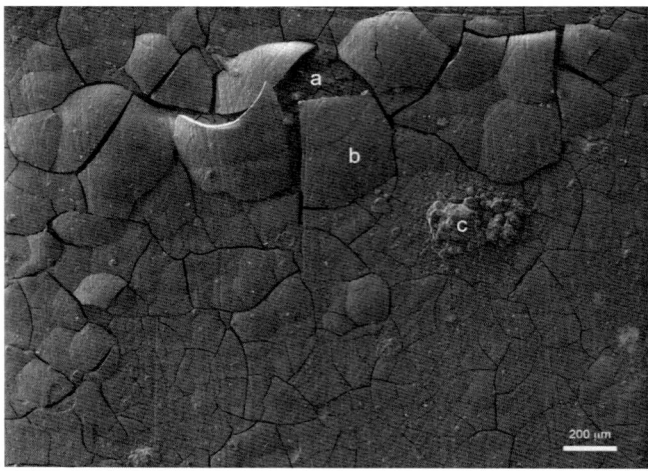

Fig. 1.8 A scanning electron micrograph of the surface of pure Mg corroded *in vitro* for 21 days in Earle's balanced salt solution (EBSS) before cleaning. Similar results were observed in all solutions. (**a**) Indicates the deep corrosion layer. (**b**) Indicates the superficial corrosion layer. (**c**) Indicates corrosion product. Image reproduced from Walker et al. [51] by permission of John Wiley and Sons

retards or enhances corrosion [51]. Of solution that have been tested, Earle's balanced salt solution (EBSS) using a 26 mM sodium bicarbonate [$NaHCO_3$] buffer has shown to provide an environment that is only slightly more corrosive than the *in vivo* environment by way of corrosion rate and loss of material in mm/year [51]. Nevertheless, potentiodynamic (varying voltage) polarization tests along with electrochemical understanding of *in vivo* corrosion do not currently exist. Simply, the relationship between physiological relevance of the corrosion medium and the resulting corrosion behavior of Mg alloys is not clearly understood and significant testing remains necessary [51].

1.5 Biocompatible Coatings

Improvement of biocompatibility is essential for the creation of more resilient biodegradable and permanent implants. Coatings are designed to alter surface characteristics, such as roughness or hemo-/

osteo-/biocompatibility, while preserving desirable bulk properties. Many biocompatible coatings and techniques exist including, but not limited to, increasing the thickness and/or changing the morphology of an oxide layer by anodizing the surface [53]; coating the surface with another material by galvanization [54], which includes electroplating, electroless deposition, and dipping in a molten metal bath; the formation of polymer and biopolymer coatings, such as polyurethane, by a solvent-casting technique or dip coating techniques [55, 56]; and plasma-spray coatings where a material is sprayed on the surface bulk surface [57]. Of these techniques anodized coatings are often used on the surfaces of orthopedic implants, whereas polymer coatings appear more in vascular applications, and techniques such as galvanization are more widely used for depositing metallic coatings.

1.5.1 Coatings for Biocompatibility of Stents

The principle reason for the coating of stents is to reduce the possible re-narrowing of a blood vessel, restenosis, following healing, as well as the thickening of the innermost layer of a blood vessel, neointimal hyperplasia, following the mechanically widening of a narrowed or obstructed blood vessel, coronary angioplasty. Aside from material characteristics such as flexibility, strength, and expandability, stents are required to be radiopaque and/or compatible with magnetic resonance imaging, resist the thrombosis formation, and ideally be able to deliver/release medications. Materials used for stents include 316L stainless steel (316L SS), platinum-iridium [Pd-Ir] alloy, tantalum [Ta], nitinol [Ni-Ti] alloy, cobalt-chromium [Co-Cr], titanium [Ti], pure iron [Fe], magnesium [Mg] alloys, polyethylene terepthalate (PET), poly-L-lactic acid (PLLA), and poly-L-glycolic acid (PLGA) with coatings used to improve biocompatibility, allow for the delivery of medications, as well as provide hemocompatibility to prevent restenosis or thrombosis [26, 36]. Notably absent from the list are common metals such as copper [Cu] and zinc [Zn] as they along with nickel [Ni], silver [Ag], chromium [Cr], cobalt [Co], including some of their alloys, produce hydrogen peroxide [H_2O_2], a strongly oxidizing agent that is harmful to arteries and can cause inflammation, as they corrode [26]. It should be noted that little H_2O_2 is produced from stainless steel unlike its alloying metals [26]. Additionally, despite the use of some polymer-based stents, the majority of stents continue to be metallic largely due to the lack of radiopacity as well as elastic modulus of biomedical polymers, which usually lies in the range of 1–5 GPa [26]. Compared to polymer-based stents, metallic stents are, with the exception of Mg which cannot be imaged using X-rays, radiopaque and typically have elastic moduli of around 100–200 GPa [26]. Nevertheless, it has been shown that the radial pressure exerted by some stents, such as PET braided mesh stents, is comparable to that of stainless steel stents [26]. Other materials investigated for coronary stents include copolymers such as PLGA and polyurethanes, though the use of biodegradable PLGA stents has been investigated more for urological than coronary applications.

The coating of stents is most broadly classified into inorganic materials, such as metals, and organic polymers including, biostable, biodegradable, and biological. Stent materials commonly used as coatings include Ti and Ta metals, biostable PET and polyurethanes along with biodegradable PLLA and PLGA, which breakdown by means of hydrolysis and the Krebs cycle, are some of the wide array of polymers and copolymers that have been used to coat stents [26]. Despite the use of polyurethanes in many cardiovascular devices, such as pacemaker lead wires and inner surface coatings for artificial hearts, some studies have shown that polyurethane stents produce an extensive inflammatory reaction and that coatings used in other biocompatible applications may not be beneficial for stents [26].

Commonly used metallic stent coatings, outside of base stent material, include gold [Au], silicon-carbide [SiC], iridium oxide [IrO_2], and diamond-like carbon. The utility in Au coatings lies largely with its excellent corrosion resistance and the significant radiopacity, six times greater than steel; a 5 μm coating on each side of an 80 μm thick steel stent doubles its radiopacity [26]. The smooth surface and low impurities of thermally process Au coatings have been found to reduce neointimal hyperplasia and inflammation compared to standard gold coatings; however, the human clinical trials on gold-coated stents were found not to be satisfactory as neointimal proliferation was greater in those receiving gold-coated stents compared to other stenting options [26]. Despite the undesirable characteristics of Au coatings, the deposition of an Au interlayer provides a means to enhance radiopacity. The highly biocompatible, inert, ceramic coatings provided by IrO_2 have been shown to reduce neointimal thickness compared to uncoated, bare 316L SS stents [26], though work remains regarding the behavior of the stent *in vivo*, particularly regarding potential conversion of H_2O_2 into H_2O and O_2. As for SiC and diamond-like carbon, both show signs of reducing thrombosis and restenosis, though further study remains required [26].

Biological polymers that have extensively explored for coating stents include phosphorycholine (PC), hyaluronic acid (HA), and fibrin. Phospholipid PC, an essential part of the red blood cell membrane, is structurally composed of both hydrophilic and hydrophobic components and has been coated on metallic stents mainly to prevent the adhesion of coagulation-inducing cells [26]. This coating, which has been reported stable up to 6 months, has been found to have antithrombogenic properties and decrease restenosis rates in addition to the ability to deliver medications as a drug-eluting stent, which has been clinically tested [26]. Similarly, though less literature is available on HA, HA, a linear polysaccharide non-sulfated glycosaminoglycan present in various tissues of the body, has been found to improve the thrombo-resistance and decrease the inflammatory response of stents [26]. The biopolymer fibrin, an insoluble protein produced during the coagulation of blood, is well known for its biocompatible, biodegradable, and viscoelastic properties and has demonstrated its potential as a coating with markedly less vessel occlusion, sudden blockage of a blood vessel, or foreign body reaction than PU and PET coatings [26]. Given positive results of biological polymers as stent coatings, the investigation of biological coatings to replace metallic coatings such as Ta and Ti as well as the more commonly used nonmetal polymers.

Regardless of the coatings type, the mechanics of stent use must also be considered. Coatings are applied on stent surfaces in the crimped state when the surface area occupied by the stent is minimal. Expansion of the stent from its crimped state can result in fissures, cracks, and pores on the coating as the area occupied by the stent increases [26]. Damaged coatings can release particulate debris as well as cause a burst in the release of medications rather than a gradual one, increasing the chance of restenosis [26]. Finally, chemical modifications must be thoroughly characterized as even traces of chemicals used in coating processes can be non-biocompatible and lead to erroneous conclusions regarding a coating [26].

1.5.2 Coatings for Better Osseointegration

As previously stated, certain metals, such as Ti, provide for better osseointegration compared to stainless steel or Co-Cr alloys. Coatings such as Zr and/or Ti on Co-Cr screws have been found to considerably improve osseointegration and bone-implant contact over uncoated implants; providing similar results to Ti screws [27]. Plasma-spray-coatings, where material is melted and propelled towards a substrate and the molten droplets flatten, rapidly solidify and form a deposit, are another method of including the biocompatibility and osseointegration of Ti [57, 58] and are presently used on FDA-approved Co-Cr-Mo implants [57].

Another type of Ti surface coating currently being studied is the formation of TiO_2 nanotube layers fabricated by electrochemical anodization [53, 59, 60]. As mentioned in Sect. 1.4.2, anodizing is an electrolytic passivation process used to increase the thickness of the natural oxide layer on a metal surface by setting the part as the anode and passing a current through the system. Specifically in the case of TiO_2 nanotube growth, the Ti anode was situated 1 cm away from the platinum [Pt] cathode in a room temperature bath containing hydrofluoric acid [HF] [53]. The anodizing conditions for TiO_2 nanotube growth of 20 min in a 0.5% HF solution have been found at potentials of 15 and 20 V [53, 59] with potentials at, or less than, 10 V resulting in the formation of nanoparticles and potentials at, or above, 25 V resulting in a disordered nanotube structure [53]. Anodizing for 20 min within a 1.5% HF solution indicated that particulate features appeared only at 5 V, while tube-like structures appeared at 10, 15, and 20 V [53]. The value in the vertical TiO_2 nanotube array structure is that adhesion/propagation of the osteoblast cells appear significantly improved by the topography of the nanotubes [59] and evidence exists that enhanced calcium deposition by osteoblasts occurs on nanotube-like structures, compared with both unanodized Ti and anodized nanoparticle Ti structures [53], potentially resulting in more rapid osseointegration.

Other options for better osseointegration of metallic implants include increasing the porosity of the implant surface or the deposition of hydroxyapatite coatings. Placed adjacent to existing bone, porous surfaces have been seen to promote the ingrowth of new bone in a process that is largely independent of chemistry and controlled by the morphology of the pores [28]. In cases of the porosity being interconnecting, and minimum pore size is in the region of 150 µm, bone is able to grow into the surface area and promote attachment [28]. Despite the lack of strength noted in calcium hydroxyapatite [$Ca_{10}(PO_4)_6(OH)_2$] for reconstructive purposes, synthetic hydroxyapatite ceramic coatings applied by air-plasma spraying are designed to enhance bone growth at the implant–bone interface [28, 61–64]. These deposits provide good matching and osseointegration due to the structural and chemical resemblance to the inorganic constituent of bone, carbonated hydroxyapatite [28, 38, 61, 62]. Such deposits allow for a highly crystalline deposit of controlled porosity and thickness [62] that bonds to bone and not soft tissue, unlike some bioactive glasses, such as 4555 bioactive glasses [46.1 mol% SiO_2—24.4 Na_2O—26.9 CaO—2.6 P_2O_5], which bond to both [64]. Hydroxyapatite coatings have been shown to increase bone contact provide better bone-implant integration on CoCr implants than their uncoated counterparts [63] as well as provide superior bone-implant contact on pre-deposit machine patterned Ti-6Al-4V implants compared to planar structures [62]. Additionally, the control of the porous matrix of the coatings allows for the inclusion, and sustainable release, of medications such as ceftriaxone sodium, an antibiotic, and sulbactam sodium for the treatment of osteomyelitis, inflammation of the bone marrow [65]. In this connection, it has been found that lower pore percentage with a distribution of mainly micropores was superior to the higher pore percentage both *in vitro* and *in vivo* animal trials [65].

In addition to biocompatibility and osseointegration, infection is a key concern in total joint arthroplasties as it can also lead to implant failure. One means by which to improve success rates and combat infection is the inclusion of silver [Ag] and Ag-oxide [Ag_2O] within the implant coating to reduce bacterial proliferation [66, 67]. Silver is cytotoxic to many bacterium including many strains of staphylococcus [68]. The inclusion of Ag may be within the coating itself [68] or simply as a surface coating [69, 70].

The inclusion of Ag within plasma sprayed hydroxyapatite coatings of Ti substrates was found to significantly reduce the adhesion of the *pseudomonas aeruginosa* bacteria in a concentration-dependent manner [68]. Although 6 wt.% Ag incorporation showed highest antimicrobial properties, it also reduced human fetal osteoblast cell proliferation, necessitating a balance between good cell proliferation and antimicrobial property, which has been presented with the inclusion of 2 and 4 wt.% Ag in the hydroxyapatite coatings [68]. The use of Ag has also been explored as surface coatings on external bone fixing elements, elements that form an interface between the surface of the skin, which is normally colonized by bacteria, and the bone tissue, the infection of which provides the most significant complication [71]. Coatings of Ag on 316L SS by means of vapor-phase ion-beam-assisted deposition for infection prevention of external bone fixation devices have shown positive results in the prevention of bacteria proliferation, while raising concerns over a potentially high release of Ag^+ ions from surface coatings, confirmed by its increase in blood concentration *in vivo* [69, 70]. Nevertheless, the ability of Ag to inhibit the initial stages of bacteria colonization along with the potential for long-term host defense against bacterial adhesion are positive results that provide for the continued research into Ag inclusion. A recent study has shown that an organ/tissue-specific cytotoxicity response appears to exist with respect to Ag nanoparticles [72] and provides further evidence for the eventual inclusion of Ag within certain implant coatings. Further study in understanding the release of Ag^+ ions from implants and resulting effects within the body is required.

1.5.3 Future Prospects for Biocompatible Coatings

Electrochemical coating and analysis techniques play a vital role in the understanding of any implanted material within the body and are of serious ongoing study. The provision of a good *in vitro* facsimile for *in vivo* conditions is essential for expanded electrochemical testing of implantable materials [9, 51]. Given the many coating techniques available for implants, the coating of traditionally active materials, such as Mg and its alloys, for *in vivo* use is something that will inevitably be further explored. Additionally, the variety of potential applications in orthopedic, vascular, and other concentrations requires further study given the potential of tissue-specific cytotoxic responses. The advent of thin film cladding of Mg alloys [73, 74] poses an interesting question as to the future of Mg implant use, especially within osteo- and biodegradable applications. The incorporation of Ag [68, 72] and other metals within implants also requires better electrochemical understanding and will be a topic of ongoing research.

References

1. Reclaru L, Lerf R, Eschler P-Y, Meyer J-M. Corrosion behavior of a welded stainless-steel orthopedic implant. Biomaterials. 2001;22:269–79.
2. Reclaru L, Lerf R, Eschler P-Y, Blatter A, Meyer J-M. Pitting, crevice and galvanic corrosion of REX stainless-steel/CoCr orthopedic implant material. Biomaterials. 2002;23:3479–85.
3. Zardiackas L, Roach M, Disegi J (2005) Galvanic corrosion of cobalt-base and titanium-base implant material couples. Medical device, materials II: proceedings from the materials & processes for medical devices conference 2004, p 398–402
4. Sawyer PN, Pate JW. Electrical potential differences across the normal aorta and aortic grafts of dogs. Am J Physiol. 1953;175:113.
5. Sawyer PN, Dennis C, Wesolowski SA. Electrical hemostasis in uncontrollable bleeding states. Ann Surg. 1961;154:556.
6. Sawyer PN. Application of electrochemical techniques to the solution of problems in medicine. J Electrochem Soc. 1978;125(10):419C–36C.
7. Alvarez E, Vinciguerra J, DesJardins JD. The effect of a novel CoCr electropolishing technique on CoCr-UHMWPE bearing frictional performance for total joint replacements. Tribol Int. 2012;47:204–11.
8. Witte F. The history of biodegradable magnesium implants: a review. Acta Biomater. 2010;6:1680–92.
9. Xue D, Yun Y, Tan Z, Dong Z, Schulz MJ. In vivo and in vitro degradation behavior of magnesium alloys as biomaterials. J Mater Sci Tech. 2012;28(3):261–7.
10. An YH, Woolf SK, Friedman RJ. Pre-clinical in vivo evaluation of orthopaedic bioabsorbable devices. Biomaterials. 2000;21:2635–52.
11. Ylikontiola L, Sundqvuist K, Sàndor GKB, Törmälä P, Ashammakhi N. Self-reinforced bioresorbable poly-L/DL-lactide [SR-P(L/DL)LA] 70/30 miniplates and miniscrews are reliable for fixation of anterior mandibular fractures: a pilot study. Oral Surg Oral Med Oral Pathol. 2004;97(3):312–7.
12. Suzuki H. Advances in the microfabrication of electrochemical sensors and systems. Electroanalysis. 2000;12(9):703–15.
13. Skladal P. Advances in electrochemical immunosensors. Electroanalysis. 1997;9(10):737–45.
14. Ahmed MU, Hossain MM, Tamiya E. Electrochemical biosensors for medical and food applications. Electroanalysis. 2008;20(6):616–26.
15. Qi Y, McAlpine MC. Nanotechnology-enabled flexible and biocompatible energy harvesting. Energ Environ Sci. 2010;3(9):1275–85.
16. Häsler E, Stein L, Harbauer G. Implantable physiological power supply with PVDF film. Ferroelectrics. 1984;60:277–82.
17. Park KI, Xu S, Liu Y, Hwang GT, Kang SJ, Wang ZL, Lee KJ. Piezoelectric $BaTiO_3$ thin film nanogenerator on plastic substrates. Nano Lett. 2010;10:4939–43.
18. Li Z, Zhu GA, Yang RS, Wang AC, Wang ZL. Muscle-driven in vivo nanogenerator. Adv Mater. 2010;22(23):2534–7.
19. Zhu G, Yang R, Wang S, Wang ZL. Flexible high-output nanogenerator based on lateral ZnO nanowire array. Nano Lett. 2010;10:3151–5.

20. Gilanyi M, Ikrényi C, Fekete J, Ikrényi K, Kovach AGB. Ion concentrations in subcutaneous interstitial fluid: measured versus expected values. Am J Physiol. 1988;255(3):F513–9.
21. Gotman I. Characteristics of metals used in implants. Journal of Endourology 1997;11(6):383-389
22. Biesiekierski A, Wang J, Abdel-Hady Gepreel M, Wen C. A new look at biomedical Ti-based shape memory alloys. Acta Biomater. 2012;8:1661–9.
23. Walkowiak-Przybyło M, Klimek L, Okrój W, Jakubowski W, Chwiłka M, Czajka A, Walkowiak B. Adhesion, activation, and aggregation of blood platelets and biofilm formation on the surfaces of titanium alloys Ti6Al4V and Ti6Al7Nb. J Biomed Mater Res A. 2012;100A(3):768–75.
24. Aguilar Maya AE, Grana DR, Hazarabedian A, Kokubu GA, Luppo MI, Vigna G. Zr–Ti–Nb porous alloys for biomedical application. Mater Sci Eng C. 2012;32:321–9.
25. Lecoeur J, Viegas MF, Abrantes LM. Metal materials biodegradation: a chronoamperometric study. J Mater Sci Mater Med. 1990;1:105–9.
26. Mani G, Feldman MD, Patel D, Agrawal CM. Coronary stents: a materials perspective. Biomaterials 2007;28:1689–1710
27. Plecko M, Sievert C, Andermatt D, Frigg R, Kronen P, Klein K, Stübinger S, Nuss K, Bürki A, Ferguson S, Stoeckle U, von Rechenberg B. Osseointegration and biocompatibility of different metal implants—a comparative experimental investigation in sheep. BMC Musculoskelet Disord. 2012;13:32.
28. Williams DF. Titanium and titanium alloys. In: Williams DF, editor. Biocompatibility of clinical implant materials, vol. II. Boca Raton, FL: CRC Press; 1981. p. 9–44.
29. Williams DF. Tissue-biomaterial interactions. Journal of Materials Science. 1987;22:3421–3445
30. Assad M, Lemieux N, Rivard CH, Yahia L'H. Comparative in vitro biocompatibility of nickel-titanium, pure nickel, pure titanium, and stainless steel: genotoxicity and atomic absorption evaluation. Bio-Med Mater Eng. 1999;9:1–12.
31. Nygren H, Tengvall P, Lundström I. The initial reactions of TiO_2 with blood. J Biomed Mater Res. 1997;34:487–92.
32. Bai Z, Filiaggi MJ, Dahn JR. Fibrinogen adsorption onto 316L stainless steel, Nitinol and titanium. Surf Sci. 2009;603:839–46.
33. Gettens RTT, Gilbert JL. Fibrinogen adsorption onto 316L stainless steel under polarized conditions. J Biomed Mater Res. Part A 2008;85(1): 176–187
34. Kereiakes DJ, Cannon LA, Ormiston JA, Turco MA, Mann T, Mishkel GJ, McGarry T, Wang H, Underwood P, Dawkins KD. Propensity-matched patient-level comparison of the TAXUS Liberté and TAXUS element (ION) paclitaxel-eluting stents. Am J Cardiol. 2011;108(6):828–37.
35. Duda SH, Wiskirchen J, Tepe G, Bitzer M, Kaulich TW, Stoeckel D, Claussen CD. Physical properties of endovascular stents: an experimental comparison. J Vasc Intervent Radiol. 2000;11(5):645–54.
36. Gertner ME, Schlesinger M. Electrochem and Solid-State Letters, 2003;6(4): J4–6.
37. Sàndor GKB. Branemark's remarkable implant: dentistry's quantum leap—the surgical perspective. Geriatr Med Quart. 1989;1:162–7.

38. Reiner T, Kababya S, Gotman I. Protein incorporation within Ti scaffold for bone ingrowth using Sol–gel SiO_2 as a slow release carrier. J Mater Sci Mater Med. 2008;19:583–9.
39. Malluche HH. Aluminium and bone disease in chronic renal failure. Nephrol Dial Transplant. 2002;17(2):21–4.
40. Matsuno H, Yokoyama A, Watari F, Uo M, Kawasaki T. Biocompatibility and osteogenesis of refractory metal implants, titanium, hafnium, niobium, tantalum and rhenium. Biomaterials. 2001;22:1253-1262
41. Niinomi M. Recent metallic materials for biomedical applications. Metall Mater Trans A. 2002;33(3):477–86.
42. Seeherman H. The influence of delivery vehicles and their properties on the repair of segmental defects and fractures with osteogenic factors. Journal of Bone & Joint Surgery. 2001;83(Supp 1):S79.
43. Liu Y, Li JP, Hunziker EB, De Groot K. Incorporation of growth factors into medical devices via biomimetic coatings. Phil Trans R Soc A. 2006; 364(1838):233–48.
44. Serlo WS, Ylikontiola L, Vesala A-L, Kaarela OI, Iber T, Sàndor GKB, Ashammakhi N. Effective correction of frontal cranial deformities using biodegradable fixation on the inner surface of the cranial bones during infancy. Childs Nerv Syst. 2007;23:1439–45.
45. Poncin P, Proft J. Stent Tubing: Understanding the Desired Attributes. 2004. Materials & processes from medical devices conference, Anaheim, CA, USA, 8–10 Sept 2003, p 253–259
46. Staiger MP, Pietak AM, Huadmai J, Dias G. Magnesium and its alloys as orthopedic biomaterials: a review. Biomaterials. 2006;27:1728–34.
47. Song GL, Song S. A possible biodegradable magnesium implant material. Adv Eng Mater. 2007;9(4):298–302.
48. Song GL. Control of biodegradation of biocompatable magnesium alloys. Corros Sci. 2007;49:1696–701.
49. Yun YH, Dong ZY, Yang D, Schulz MJ, Shanov VN, Yarmolenko S, Xu Z, Kumta P, Sfeir C. Biodegradable Mg corrosion and osteoblast cell culture studies. Mater Sci Eng C. 2009;29:1814–21.
50. Gu XN, Zheng YF, Cheng Y, Zhong SP, Xi TF. In vitro corrosion and biocompatibility of binary magnesium alloys. Biomaterials. 2009;30(4):484–98.
51. Walker J, Shadanbaz S, Kirkland NT, Stace E, Woodfield T, Staiger MP, Dias GJ. Magnesium alloys: predicting in vivo corrosion with in vitro immersion testing. J Biomed Mater Res B Appl Biomater. 2012;100B(4):1134–41.
52. Zhang J, Gu Y, Guo Y, Ning C. Electrochemical behavior of biocompatible AZ31 magnesium alloy in simulated body fluid. J Mater Sci. 2012;47: 5197–204.
53. Yao C, Slamovich EB, Webster TJ. Enhanced osteoblast functions on anodized titanium with nanotube-like structures. J Biomed Mater Res A. 2008;85A(1):157–66.
54. Hehrlein C, Zimmermann M, Metz J, Ensigner W, Kubler W. Influence of surface texture and charge on the biocompatibility of endovascular stents. Coron Artery Dis. 1995;6(7):581–6.
55. Severini A, Mantero S, Tanzi MC, Cigada A, Salvetti M, Cozzi G, Motta A. Polyurethane-coated, self-expandable biliary stent: an experimental study. Acad Radiol. 1995;2(12):1078–81.

56. Whelan DM, van der Giessen WJ, Krabbendam SC, van Vliet EA, Verdouw PD, Serruys PW, van Beusekom HMM. Biocompatibility of phosphorylcholine coated stents in normal porcine coronary arteries. Heart. 2000;83:338–45.
57. Giannakopoulos HE, Sinn DP, Quinn PD. Biomet microfixation temporomandibular joint replacement system: a 3-year follow-up study of patients treated during 1995 to 2005. J Oral Maxillofac Surg. 2012;70(4):787–94.
58. Vercaigne S, Wolke JGC, Naert I, Jansen JA. Histomorphometrical and mechanical evaluation of titanium plasma-spray-coated implants placed in the cortical bone of goats. J Biomed Mater Res. 1998;41(1):41–8.
59. Oh S, Daraio C, Chen L-H, Pisanic TR, Fiñones RR, Jin S. Significantly accelerated osteoblast cell growth on aligned TiO_2 nanotubes. J Biomed Mater Res A. 2006;78A(1):97–103.
60. Yu W, Zhang Y, Xu L, Sun S, Jiang X, Zhang F. Microarray-based bioinformatics analysis of osteoblasts on TiO_2 nanotube layers. Colloids Surf B Biointerfaces. 2012;93:135–42.
61. De Lange GL, Donath K. Interface between bone tissue and implants of solid hydroxyapatite or hydroxyapatite-coated titanium implants. Biomaterials. 1989;10(2):121–5.
62. Ripamonti U, Roden LC, Renton LF. Osteoinductive hydroxyapatite-coated titanium implants. Biomaterials. 2012;33:3813–23.
63. Grandfield K, Palmquist A, Gonçalves S, Taylor A, Taylor M, Emanuelsson L, Thomsen P, Engqvist H. Free form fabricated features on CoCr implants with and without hydroxyapatite coating in vivo: a comparative study of bone contact and bone growth induction. J Mater Sci Mater Med. 2011;22:899–906.
64. Hench LL. Biomaterials: a forecast for the future. Biomaterials. 1998;19:1419–23.
65. Kundu B, Soundrapandian C, Nandi SK, Mukherjee P, Dandapat N, Roy S, Datta BK, Mandal TK, Basu D, Bhattacharya RN. Development of new localized drug delivery system based on ceftriaxone-sulbactam composite drug impregnated porous hydroxyapatite: a systematic approach for in vitro and in vivo animal trial. Pharm Res. 2010;27:1659–76.
66. Gravens DL, Margraf HW, Butcher HR, Ballinge WF. The antibacterial effect of treating sutures with silver. Surgery. 1973;73(1):122–7.
67. Pollini M, Paladini F, Licciulli A, Maffezzoli A, Nicolais L, Sannino A. Silver-coated wool yarns with durable antibacterial properties. J Appl Polymer Sci. 2012;125(3):2239–44.
68. Roy M, Fielding GA, Beyenal H, Bandyopadhyay A, Bose S. Mechanical, in vitro antimicrobial, and biological properties of plasma-sprayed silver-doped hydroxyapatite coating. ACS Appl Mater Interfaces. 2012;4:1341–9.
69. Massè A, Bruno A, Bosetti M, Biasibetti A, Cannas M, Gallinaro P. Prevention of pin track infection in external fixation with silver coated pins: clinical and microbiological results. J Biomed Mater Res. 2000;53(5):600–4.
70. Bosetti M, Massè A, Tobin E, Cannas M. Silver coated materials for external fixation devices: in vitro biocompatibility and genotoxicity. Biomaterials. 2002;23:887–92.
71. Mahan J, Seligson D, Henry S, Hynes P, Dobbins J. Factors in pin tract infections. Orthopedics. 1991;14:305–8.

72. Martínez-Gutierrez F, Thi EP, Silverman JM, de Oliveira CC, Svensson SL, Vanden Hoek A, Sánchez EM, Reiner NE, Gaynor EC, Pryzdial ELG, Conway EM, Orrantia E, Ruiz F, Av-Gay Y, Bach H. Antibacterial activity, inflammatory response, coagulation and cytotoxicity effects of silver nanoparticles. Nanomedicine. 2012;8:328–36.
73. Petro R, Schlesinger M. Direct electroless deposition of nickel boron alloys and copper on aluminum containing magnesium alloys. Electrochem Solid State Lett. 2011;14(4):D37–40.
74. Petro R, Schlesinger M. Direct electroless deposition of low phosphorous Ni-P films on AZ91D Mg alloy. J Electrochem Soc. 2012;159(7):D455–61.

Chapter 2
Applications of Electrochemistry in the Design and Development of Medical Technologies and Devices

Roy Asaf and Shany Blum

2.1 Introduction

Medicine comprise a vast field of research that has crossed the demarcation lines between disciplines over the past decades becoming a multidisciplinary milieu that encompasses a variety of fields from basic sciences (e.g., mathematics and physics) to engineering, and even social sciences and epidemiology. All of these directly contribute to the evolution of medical technologies. In other words, multidisciplinary is a key concept in the evolution of modern medical devices and technologies. It is clear now that the development of biocompatible and durable medical devices rests upon implementation of discoveries from all basic sciences rather than major breakthrough in one field. This will be the scarlet thread in the following sections of this chapter.

During the last century it became understood and accepted that hard and fast rules could not easily be applied in medicine. Apparently, this is largely due to population-related genetic diversity, but there is no

---

R. Asaf (✉)
The Edmond and Lily Safra Children's Hospital,
Sheba Medical Center at Tel Hashomer,
Ramat-Gan, Israel
e-mail: royasaf@gmail.com

S. Blum
Laboratory of Vascular Medicine, Faculty of Medicine,
Technion—Israel Institute of Technology, Haifa, Israel
e-mail: blum.shany@gmail.com

doubt that other factors are reflected here as well. Undisputed agreement of opinions is hardly ever achieved in medicine, even after a medical technology has made it to broad clinical acknowledgement. However, not consent; a medical technology is used on the primary basis of extensive understanding of its chemical and physical properties and, for most cases, after showing efficacy and safety in large population trials.

The field of electrochemistry has greatly evolved in the past decades, and its applications to medical devices have involved various medical disciplines. The surface-tissue interface level has especially drawn the attention of electrochemical research in the medical arena. In fact, the impact of research in this aspect on the biological (and clinical) results of a treatment with a certain medical device is imperative.

The main rational of writing this present contribution is to provide the reader with an overview of the various capabilities of electrochemistry for the advancement of medical treatment. In respect to the authors' clinical perspective, the various applications discussed in this review are categorized in three topics which are: (1) bio-stimulating devices, (2) biosensors, and (3) biomechanical support devices.

2.2 Applications of Electrochemistry in the Development of Electrical Stimulation Devices

The application of stimulating electrodes in modern medicine is familiar mainly from its wide usage in the field of cardiology. The first implantable cardiac pacemaker for arrhythmia introduced in 1958 has undergone a great deal of variations. It has evolved into the small size, programmable, and chargeable device with multiple synchronized electrodes commonly used nowadays. In recent years a new category of medical stimulating devices has made its appearance to become of scientific interest. These stimulating devices are applied to the brain for treatment of neurodegenerative diseases and other neurologic or psychiatric disorders. Herein we focus the discussion on electrochemical aspects of the ongoing development of deep brain stimulation (DBS) systems.

Parkinson's disease (PD) is characterized by a progressive degeneration of dopaminergic neurons in a distinct zone of the midbrain, called substantia nigra, which is involved in regulation of motor neuron activity. Well-known symptoms like tremor, rigidity, and abrasive movements are also accompanied by cognitive impairment that altogether severely debilitate patients and eventually shorten their life span. The administration of L-Dopa has become the best medical treatment for patients with PD; however, controversy still exists regarding the timing of treatment initiation, and great efforts are still made to overcome issues of tolerance to the drug and its adverse effects. The use of DBS whose seeds were first planted back at the mid-twentieth century was heralded by studies of Benabid and Pollack in the late 1980s. DBS for tremor preceded trials in Parkinson's disease and other neuropsychiatric disorders [1]. Nevertheless, PD has set the framework to the development of this cutting-edge technology as a promising therapy.

DBS is an evolving treatment of PD which is based on placement of a pulse electrode into deep structures of the brain. For PD patients, the electrode is inserted into the subthalamic nucleus (STN) of the basal ganglia. This area regulates release of the neurotransmitter dopamine from dependent neurons in the substantia nigra, which becomes degenerated in PD patients' brain. Stimulation of the STN in PD patients has been shown to alleviate the symptoms related to PD more effectively than the common medical treatment alone and it also enables the reduction of the required medical treatment [2, 3]. The stimulating electrode is made to act as a cathode (that is a reducing rather than oxidizing element) with respect to the vicinity of its position in the STN and induces chemical and biologic reactions in a yet to be precisely determined mechanism.

Biomedical engineering is called upon to tackle issues related to the electrode-tissue interface. Biocompatibility questions relating to both short- and long-term time-scale must be resolved to ensure the reliability of data acquisition on the one hand and to increase durability of the stimulating probe on the other. Controlling local inflammatory response and remodeling processes is undoubtedly one of the ultimate hindrances in the development of future stimulating devices. It is also common place to try and adjust the physical properties of an implanted device (i.e., diameter, elasticity, strength) to the target tissue. Altogether, this developing niche entails a great deal of multidisciplinary challenges for the medical practitioner.

There exist a variety of manufacturing methods of a reliable and biocompatible probe for DBS. The selection of suitable materials is a fundamental issue to any implanted medical device and therefore has been the focus of a large number of research projects in the past few decades. Selection is driven mainly but not solely by considerations of biocompatibility and endurance for all that entails. That is so in particular with respect to stimulating electrodes. The interface of the electrode and its biologic bed is surprisingly dynamic and has a profound impact on the electrical conductivity of the device. Therefore, it is the electrode-tissue impedance factor that serves as a unique criterion in the design of electrical stimulation devices. The science of biomaterials is constantly introducing novel materials custom made to improve biocompatibility and longevity of implanted devices. Currently, the experience in animal and clinical studies has shown superiority of few metals and alloys used for electric stimulation devices. These include Platinum (Pt), Platinum\Iridium (Pt\Ir), and gold, all share overall beneficial characteristics in terms of biocompatibility, impedance, allergy, and radiographic visibility [4]. The corrosion resistance properties of a given material are also crucial when designing implanted devices. This issue as well as the revolutionary shape reserving alloy Nitinol will be elaborated in Sect. 4. With recent advances in insulating materials and the application of more stimulating techniques (i.e., capacitive electrodes), the material choice may become less cardinal in some circumstances; however, for DBS it will probably stay so owing to the unacceptable and irreversible risk involved in neurosurgery.

Advances in the semiconductor industry provided new tools for fabrication of micro-scale probes. Electrodeposition techniques have allowed decreasing the size of a stimulating probe while maintaining its physical strength. These techniques have greatly contributed to the reduction of local tissue damage during the insertion of a probe. The correlated inflammatory response around the stimulating electrode forms an in-excitable fibrotic tissue that increases the impedance and the stimulating threshold of the electrode. In the case of cardiac pacemakers, it was the introduction of steroid eluting electrodes that made it possible to overcome this drawback [5]. It is the development of electrochemical methods that will ultimately allow a continuous deposition of an anti-inflammatory substance from the stimulating probe to its local environment. This has been demonstrated mainly through modern vascular stents (discussed below). Low electrode-tissue impedance was also achieved by the use of Pt and to an even greater extent with Pt\Ir. In addition, roughening the surface area of the electrode can further decrease the electrode impedance and the first method reported to achieve this was electrochemical deposition of finely dispersed platinum upon a platinum electrode [6]. These principles are still implemented in the cardiac electrode arena and will probably play an important role in the development of DBS electrodes as well.

Electrodeposition of metals such as Iridium (Ir), Platinum, Nickel (Ni), and Gold has been the center pieces of the development process of electric stimulation micro-devices. These metals are applied at the electrode sites, as interconnectors, and as the supporting mechanical layer. Platinum has been shown in numerous instances to have high biocompatibility. Yet, high melting point of same presents difficulties when attempting at connecting (soldering) the probe to the electrical microwires. Ge et al. [7] have suggested a novel technique to connect microwires to the stimulating probe using Pt electroplating. They have demonstrated improved electrochemical stability of the probe with this method as well [7]. A novel application of electroplating was described in 2005 by Motta and Judy [8] aimed at the trade-off between probe strength and tissue damage. Using a pre-patterned seed layer it was made possible to control the diameter of the probe during the Ni electroplating process. That, in turn, resulted in a probe with tapered geometry that has relatively high strength yet lower insertion tissue damage. The use of a final thin gold layer electroplated onto the probe to encapsulate the potentially toxic Ni has substantially attenuated the inflammatory response of the tissue in contact with the probe [8]. The concept of electrochemical deposition has spread much beyond metal deposition with the discovery of conducting polymers like polypyrrole and PEDOT. Not only that these polymers can be synthesized by electrochemical methods, but accurate electrodeposition onto micro-electrodes roughens the electrode surface area. This improves integration of the electrode into the tissue, decreases the inflammatory response at the electrode site, and decreases the impedance at the electrode-tissue interface [9]. Significantly, the later will also extend battery life due to lower power requirements.

It is commonly accepted that a key role in the therapeutic strategy of DBS is played by the precise positioning of the stimulating electrode. The elaborate basic research in this field and the pursuit of significant clinical effect obviate the need for an accurate microscale real-time neurochemical measurements. These must also correlate with advanced imaging modalities used in DBS operations. One of the newest developments in this field addresses this need. Lee and colleagues [10] have engineered such a micro-scale system that can measure discrete changes in the concentration of neurotransmitters. This system can use carbon-fibered microelectrode or enzyme-based microsensor to measure temporal and spatial changes in currents. These changes reflect the distinct signature of oxidation-reduction for a given substance (i.e., dopamine, adenosine) and can also be plotted onto a 3D scale. This method may be used in real-time in conjunction with stereotactic guidance to surgically place the DBS electrode [10]. Moreover, this system enables the physician to continuously adjust the electrical stimuli in order to achieve a satisfying biologic response in correlation with the clinical response. This new breakthrough will ultimately lead the way to smart closed-loop systems.

Despite this advanced technology, the exact mechanism of DBS is still unclear. Several hypotheses have been suggested based on electrophysiological studies and measurements of neurotransmitters release in various areas of the brain. In recent years a new side of the DBS paradigm has been elucidated. The astonishing discovery of a harmonized crosstalk between neurons and glia cells, the principal supporting cells of neurons in the brain, was rapidly followed by studies showing that not only DBS interferes with neuronal activity, but it also affects the activity of glia cells and amplifies this crosstalk. Presumably this phenomenon is responsible for the attenuation of clinical symptoms in DBS for PD [11].

The field of electric stimulation is continuously evolving. With the need for multidisciplinary advances, it is clear now that electrochemistry principles are essential in the design and development of stimulating electrodes. Appreciation of the surface electrochemistry and utilization of basic electrochemical methods in probe fabrication is paramount. It will no doubt advance the early initiation of treatment with this novel modality in PD patients and in other diseases as well.

2.3 Applications of Electrochemistry in the Development of Biosensors

Biosensors are used in a variety of fields. It extends from medicine for the detection of various biochemical substances to the detection of drugs and explosives in the framework of homeland security. Modern biosensors are fondly called today "electronic noses" reflecting their capabilities to detect the fingerprint of substances that are undetectable by the human senses. Extensive research is conducted to develop these sensors that will be able to detect the presence of a disease or a pathologic process in the human body as early as possible, and presumably to assist in preventing the disease altogether. Biosensors are based on various factors, including electrochemical, photometric, and piezoelectric ones. This section will review the medical evolution of biosensors and the electrochemical principles utilized in their development with glucose sensor as a market defining prototype.

Diabetes is a worldwide growing epidemic. Two-hundred and eighty-five million diabetes patients are currently diagnosed in the world, with an alarming growth rate of additional seven million new patients are diagnosed every year. Diabetes is a state of increased blood glucose (Sugar) levels known as hyperglycemia. Insulin, a peptide secreted from specific cells (Langerhans cells) in the pancreas, is the hormone which regulates normal physiologic glucose levels in the blood. Two classic types of diabetes are known. Type 1 (also known as juvenile diabetes) is commonly diagnosed early in puberty and featured in decrease production and secretion of insulin from the pancreas, due to autoimmune destruction of the Langerhans cells. Type 2 diabetes commonly develops later in adulthood and related to the metabolic syndrome (dyslipidemia, obesity, and insulin resistance). Diabetes is believed to have a multi-factorial pathogenesis, presumably involving numerous genetic factors. Both types of diabetes share the feature of hyperglycemia. Hyperglycemia, via various oxidative-based mechanisms, results in detrimental cardiovascular complications including myocardial infarction (MI), stroke, and slowly debilitating microvascular complications (e.g., retinopathy and nephropathy). Cardiovascular complications lead to 80% of death cases in diabetic patients and to a significant health expenditure of up to 400 billion US$ in the USA and Europe (combined). Tight glucose control not only prevents imminent short-term complications such as hypoglycemic coma, but also slows down the progression of all types of diabetic complications. Hence, adequately and accurately monitoring glucose levels is essential in order to allow the adjustment of treatment [12].

The basis for glucose measurement is glucose electro-oxidation [13]. There are two main families of enzymes which are used most widely for the electro-oxidation of glucose. These are the glucose oxidases (GOx) and the PQQ-glucose dehydrogenases (PQQ-GDH). Though originally derived from yeast, these enzymes are replaced today by engineered forms in order to increase enzyme yield, facilitate enzyme purification, improve the enzyme stability, increase specific activity, and enhance selectivity for glucose [14] (see also Chap. 5).

The dependence of platinum electrode potential on O_2 partial pressure as well as the electro-oxidation of H_2O_2 on platinum was known since 1939. This, in fact, was the basis for designing GOx-based electrodes that monitored glucose concentration potentiometrically by measurement of the decrease in O_2 partial pressure upon its consumption in the reaction. Glucose concentration could also be measured amperometrically, by the electro-oxidation of the H_2O_2 which is produced. Two decades after these electrodes were first built, an electrochemical glucose assay was described by Malmstadt and Pardue [15, 16]. Updike and Hicks simplified the electrochemical glucose assay by immobilizing and thereby also stabilizing GOx [17, 18]. In order to allow for home glucose monitoring devices, the redox potential had to be increased. This was done by introducing redox mediators into the reaction. These mediators were both organic (e.g., quinoid dyes, quinones), inorganic (e.g., hexacyano-complexes of iron, cobalt, and ruthenium), and metal-organic (e.g., pentacyanoferrate (III) with pyridine, pyrazole, imidazole, histidine) [19, 20]. Most home glucose monitoring devices use the amperometric method, like other biosensors used to measure different organic substances such as lactate (which is clinically important in intensive care medicine and in sport medicine) [21].

Exactech was the first electrochemical home blood glucose monitor, manufactured by Genetics International/Medisense which utilized PQQ-GDH and a ferrocene-derivative as electron-shuttling mediator between both PQQ-GDH and electrodes [22]. The ferrocene-derivative mediator was introduced by researchers at the University of Oxford and the strip development was contracted partially out to Cranfield University. Due to the advantage of the electrochemical assays, the other manufacturers also began to develop electrochemical assays, all of which used enzyme-catalyzed oxidation of glucose, reduction of an oxidized redox mediator by the glucose-reduced enzyme, and electro-oxidation of the resulting reduced mediator. The glucose concentration was obtained by monitoring the steady-state or decay of the current associated with reduced mediator electro-oxidation.

Coulometry is not widely employed in medicine, but rather in different applications such as the analysis of very low oxygen concentration within a gas mix. Its use in glucose monitoring, however, was first proposed in 1996 and today it is in the basis of the freestyle home glucose monitor [14]. Coulometry has the advantage over the amperometric and chronoamperometric methods for several reasons. First, in the coulometric assay, the measured electrical signal varies more linearly with the glucose concentration than it does in amperometry under the constraints of the cell volume. Second, the outcome of coulometry is only modestly affected by variables like temperature, glucose electro-oxidation-catalyzing enzyme activity, blood viscosity, etc. The small interelectrode separation together with the fast redox mediator/enzyme chemistry made it possible to complete coulometric oxidation of glucose in a very short time.

Electron conducting redox hydrogels serve to electrically connect the redox centers of enzymes to electrodes. This way they enable their use in cases when leaching of electron-shuttling diffusional redox mediators must be avoided. An example of such a case is glucose level monitoring electrodes which are implanted in the subcutaneous tissue of diabetic patients for continuous glucose levels monitoring. One example for a continuous glucose monitor is the FreeStyle Navigator of Abbott Diabetes Care of Alameda, CA. This sensor measures and transmits the glucose level readings in the extra cellular fluid to a PDA-like device about once a minute. The core component of this device is a disposable miniature amperometric glucose sensor which is implanted in the subcutaneous tissue and comprises redox hydrogel-wired GOx. [23].

The home monitoring of biochemical substances is a major advantage in the management of chronic diseases. Similar to glucose control in diabetes, the need for chronic anti-coagulant treatment in the management of medical disorders that involve hyper-coagulability states requires close monitoring. Home monitoring for coagulation status (measurement of clot formation time) is available in several devices utilizing different analysis methods, two of which are based on electrochemical principles. The CoaguCheck of Roche Diagnostics involves paramagnetic iron oxide particles which are mixed with thromboplastin reagent (that initiates the clotting cascade in reaction to a drop of blood) and the movement of the particles in response to an electromagnet which is restricted as a clot forms. Another device, the INRatio Prothrombin Time Monitoring System, detects an impedance difference between electrodes during clot formation [24].

Medical practices develop nowadays more and more towards a world of minimal invasive procedures. From advanced imaging techniques and noninvasive treatments using Ultra Sound to destroy tumors and radiation as a replacement for the surgeon's blade, to minimal invasive procedures using surgery robotics. Therefore, it is only natural that the field of diagnostics and bio-sensing is also becoming minimal or noninvasive in the future. Utilization of electrochemical principals in medical biosensors will continue to be fundamental for their development.

2.4 Applications of Electrochemistry in the Development of Implanted Biomechanical Support Devices

As we have shown in the present chapter, a true appreciation of surface electrochemistry and utilization and development of electrochemical methods are imperative for the design of biocompatible medical devices. Where the previous sections dealt with devices that support a failing biologic system by working at the cell level to replace or augment its physiologic capacity, this section will focus on a large group of medical devices whose purpose, as was originally defined, is to reconstruct and support the natural form of a biologic tissue or an organ. This group of devices has aided the health of many millions of patients to date. The vascular stent is a paramount supplement to the armamentarium against cardiovascular diseases (CVD) and its use has become a common practice in western medicine. In addition, orthopedic implants have greatly evolved over the past decades and it is estimated that, in the near future, several millions worldwide will need orthopedic fixation implants and some millions will have prosthetic joints. These seemingly two distinct categories of medical devises do share a few common features bringing us to discuss them both under the scope of this section.

One of the ultimate goals of orthopedic implants is an optimal integration of the implant in the surrounding native environment. This is hindered by several factors. Firstly, by its nature, an orthopedic implant is subjected to and reacts to constant stresses and forces varied in direction and/or amplitude. Shear forces and micro-movements result in progressive dislocation of the implant; moreover, stress and poor mechanical stability lead to abrasion of the thin passivating film that protects the implant from the formation of corrosion (see below) and this further contributes to implant dislocation. Secondly, any foreign body placed in a living tissue initiates a unique inflammatory reaction which assumed goal is to isolate it from the native environment. In electric stimulation devices this consequently impacts the electrode-tissue impedance posing a challenge to research and development in this field. In orthopedic implants the foreign body type inflammatory reaction at the vicinity of the implant results in a weak encapsulating layer that inhibits direct contact between the implant and the growing bone tissue in its bed. This directly affects the implant's mechanical support capacity. Thirdly, as post mortem analysis and revision operations have shown, corrosion is a common phenomenon in orthopedic implants especially multi-module implants and in particular artificial joints. Galvanic corrosion usually develops when two different metals come in contact in an electrolytic solution. Electrons migration from the active metal to the noble one results in corrosion of the active metal (anodization). This process is likely to start when two modular parts of an orthopedic implant come in contact or when stainless steel wires come in contact with other metallic part of the implant, even if the alloys are very similar in composition. A unique phenomenon called crevice corrosion is a corrosive attack on a single metal implant (e.g., stainless steel screw or plates) where a crevice electrolytic environment becomes distinct from the electrolyte outside the crevice resulting in increased chemical potential difference which promotes corrosion at the crevice. This phenomenon is seen in a substantial portion of internal fixation devices [25]. Implant corrosion leads to bone loss, systemic and local toxicity, and implant fracture. Metal fragments of a corroding implant increase the corrosive surface area and distant precipitates may ignite a complex inflammatory response that impacts the adjacent tissue and remote structures. In several reported cases this impact was described as sharing the nature of malignancy [26]. Taken together these put obstacles on the road to a major breakthrough in the orthopedic implants arena. It is preventing the discovery and utilization of innovative methods to fix and integrate an implant in its positioned environment.

Nitinol (NiTi), a nearly equiatomic alloy of Ni and Ti, has established its place of honor in medicine due to its shape memory quality. NiTi-fabricated vascular stents have revolutionized interventive medicine in the field of cardiology and significantly contributed to medical treatments in other fields as well. In recent years, robust clinical experience has forced extensive research on the long-term behavior of NiTi devices under physiologic conditions. NiTi corrosion resistance greatly depends on the properties of its self-passivating oxide film (capacity that decreases after abrupt damage). Electrochemical methods along with other surface modification techniques are used in many studies in an attempt in improving NiTi corrosion resistance, decrease Ni release, and to abolish the unwanted biologic responses such as thrombus formation and smooth muscle cell proliferation that eventually present as restenosis and failure of the stent. [27].

Nobel metals such as gold and platinum do not tend to corrode under normal physiologic conditions due to their positive equilibrium potential. However, some metals that are more often used in medical implanted devices with a negative equilibrium potential (e.g., Titanium) have a higher driving force for oxidation and therefore must be passivated prior to implantation. Passivation is commonly achieved through cladding the desired metal with a thin oxide film that serves as a kinetic barrier to electron migration from the metal to the environment. Ti alloys were found to have the desired strength and biocompatibility in long-term implanted medical devices [28]. One of the unique features of Ti is that in oxygen-rich environment self-passivation occurs through the formation of the corrosion-resistant Titanium-oxide layer (TiO_2). This feature of Ti is affected by the electrochemical conditions in the implant's bed in a manner that attenuates TiO_2 aging, but may also hasten Ti dissolution into the environment [29]. Nevertheless, this may necessitate the preimplantation coating of Ti and Ti alloy devices with another component to maintain biocompatibility and to promote bio-integration.

Contemporary developments in nano-technology have offered a practical approach to a thorough analysis of the tissue-implant interface. The enormous significance of the implant's interface phase was long ago recognized. It has been tackled from different angles resulting in an impressive spectrum of coatings and fabrication techniques designed to meet specific requirements. The various techniques include the passive metal-oxide layer and coating layers co-deposited with specific biomolecules to enhance tissue incorporation of the implant. Recently, the application of the revolutionary concept of pharmaco-active layers of implants has accelerated [30]. Undoubtedly, electrochemistry plays a key role in the development of coated biological implants.

In the literature there are identified two fundamental principles which dictate the design and development of implants' coating. The first principle is to cover the entire surface area of the implant to minimize corrosion and the second is to promote cell growth and incorporation of the implant in its site. The former target is difficult to achieve with standard methods such as the commonly used plasma spray coating of Hydroxyapetite (HA), a calcium–phosphate (Ca–P) crystal ubiquitous in bones. However, other methods like biomimetic nucleation and electrodeposition coating of HA do allow a complete coverage of the implant. This is important to the biocompatibility and durability of an implant. A remarkable character of biomimetic nucleation of HA is the great porosity and unique geometry of the coating. The increased surface area of the coating allows a better interaction between the reforming bone and the implant. HA coating was also studied in association with vascular stents in light of its biocompatibility properties and drug delivery capacity [31]. In addition, when HA coating is applied to a shape memory alloy based on NiTi, it preserves its integrity after changes in temperature and shape of the coated plate. These changes resemble the process of opening a stent within the lumen of a blood vessel [32]. The effectiveness of biomimetic nucleation techniques is enhanced when the surface to be plated is activated by electrochemical methods [33]. Shirkhanzadeh [34] described the formation of HA by electrochemical deposition and others have described modification of this method aimed to control the geometry and properties of the coating [35]. The advantage achieved by this method is a leveled coating of the desired metal which is derived by a negative feedback of current density from the forming HA coating that shifts deposition to bare areas of the implant.

A large body of evidence has clearly shown that osteogenesis is not merely a simple mineralization process, but rather a complex interaction between cellular components and noncellular components of the boney milieu [36]. An outstanding strategy in the field of coatings deals with the idea of incorporating biologic molecules into the coating of a metal implant. Research has focused on several candidate molecules that are apparently essential to osteoinduction.

The combination of these molecules with a porous and stress-proof HA coating may enhance cell interaction with the implants' surface and promote bone growth into it. Collagen is a large protein that exists in a few biologic forms and orders in different tissues. It is ubiquitous in bones and its involvement in bone mineralization and formation is well recognized and described [36]. In recent years the added value of collagen to HA coatings of metal implants has been thoroughly investigated. Ti bodies were coated by different methods, including biomimetic nucleation and spin coating assays. Repeatedly, collagen containing coatings were found to empower cellular response and bone-implant interaction. [37]. Electrochemical methods may further potentiate biomimetic coating techniques and therefore will play a role in future development of biological coating of medical implants [33].

The next stage in the evolution of implants was designing devices that were able to actively modify tissue remodeling by eluting certain pharmaceutical agents pre-loaded on their surface. Such medical devices, mostly familiar from the vascular stent arena, are classified by the Food and Drug Administration (FDA) as "combination devices." These devices offer the advantage of judiciously administering therapeutic drug concentrations alongside with avoiding its systemic toxicity. The drug-eluting stent is an excellent example for the above described paradigm. Its engineering was a multidisciplinary attempt to resolve the most dreaded drawback of bare-metal stents, that is, in-stent restenosis. Rapidly growing understanding of the mechanisms underlying restenosis in bare-metal stents has led to the emergence of drug-eluting stents. The therapeutic agent arsenal used in stents was originally inspired by the similarity to malignant processes. Later, this has encompassed other pharmacologic groups as well after the publication of long-term outcomes in clinical trials [38].

A large range of coatings and drug loading techniques were previously reported. However, electrochemical methods seem to raise the bar for the development of this group of devices. In their paper from 2003, Gertner and Schlesinger described the superiority of an electroless deposition of Ni solution containing pharmaceuticals on different metallic substrates. This novel assay enables control over the quantities of the therapeutic agent stored on the device and its rate of delivery [39]. Electrolytic deposition of bisphophonate, a widely used osteoclastic inhibitor, onto a Ti plate was also characterized. Using this method, the researchers achieved a continuous coating with a prolonged and quantifiable drug elution which contributed to successful integration of the implant [40]. By contrast with methods based on co-deposition of the drug and the coating, porous coatings that are first formed and only subsequently loaded with the pharmaceutical agent still offer a measurable drug concentration and controlled elution and are now used in various medical applications. Electrochemistry is utilized here to create nanopores on aluminum coatings, and for self-ordering of Ti nanotubes [41]. The unique geometry of these surface coatings is apt for efficient drug loading and it has also decreased fatigue strength. One significant value of these coating methods is the ability to easily determine the concentration of a drug and its elution kinetics by changing the diameter, density, and length of the pores (or tubes) during the primary coating or by secondary modification techniques [42]. As mentioned earlier, this quality has a crucial effect on the clinical outcomes with these devices.

However still debated, the idea of multifunctioning implanted devices (i.e., passively support a tissue and actively promote its repair) is fascinating and promising. The advantages of such devices, some of which are discussed in this section, are obvious and thus lead to a rapidly growing research in this field. Notably, the importance of biocompatibility is underscored in medical devices that ought to stay in the body for many years. For this matter, the use of degradable materials also poses challenges and may define new frontiers in medical treatment in the vascular arena and possibly in the field of orthopedics [43]. Furthermore, the emerging technologies of biologic scaffolds in tissue remodeling and stem cell targeting may be used in the future with implanted medical devices.

2.5 Future Prospective

A fairly logical conclusion that may be deduced thus far is that the progression and evolution of the technologies reviewed here (and many other technologies) is aimed at minimizing tissue damage and increasing lifetime span of the device and its intended functioning. Naturally, all the above mentioned necessitate the embracement of some principles in the process of designing a medical device such as multi-functionality, biocompatibility, and durability. Reflected in all of the above is the fundamental understanding of surface electrochemistry as we sought to illuminate in this chapter.

The Artificial Pancreas, a Juvenile Diabetes Research Fund (JDRF) project, represents a new frontier in medical device development. It brings together diabetes researchers, mathematicians, and engineers all aimed at the development of a device that will combine the continuous glucose sensor and the insulin delivering pump, which will be fully automated and auto-regulated. Closed circle neurostimulator and sensor devices that are capable of self-adjusting are already under development and bring closer the day in which nerve reconstruction and function rehabilitation will make its giant leap. Moreover, fascinating breakthroughs in engineered scaffolds and stem cell research lead the way to sophisticated treatments that are applied in a tissue-specific and timely manner. In this way, a small scale implant would deliver several therapeutic agents (whether a drug, a cellular component, or another mediator) to the desired tissue and allow their synchronized activation.

The authors wish to state that though not discussed under the scope of this chapter, the significant field of electrochemical power sources is not overlooked. Electrochemistry exerts a profound effect on the development of electrical devices in medicine and in particular on the field of cardiac pacemakers. Fuel cells make an intriguing area for research and its application in the medical environment will bring far-reaching change to future development of medical devices.

2.6 Summary

Medicine encounters new and complex pathological entities as lifespan continuous to rise in developed countries. The horrific consequences of heart diseases, diabetes, and osteoporosis on human life expectancy and quality are so inherent in people's mind today it is nearly unbelievable that few decades ago their impact on public health had been merely noticeable. Hopefully, this chapter has enlightened the reader about the infinite possibilities that still exist in the pursuit of medical therapy and human health.

References

1. Benabid AL, Pollak P, Louveau A, Henry S, de Rougemont J. Combined (thalamotomy and stimulation) stereotactic surgery of the VIM thalamic nucleus for bilateral Parkinson disease. Appl Neurophysiol. 1987;50:344–6.
2. Weaver FM, Follett K, Stern M, Hur K, Harris C, Jr Marks WJ, et al. Bilateral deep brain stimulation vs. best medical therapy for patients with advanced Parkinson disease: a randomized controlled trial. JAMA. 2009;301:63–73.
3. Kleiner-Fisman G, Herzog J, Fisman DN, Tamma F, Lyons KE, Pahwa R, et al. Subthalamic nucleus deep brain stimulation: summary and meta-analysis of outcomes. Mov Disord. 2006;21 Suppl 14:S290–304.
4. Geddes LA, Roeder R. Criteria for the selection of materials for implanted electrodes. Ann Biomed Eng. 2003;31:879–90.
5. Stokes KB, Bornzin GA, Weabusch WA. A steroid-electing, low-threshold, low polarizing electrode. In: Steinkoff D, editor. Cardiac pacing. Darnstadt: Steinkoff; 1983. p. 369.
6. Schwan HP. Determination of biological impedances. In: Nastuk WL, editor. Physical techniques in biological research. New York: Academic; 1963.
7. Ge Q, Yaxiong L, Hongzhong L, Yucheng D, Xiaping Q, Rukun D. Fabrication of bio-microelectrodes for deep-brain stimulation using microfabrication and electroplating process. Microsyst Technol. 2009;15:933–9.
8. Motta PS, Judy JW. Multielectrode microprobes for deep-brain stimulation fabricated with a customizable 3-D electroplating process. IEEE Trans Biomed Eng. 2005;52(5):923.
9. Cui X, Lee VA, Raphael Y, Wiler JA, Hetke JF, Anderson DJ, et al. Surface modification of neural recording electrodes with conducting polymer/biomolecule blends. J Biomed Mater Res. 2001;56:261–72.
10. Lee KH, Blaha CD, Garris PA, Mohseni P, Horne AE, Bennet KE, et al. Evolution of deep brain stimulation: human electrometer and smart devices supporting the next generation of therapy. Neuromodulation. 2009;12(2):85–103.
11. Perea G, Araque A. GLIA modulates synaptic transmission. Brain Res Rev. 2010;63:93–102.
12. King H, Aubert RE, Herman WH. Global burden of diabetes, 1995–2025: prevalence, numerical estimates, and projections. Diabetes Care. 1998;21:1414–31.
13. Iswantini D, Kano K, Ikeda T. Kinetics and thermodynamics of activation of quinoprotein glucose dehydrogenase apoenzyme in vivo and catalytic activity of the activated enzyme in Escherichia coli cells. Biochem J. 2000;350(Pt 3):917–23.
14. Heller A, Feldman B. Electrochemical glucose sensors and their applications in diabetes management. Chem Rev. 2008;108:2482–505.
15. Malmstadt HV, Pardue HL. Specific enzymatic determination of glucose in blood serum or plasma by an automatic potentiometric reaction-rate method. Clin Chem. 1962;8:606–15.
16. Pardue HL, Simon RK. Automatic amperometric assay of glucose oxidase. Anal Biochem. 1964;9:204–10.
17. Updike SJ, Hicks GP. The enzyme electrode. Nature. 1967;214:986–8.

18. Updike SJ, Hicks GP. Reagentless substrate analysis with immobilized enzymes. Science. 1967;158:270–2.
19. Kulys J, Tetianec L, Ziemys AJ. Probing Aspergillus niger glucose oxidase with pentacyanoferrate (III) aza- and thia-complexes. Inorg Biochem. 2006;100:1614–22.
20. Kulys JJ, Samalius AS, Svirmickas GJ. Electron exchange between the enzyme active center and organic metal. FEBS Lett. 1980;114:7–10.
21. Nikolaus N, Strehlitz B. Amperometric lactate biosensors and their application in (sports) medicine, or life quality and wellbeing. Microchim Acta. 2007;160(1–2):15–55.
22. Matthews DR, Holman RR, Bown E, Steemson EJ, Watson A, Hughes S, et al. Pen-sized digital 30-second blood glucose meter. Lancet. 1987;1:778–9.
23. Heller A. Integrated medical feedback systems for drug delivery. AIChE J. 2005;51:1054–66.
24. Yang DT, Robetorye RS, Rodgers GM. Home prothrombin time monitoring: a literature analysis. Am J Hematol. 2004;77:177–86.
25. Cook SD, Thomas KA, Harding AE, Thomas KA, Harding AE, Collins CL, et al. The in vivo performance of 250 internal fixation devices: a follow-up study. Biomaterials. 1987;8:177–84.
26. Jacobs JJ, Gilbert JL, Urban RM. Current concepts review—corrosion of metal orthopaedic implants. J Bone Joint Surg Am. 1998;80:268–82.
27. Shabalovskaya S, Anderegg J, Van Humbeeck J. Critical overview of Nitinol surfaces and their modifications for medical applications. Acta Biomater. 2008;4:447–67.
28. Long M, Rack HJ. Titanium alloys in total joint replacement—a materials science perspective. Biomaterials. 1998;18:1621–39.
29. Alkhateeb E, Virtanen S. Influence of surface self-modification in Ringer's solution on the passive behavior of titanium. J Biomed Mater Res A. 2005;4:934–40.
30. Duan K, Wang R. Surface modifications of bone implants through wet chemistry. J Mater Chem. 2006;16:2309–21.
31. Rajter A, Kaluza GL, Yang Q, Hakimi D, Liu D, Tsui M, et al. Hydroxyapatite-coated cardiovascular stents. Eurointervention. 2006;2:113–5.
32. Choi J, Bogdanski D, Köller M, Esenwein SA, Müller D, Muhr G, et al. Calcium phosphate coating of nickel–titanium shape-memory alloys. Coating procedure and adherence of leukocytes and platelets. Biomaterials. 2003;24:3689–96.
33. Zhang Q, Leng Y. Electrochemical activation of titanium for biomimetic coating of calcium phosphate. Biomaterials. 2005;26:3853–9.
34. Shirkhanzadeh M. Direct formation of nanophase hydroxyapatite on cathodically polarized electrodes. J Mater Sci Mater Med. 1998;9:67–72.
35. Lin S, LeGeros RZ, LeGeros JP. Adherent octacalciumphosphate coating on titanium alloy using modulated electrochemical deposition method. J Biomed Mater Res. 2003;66A(4):819–28.
36. Palmer LC, Newcomb CJ, Kaltz SR, Spoerke ED, Stupp SI. Biomimetic systems for hydroxyapatite mineralization inspired by bone and enamel. Chem Rev. 2008;108:4754–83.

37. Shu-Hua T, Eun-Jung L, Chee-Sung P, Won-Young C, Du-Sik S, Hyoun-Ee K. Bioactive nanocomposite coatings of collagen/hydroxyapatite on titanium substrates. J Mater Sci Mater Med. 2008;19:2453–61.
38. Fattori R, Piva T. Drug-eluting stents in vascular intervention. Lancet. 2003;361(9353):247–9.
39. Gertner ME, Schlesinger M. Drug delivery from electrochemically deposited thin metal films. Electrochem Solid-State Lett. 2003;6(4):4–6.
40. Duan K, Fan Y, Wang R. Electrolytic deposition of calcium etidronate drug coating on titanium substrate. J Biomed Mater Res B Appl Biomater. 2005;72:43–51.
41. Ghicov A, Tsuchiya H, Macak JM, Schmuki P. Titanium oxide nanotubes prepared in phosphate electrolytes. Electrochem Commun. 2005;7:505–9.
42. Gultepe E, Nagesha D, Sridhar S, Amiji M. Nanoporous inorganic membranes or coatings for sustained drug delivery in implantable devices. Adv Drug Deliv Rev. 2010;62:305–15.
43. Zilberman M, Eberhart RC. Drug-eluting bioresorbable stents for various applications. Annu Rev Biomed Eng. 2006;8:153–80.

Chapter 3
Intracoronary Stents: Medical Devices at the Interface of Biology and Electrochemistry

Michael B. Boffa and Marlys L. Koschinsky

3.1 Introduction

Despite the many significant advances in the diagnosis and treatment of cardiovascular diseases that have occurred over the past three decades, these disorders remain a primary cause of both morbidity and mortality worldwide. In fact, ischaemic heart disease (essentially, heart attacks and their sequelae) are the single leading cause of death, accounting for almost 13% of deaths worldwide [1]. Since their introduction nearly 20 years ago, intracoronary stents have become an indispensable tool in the arsenal of interventional cardiologists for the prevention and treatment of ischaemic heart disease. These devices, generally consisting of a tubular metal scaffold with some coating to ensure biocompatibility and reduce complications (Fig. 3.1), serve to restore patency to occluded arteries and to stave off their reocclusion. What is fascinating about stents is that they stand at the interface of the fields of engineering, chemistry (particularly materials chemistry and electrochemistry), biology, and medicine. In this chapter, we will review some of the disease biology of atherosclerosis—the pathological process that ultimately gives rise to ischaemic heart disease—before reviewing the history

M.B. Boffa (✉) • M.L. Koschinsky
Department of Chemistry and Biochemistry, University of Windsor,
Windsor, ON N9B 3P4, Canada
e-mail: mboffa@uwindsor.ca

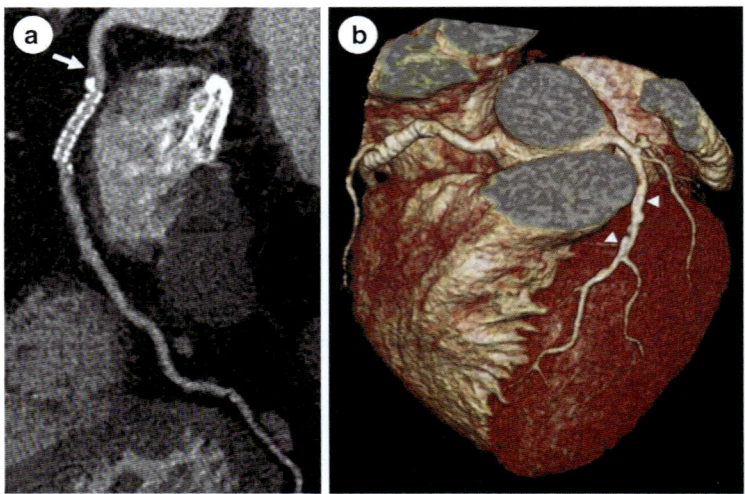

Fig. 3.1 Coronary computed tomography (CT) angiogram in a subject with previous stent implantation in the right coronary artery. The acquisition technique utilized involved prospective electrocardiogram-triggering; a curved multiplanar reformation of the CT data is shown in (**a**) and a volume rendered image is shown in (**b**). Stent patency is observed, along with mixed plaque proximal to the stent causing mild coronary stenosis (*arrow*). Calcified plaques are also observable in the left anterior descending coronary artery (*arrowheads*). Reprinted from Semin Thorac Cardiovasc Surg, Vol 20, Bastarrika G and Schoepf UJ, Evolving CT applications in ischemic heart disease, pp. 380.e1–380.e14, Copyright 2008, with permission from Elsevier

of stent development and some of the common features of these devices. The heart of this chapter, however, will focus on the ongoing research in the area of electrochemistry that is aimed at optimizing the properties of intracoronary stents.

3.2 The Process of Atherosclerosis

Cardiovascular diseases arise primarily from the process of atherosclerosis in the arterial wall [2]. Only certain arteries are susceptible—principally the large muscular arteries—and the disease is most prevalent in areas of disturbed or turbulent blood flow. Lesions in the diseased arteries lead to significant narrowing of the artery wall which impedes blood flow. An even more clinically catastrophic consequence of the atherosclerotic lesions is the sudden formation of an occlusive thrombus: a blood clot that blocks the artery. In coronary arteries, this results in myocardial infarction (heart attack), and in the carotid or cerebral arteries this results in ischaemic stroke. Intracoronary stents come into the clinical picture after a myocardial infarction or after vessel occlusion is detected by visualization of the coronary circulation using a radiological procedure called an angiogram. In the ensuing paragraphs we provide an overview of the pathophysiology that underlies atherosclerosis, an understanding of which is necessary to appreciate the various challenges involved in stent design and manufacture.

Arteries contain distinct concentric cellular layers that underlie their function (Fig. 3.2). A single layer of endothelial cells lines the vessel and is in contact with the flowing blood. Under the endothelial cells is found the intima layer (which normally contains elastic fibers and some connective tissue), the media (which contains primarily contractile smooth muscle cells, SMCs), and the adventitial layer which contains elastic and connective tissue. In a non-diseased artery the intima occupies a very small proportion of the arterial cross-section (Fig. 3.2).

The steps in the development of atherosclerotic lesions have been well-characterized (Fig. 3.2). The most accepted hypothesis for the initiation of atherosclerosis is the so-called response to injury [2]. In response to various injurious stimuli including turbulent (non-laminar) blood flow [3] or oxidative stress from factors such as smoking, endothelial cells that line the arteries initially become altered in their function [4–6]. They develop a phenotype that promotes blood clot formation, attracts circulating immune and inflammatory cells, and exhibits leakiness of the normally non-permeable intact endothelial cell barrier that separates the flowing blood from the interior of the vessel. As a consequence, members of a class of white blood cells called monocytes adhere to the endothelium and penetrate into the sub-endothelial space [2]. Here, they differentiate into macrophages which under normal circumstances are cells that destroy and ingest microbial invaders or cancer cells. In an artery wall, however, these cells voraciously accumulate lipids from sources such as oxidized LDL to form enormous fat-laden bodies called foam cells [7, 8]. The foam cells promote the migration and proliferation of SMCs from the medial layer of the artery to the intimal layer and result in continued monocyte recruitment to the endothelial cells [2, 7, 8]. Within the

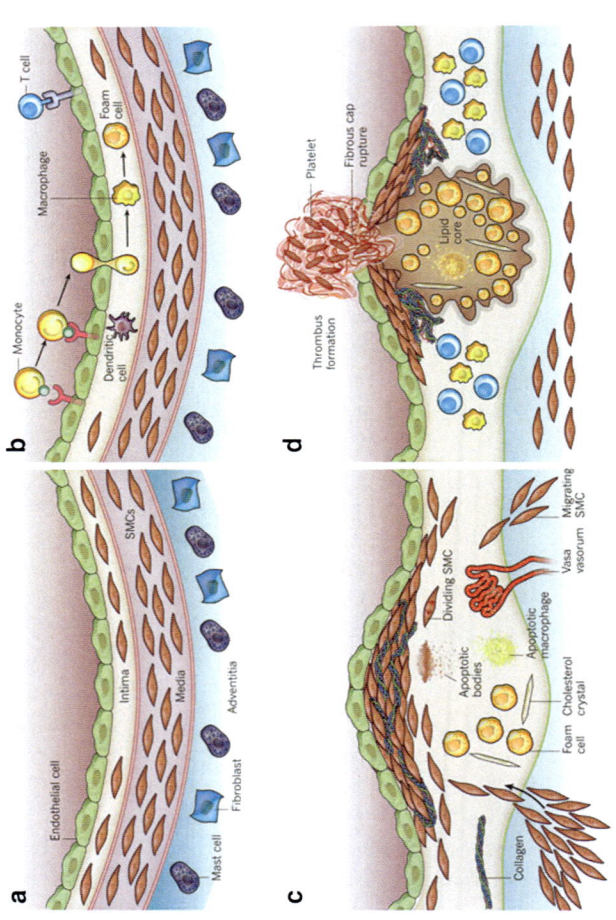

Fig. 3.2 The development of atherosclerosis. Schematic diagram showing the progressive stages of atherosclerosis: normal, healthy artery (**a**); early lesion development leading to the formation of fatty streaks (**b**); complex lesion formation (**c**); and plaque rupture leading to thrombosis (**d**). Reprinted by permission from Macmillan Publishers Ltd [Nature]: Libby P, Ridker PM, Hansson GK. Progress and challenges in translating the biology of atherosclerosis. Nature 2011;473:317–25. Copyright 2011

intima, the SMCs change in their properties to assume a non-contractile, proliferative, and extracellular matrix-synthesizing phenotype, thereby initiating the process of intimal thickening which is a key step in the formation of an atherosclerotic lesion (Fig. 3.2). As a consequence of intimal thickening, the lumens of arteries become narrowed, thereby compromising arterial blood flow to critical organs. A mature atherosclerotic plaque consists of a necrotic core of foam cells and fatty material released from dead foam cells, varying extents of intimal thickening due to SMC proliferation and synthesis of extracellular matrix, and a fibrous cap of SMCs and matrix that separates the necrotic core from the flowing blood (Fig. 3.2) [2].

As mentioned above, the developing atherosclerotic lesion is at its most dangerous when it promotes the formation of blood clots or occlusive thrombi within the vessel. These clots may be transient, giving rise to the episodic and unprovoked chest pain of unstable angina, or they may result in more stable clots that occlude the vessel completely and cause a heart attack. These clotting events are precipitated by rupture of the plaque [9]. This occurs when the fibrous cap is somehow compromised and the flowing blood is exposed to the fatty material in the necrotic core of the plaque. This fatty material is very potent at stimulating blood clotting [10]. A relatively recent concept in the characterization of atherosclerotic lesions is assessment of the extent to which the plaques are vulnerable to rupture [9]. Vulnerable plaques tend to have a thin fibrous cap covering the lesion as well as a high lipid and macrophage content. Conversely, more stable plaques, which in fact can often result in a greater extent of narrowing of the artery than vulnerable plaques, have a smaller necrotic core and a thicker fibrous cap. In theory, identifying individuals having vulnerable plaques could lead to interventions that would prevent a large number of heart attacks. In reality, it is very difficult if not impractical to identify these individuals in the population. No blood test exists to identify them. Moreover, the anatomy of these plaques allows them to escape detection by traditional angiographic techniques for plaque detection.

Intracoronary stents are prescribed to patients after a heart attack, if regions of compromised flow are found to have persisted, or upon angiographic examination of patients referred for angina symptoms. As we discuss below, the development of these devices has been a major leap forward in the management of acute coronary syndromes.

3.3 Development of Mechanical Treatments for Occlusive Atheroslcerotic Lesions: The Rise of Interventional Cardiology

Percutaneous transluminal coronary angioplasty (PTCA) is a minimally invasive procedure designed to open up blocked coronary arteries, thereby allowing blood to circulate unimpeded to the heart muscle [11]. It is a considerably less invasive procedure than the alternative: coronary artery bypass grafting (CABG). In CABG, a vessel from elsewhere in the patient's body (most commonly the saphenous vein or the mammary artery) is grafted onto the coronary circulation from the aorta (the major artery by which oxygenated blood is pumped from the heart to the rest of the body including the coronary circulation) to the coronary arteries beyond the blockage. CABG requires sawing through the breastbone and up to 90 min of mechanical cardiopulmonary bypass.

The first PTCA or percutaneous coronary intervention (PCI) was performed in 1977 [12]; this marked the birth of the field of interventional cardiology. In this process [11], a diagnostic catheter is inserted into the femoral artery at a site in the groin of the patient and is guided through the aorta to the site where the coronary arteries branch off from this vessel; the guide wire is then removed. Contrast dye is injected and the patient's coronary circulation is visualized by X-ray. If a treatable blockage is noted, the diagnostic catheter is exchanged for a guiding catheter. Once the guiding catheter is in place, a guide wire is advanced across the blockage, and then a balloon catheter is moved to the blockage site. The balloon is inflated for a few seconds to compress the blockage against the artery wall and then deflated. This process may be repeated several times, to maximize the restoration of blood flow, prior to removal of the catheter and completion of the procedure. By the mid 1980s, over 300,000 PTCA procedures were being performed in the USA on a yearly basis which approached the number of cardiac bypass surgeries being performed to treat coronary artery disease [13].

Restenosis (vessel re-narrowing) was noted as one of the key complications following PTCA and was reported to occur in as many as 30–40% of individuals who underwent this procedure [14, 15]. As a consequence, restenosis often necessitated a second revascularization procedure. The complication of restenosis subsequently led to the development of the intracoronary stent device. Stents are mechanical

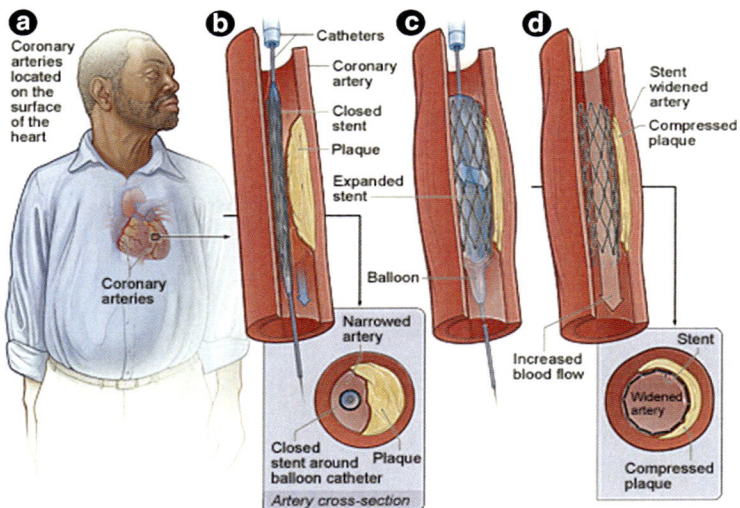

Fig. 3.3 Insertion of a coronary stent. From an insertion site in the groin, the femoral artery is catheterized and a catheter containing a collapsed stent and a balloon is advanced along a guide wire to the lesion causing impeded blood flow. The balloon is then inflated using pressurized saline causing elastic deformation of the stent which expands the artery. The balloon is then deflated and removed, leaving the stent in place. Source: National Heart, Lung and Blood Institute; National Institutes of Health; U.S. Department of Health and Human Services

devices that are approximately 3 mm in diameter and between 20 and 40 mm in length. Fig. 3.3 illustrates the placement of a stent within a coronary artery. A flexible catheter containing a collapsed stent and a balloon is placed along a guide wire to the region of interest in the vessel. The balloon is then filled with saline which generates pressure in the order of 10 atm or greater for up to 3 min. This causes elastic deformation of the stent which expands the artery. The balloon is then deflated and removed, leaving the stent in place. The first intracoronary stents were successfully implanted in 1986 [16, 17] and were self-expanding bare metal devices that reopened narrowed arteries and provided scaffolding to prevent vessel recoil and negative vessel remodeling or vessel shrinkage [18]. The use of intracoronary stents was quickly identified as a successful method to reduce restenosis following PTCA with the reduction of restenosis rates by up to 30% compared with balloon angioplasty [14, 15].

In 1989, an improved device in the form of a balloon-expandable intracoronary stent was developed [19–21]. Initial results with these Palmaz-Schatz stents were promising compared to balloon angioplasty, with a significantly lower incidence of abrupt closure of the vessel and of heart attacks in the short-term following the procedure [22]. Late restenosis rates with Palmaz-Schatz stents were also significantly improved when compared with balloon angioplasty [14, 15].

3.4 The Development of Drug-Eluting Stents

Although PCI using stenting was a considerably less invasive procedure than CABG, the former was not without its complications and, in fact, was associated with a greater need for re-intervention in the longer term [23–26]. The rates of subacute thrombosis and bleeding complications associated with deployment of the Palmaz-Schatz stents remained significant clinical issues [27]. Additionally, despite the widespread increase in the use of stents, neointimal proliferation (a biological vessel response to injury caused by stent implantation) resulting in restenosis persisted as a problem [28]. In other words, the stent itself triggered accelerated restenosis. This, in effect, created a new challenge which was especially prominent in some types of patients (those with diabetes mellitus, e.g.) and in certain types of lesions [29, 30]. These tended to include lesions close to the site of vessel bifurcation or lesions with long areas of stenosis present [31]. This subsequently led to the second phase of stent development: the drug-eluting stent (DES) [28]. Most DES are composed of stainless steel scaffolds that are coated as a means of increasing their biocompatibility as well as for providing the opportunity for local therapeutic delivery of anti-restenosis agents.

The main components of the neointima are SMCs and the collagen matrix that they secrete [2]. As such, in order to reduce the migration and proliferation of SMCs from the medial layer of the vessel, the first two drugs that were studied in the development of DES were the antiproliferative agents sirolimus (rapamycin) and paclitaxel (Taxol) [32, 33], drugs used primarily to that point as cancer chemotherapy.

Cordis (a division of Johnson & Johnson) was responsible for developing the Sirolimus-Cypher stent, a bare metal stainless steel (SS) stent containing a nonbiodegradable polymer coating that releases sirolimus over time [32]. The first study of these individuals receiving this stent during PCTI revealed an impressive lack of restenosis (0%) within the first 6 months [34]. Further trials with the Cypher stent, however, revealed that restenosis did occur in some individuals with high-risk features (such as long areas of stenosis or a history of diabetes mellitus; see above), but that the restenosis rate was significantly lower than with bare metal stents (3% compared to 35%, respectively) [35]. The Cypher stent was approved by the United States Food and Drug Administration (FDA) as the first DES for use in the general population in the USA.

Concurrent with the development of the Cypher stent, Boston Scientific started development of the Taxus stent. The Taxus stent was designed using a bare metal stent with a nonbiodegradable copolymer coating of paclitaxel that inhibits cell division. As with the Cypher stent, the first trials of the Taxus stent revealed no evidence of in-stent restenosis at 6 months after the procedure [36]. While later studies showed some restenosis, the rate was still much lower than the bare metal stent counterpart [37]. By the end of 2004, DES were used in nearly 80% of all PTCI interventions performed [38].

Currently, a second generation of DES has come on the market [39]. These include the Endeavor (Medtronic), a zotarolimus-eluting stent (zotarolimus is a sirolimus analogue) on a cobalt-based platform, and the Xience V (Abbott Laboratories) which is on a cobalt–chromium ally platform and elutes everolimus, a different derivative of sirolimus. The coating of both of these stents, like those the first-generation devices, is not bioabsorbable (see below). The second-generation devices are constructed on a different platform than the previous devices, which were made of stainless steel. The second-generation platforms also have thinner struts for enhanced flexibility and thus easier implantation with less vessel damage. Moreover, the thinner struts are more easily coated by a new layer of endothelial cells (re-endothelialization) while better preventing neointimal formation. The new generation of sirolimus analogues are also purported to have additional beneficial effects through targeting of harmful cells, such as macrophages, in addition to the SMCs that form the bulk of the neointima [28].

The fact that DES must be coated in order to elute drugs has emerged as a significant concern with respect to the safety of the devices. When DES were initially approved, durability and stability of the coating was not thought to be an issue [40]. Furthermore, preclinical studies did not show adverse effects arising from polymeric coating defects including cracking, peeling, and delamination [40]. However, in recent years the US FDA together with a number of published studies have raised several safety issues with DES [40–43] that are not shared with BMS. In this context, the relationship between coating defects and safety concerns has been raised [44, 45]. Indeed, although DES have shown promising results in the clinic, the occurrence of late-stent thrombosis (LST) associated with DES devices has been a focus of concern [46, 47]. Although LST is a rare complication, it is justifiably feared by patients as it can cause sudden death. Two contributing factors to these thrombotic complications have been identified: (1) adverse effects of polymer coatings; and (2) delayed re-endothelialization which leaves surfaces exposed for the formation of blood clots. Delayed re-endothelialization occurs as a consequence of the use of antiproliferative agents to coat stents which inhibit SMC proliferation but also reduce endothelial growth [18]. Areas lacking endothelial cells are susceptible to blood clot formation. It has also been suggested that a component of delayed endothelialization may be related to polymer coatings used in a subset of patients [48].

Clinical studies of the second-generation stents have shown them to offer an improved safety profile [28, 39], with the exception of one large trial [49]. This last study is considered to be more representative of the employment of DES in routine clinical practice, however. In addition, even larger studies with longer period of follow-up are currently ongoing [39]. Therefore, although DES have vastly improved the efficacy of PCI, there remains room for improvement of these devices to enhance their safety profile, particularly over the long term.

A new generation of DES is currently in development and undergoing clinical testing [50]. The array of drugs employed and the means of associating them with the stents continues to expand. The majority of polymeric delivery matrices now being developed are biodegradable/bioabsorbable, which is aimed at circumventing problems with biocompatibility and durability that have arisen with the existing polymer technologies. Once the polymer has disappeared (along with the drug), it is thought that re-endothelialization of the remaining bare metal occurs. Layered series of coatings have been devised to release drug in a series of stages (Supralimus and Supralimus-Core; Sahajanand Medical Technologies Pvt Ltd) [50]. Other innovations include reservoirs within the stent that can be filled with biodegradable polymer containing drug (Nevo; Cordis), and technology to specifically coat the abluminal surface of the stent (i.e. that surface that faces the vessel wall) (JACTAX; Boston Scientific Corporation) [50]. In the latter case, eluted drug passes into the vessel wall and is not released at all into the bloodstream, allowing for delivery of the drug only to where it is required and allowing lower doses to be incorporated into the stent. Another notable innovation, which has yet to be tested clinically, involves coating of the drug/polymer layer with an additional layer containing antibodies against CD34, a protein expressed on the surface of circulating endothelial progenitor cells (Sirolimus+EPC Capture; OrbusNeich Medical, Inc) [50]. It is anticipated that by capturing these cells from the blood, re-endothelialization of the stent will be accelerated.

Some novel DES technologies have eliminated polymer altogether, in theory preventing complications due to damage of the polymer and hastening re-endothelialization [50]. Examples include microspray application of drug directly onto the abluminal surface (Amazonia PAX, MINYASYS), creation using abrasives of a microstructured surface that retains the drug solution (BioFreedom; Biosensors Inc), loading of the drug into groove-like reservoirs on the abluminal surface (OPTIMA; CID S.r.l.), and absorption of the drug into a microporous and bioabsorbable surface layer of hydroxyapatite (which is chemically identical to the main structural component of bone) (VESTAsync; MIV Therapeutics) [50].

3.5 Coating of Drug-Eluting Stents: General Principles

Stents are produced using various metal alloys including cobalt chromium [51] and nitinol [52], although the majority of stents are made of stainless steel (SS) [53, 54]. The surfaces of SS stents are usually coated with an oxide layer (10—50 Å) and hydroxides [55, 56]. The oxide layer is fragile, mechanically unstable, and can be altered with exposure to physical environments [55, 57]. Therefore, methods for the coating of stents must include consideration of biocompatibility and prevention of the leeching of toxic metals into the surrounding tissue which can result in inflammatory reactions *in vivo* [58]. To improve the biocompatibility of DES while reducing thrombogenic (blood clot-promoting) characteristics, hemocompatible coatings have been used which include non-thrombogenic polymers such as heparin and phosphorylcholine-based agents [59–61]. Heparin is an anticoagulant molecule, and phosphorylcholine mimics the major membrane lipid found in red blood cells.

A drug can be incorporated during stent coating either by attaching it to the polymer functional groups or by loading it into the polymeric coating after it is generated. The matrix allows loading of different drugs or agents in a range of concentrations, resulting in diverse physicochemical properties of the DES. Most commercially available stents rely on the use of polymeric coatings containing drugs [18]; the polymers are often hydrophobic in nature to accommodate the properties of drugs used which are often hydrophobic such as paclitaxel [62]. The drug-matrix, which was biostable in the initial devices although biodegradable coatings are becoming predominant in contemporary devices [50], is often applied directly onto BMS using either spray or dip coating [62]. The durability of the drug-in-polymer matrix is critical because the stent is a permanent implant that undergoes mechanical stress during deployment. Unfortunately, the adhesion properties of the matrix to the metal are often poor owing to the sub-optimal interaction between the hydrophilic oxide metal surface and the hydrophobic drug-polymer matrix; cracking, delaminating, and peeling have been reported as a consequence, which may lead to adverse clinical effects [44, 63]. This has led to studies altering the chemistry and morphology of the stent surface to increase adhesion of the drug-matrix coating, which will be described below. The ultimate goal of such efforts is to obtain a simple, reproducible stent coating process to create uniform, thin composites on the stent surface.

Current approaches for the application of coating onto stents include the following [64]: (1) *Dipping*: this is a relatively common and simple technique in which the stent can be dipped into polymer-drug solution. Its principle disadvantage is limited controllability leading to bridging, pooling, and therefore lack of uniformity. (2) *Spraying*: this is another mechanical process which uses a fine droplet beam and is used for fabrication of the majority of DES [62]. This technique can achieve uniform continuous coating, but the thickness of the coat is greater than optimal (~8–15 μM). (3) *Chemical vapor deposition (CVD)*: this method allows uniform coatings of controlled thickness and morphology. CVD methods require environments of high temperature and pressure as well as complex equipment [65]. It is used preferentially where adhesion promotion or low frictional properties are required. (4) *Surface grafting*: in this technique, the stent is incubated in an organic solvent with a solid polymer such as polyethylene glycol, which allows grafting of the polymer onto the stent surface.

3.5.1 Problems with DES Coatings

Several reports have described problems associated with the morphology and mechanical stability of the coatings of commercially available DES [44, 63, 66–68]. In this regard, adhesion and uniformity irregularities of the coatings before and after balloon expansion can lead to thrombosis due to exposure of clot-promoting surfaces on the stent, the breaking off of very small fragments of coating, and late inflammatory reactions and growth of neointima.

Electrochemical deposition of polymers on the surface of stents has been shown to be a relatively simple method for coating of these devices [40]. Electrochemistry has wide applications and is most commonly associated with the process of depositing metallic coatings such as in electroplating. Electrochemical coating of stents requires that the stent surface is conducting. Since stents are primarily made of stainless steel or other metal alloys, it is possible to use their conducting properties to deposit the coating onto the stent [40]. Typically this involves the electrochemical oxidation of a monomer, such as pyrrole or phenol, and results in the deposition of an organic polymer. Most, but not all, electrochemical coatings result in non-covalent bonding between the layer and the metallic surface [40]. It has been shown that the application of a coating to BMS can improve the adhesion of the drug-matrix, as well as the biocompatibility and durability of the resulting stents [64]. Electrochemical deposition forms homogeneous layers, the thickness of which can be controlled [69]. The organic coating on stents can be grafted with functional groups that can improve biocompatibility and modify the mechanical and chemical properties to optimize their function [70].

The remainder of this chapter will examine progress that has been made on the use of electrochemical techniques to coat BMS and the advantages of this methodology for the generation of DES.

3.6 Electrochemical Methods for Improving the Durability and Adhesion of Drug-in-Polymer Coatings of SS Stents

3.6.1 *Surface Modification of BMS Through the Covalent Attachment of a Conducting Organic Layer*

Shaulov and colleagues [71] describe the use of a three-step process for the surface modification of stents to improve their durability through the covalent attachment of organic species. In the first step, 4-(2-bromoethyl)benzendiazonium tetrafluoroborate was electrografted onto a stainless steel stent (BrD DES). X-ray photoelectron spectroscopy (XPS) was used to confirm the formation of the organic layer. In the second step, methyl methacrylate was polymerized onto the grafted surface by atom-transfer radical polymerization (PMMA-BRD DES). The biocompatible polymer brushes were characterized by XPS, electrochemical impedance spectroscopy, and contact-angle measurements. In the last step, the stent was spray-coated with a drug-in-polymer matrix (in this case, PMBA (poly(*n*-butyl methacrylate)/poly(ethylene co-vinyl acetate)) containing paclitaxel). The stent was incubated in physiological buffer (0.1 M phosphate, pH 7.4) for 3 days at 60°C, at which point the improved durability of the matrix was documented using scanning electron microscopy (SEM). Controls in which the drug-in-matrix was coated onto bare metal or BrD electrocoated stents showed increased cracking and delamination of the coating than the 3-step modified stents under these conditions. This study showed the beneficial effects of introducing a thin organic layer as the interface between hydrophilic stainless steel and the hydrophobic drug-in-polymer matrix. The enhanced durability and stability of the drug-containing matrix corresponded to a slightly slower kinetics of drug release *in vitro* for the PMMA-BrD DES compared to BrD or bare metal DES.

A recent publication by Levy and coworkers outlines a novel approach for the increased adhesion of polymer coatings onto DES [40]. The method is based on the electrochemistry of diazonium salts which are organic compounds with the characteristic functional group $R\text{-}N_2^+\, X^-$ where R is an organic residue (e.g., alkyl or aryl), and

X is an anion (e.g., a halide). These substances undergo irreversible reduction whereby the C–N bond is cleaved to generate reactive radical species. These species bind covalently to the metallic surface which increases adhesion and durability of the coating. Diazonium salts are an attractive option for basecoating of stents since they can be tailored (by substitution of different group at the R position) to accommodate a variety of applications including adhesion promoters, active groups for chemical reactions, and protein grafting.

Accordingly, Levy and coworkers described the use of the diazonium salt 4-(1-dodecyloxy)-phenyldiazonium tetrafluoroborate (C_{12}-phenyldiazonium) to basecoat metallic stents [40]. The basecoat, deposited on the metal stent (either stainless steel or CoCr) by electrochemistry, resulted in a surface grafted with aliphatic aryl moieties. This surface was studied by cyclic voltammetry, alternating cyclic voltammetry, and electrochemical impedance spectroscopy. Additionally, to test the adhesive properties of the coated surface, a secondary drug/polymer matrix was applied by dip coating using PEVA:PMBA (poly(ethylene-vinyl acetate):polybutyl methacrylate; 3.7 weight ratio) containing paclitaxel. Following incubation under physiological conditions at 37°C, basecoated stents were examined by SEM and shown to exhibit a much lower percentage of peeling after 30 days, compared to control stents lacking the basecoating. The two stents differed somewhat in their drug release properties, as release in the initial "burst phase" is delayed slightly by several days in the basecoated stents, with significant differences with the control stents in release persisting for as long as 12 days. Both biocompatibility and cytotoxicity were measured *in vitro* using extracts from control or C_{12}-phenyldiazonium-coated stents. Neither hemolysis rates nor cell viability were found to be affected by the electrocoating of the stents. *In vitro* biocompatibility was tested by implantation of electrocoated and control stents into the iliac arteries of rabbits in the context of denudation of the endothelium by balloon injury. No differences were observed between the two groups of stents after 7 days with respect to inflammation, fibrin deposition, stenosis, and endothelialization.

In both the studies of Shaulov et al. and Levy et al. [40, 71], the data strongly suggest that the novel electrochemical coatings hold the promise of enhanced durability and improved drug release characteristics, with the newly developed surfaces otherwise being wholly biocompatible and hemocompatible. The suitability of these coating to medical devices for use in patients requires further preclinical and clinical studies, however. Perhaps more significant is the ability of the diazonium salt coatings to be specifically functionalized to give rise to even better performance *in vivo* as well as novel functional properties. These issues will certainly receive greater attention in the coming years.

The properties of copolymerized heterocyclic compounds, such as polypyrrole, have also been studied. Polypyrrole has particular advantages in terms of relatively low oxidation potential, high stability, and electrical properties [72]. Further, polypyrrole has been shown to the biocompatible and hemocompatible [73, 74]. Controlling the properties of conducting polymers has also been investigated by the synthesis and co-polymerization of substituted monomers. A study by Okner and coworkers [70] examined this approach to alter the properties of polypyrroles, which can be alkylated at their N position. Specifically, two pyrrole derivatives were examined: N-(2-carboxyethyl)pyrrole (PPA) and a butyl ester of PPA (BuOPy). A series of copolymers and respective homopolymers were deposited on stainless steel surfaces using cyclic voltammetry. Changing of the monomer solution ratio and deposition conditions resulted in polymer films with different electrical, structural, and chemical properties that related, in part, to their distinct kinetics of nucleation and growth [70].

In another study by this group [64], PPA and BuOPy were electrodeposited on a stainless steel stent. As they reported previously, the electrocoating process afforded control over the thickness of the surface, with a high level of uniformity as well as defined morphology. Polymer stability was examined *in vitro* by immersing the stent into a solution containing saline and fetal calf serum (in the presence of antibiotics) for up to 1 year. Neither delamination nor morphological changes were observed, leading to the conclusion that the polymer-coated stents were durable and stable. A similar lack of surface changes was observed following implantation of the coated stents subcutaneously in mice for 7 days. There was no evidence of histopathological changes at the site of implant: the stents were surrounded by mature connective tissue with virtually no sign of any inflammatory reaction; these findings thus suggest good biocompatibility of the coated stent [64].

Loading of paclitaxel onto the electrocoated stents was carried out by dipping and the release of the drug was monitored by HPLC [64]. For these experiments, in addition to pBuOPy, a more hydrophobic copolymer of BuOPy and PPA (BuOpy:PPA 9:1) was used for the coating in order to obtain better adhesion to the stent. Drug loading was not affected by addition of the monomer PPA. The drug diffused into the coating and was maintained there due to the rough morphology on the surface. The pBuOPy- and copolymer-coated stents could elute paclitaxel over a 3-week period. Two phases of drug release were reported for both coated stents: a rapid release within the first 24 h and a slow release for up to 3 weeks during which time all of the drug was eluted; a top coating serving as a diffusion barrier for the drug can be applied to extend the time for total drug release [75]. BMS, on the other hand, showed only very rapid release (within a few days). The polymer coating was unaffected by the drug addition and release with no changes in surface morphology observed [64].

Most recently, this group has reported the electrochemical deposition of a tricopolymer consisting of *N*-pyrrole derivatives (pN-me-mix) as a stent primer coating [69]. In addition to the use of PPA and BuOPy, they added *N*-methylpyrrole (*N*-me). The tricopolymer was used as a primer coating on BMS and was characterized using a variety of techniques including AFM and electron and optical microscopy. The adhesion of the coat to the stent as well as elasticity of the coat was measured using the nano scratch test method. A nonbiodegradable polymer (pLM) based on methacrylate derivatives MMA and LMA as a carrier of paclitaxel was deposited by dip coating. While the primer coating was found to be rough (Fig. 3.4), the drug-in-polymer coating smoothens the surface considerably and also makes for a smoother coating compared to coating directly onto BMS (Fig. 3.5). The primer layer also results in a more uniform pLM/paclitaxel coating (Fig. 3.5). Expansion of the coated stents showed no cracking or delamination at the sites of maximum expansion. Drug release was studied *in vitro* and compared release from pLM/paclitaxel on a BMS, on a p*N*-me mix primer-coated stent, and on the latter but also containing a third diffusion-controlled layer. Coating on the BMS resulted in a rapid release of 55% of the drug within the first 3 h; this is comparable to the primer-coated device although the latter device releases 20% less drug in the following 30 days. The device also containing the diffusion-controlled layer results in reduced rapid drug release (20% in the first 3 h); while this device eluted drug with kinetics similar after this initial period to the device lacking the additional layer, the reduced initial release meant that more than 50% of the drug remained on the stent after 30 days. The electrochemically applied tricopolymer primer coat, therefore, is a promising development for the creation of more durable stents with more appropriate surface properties.

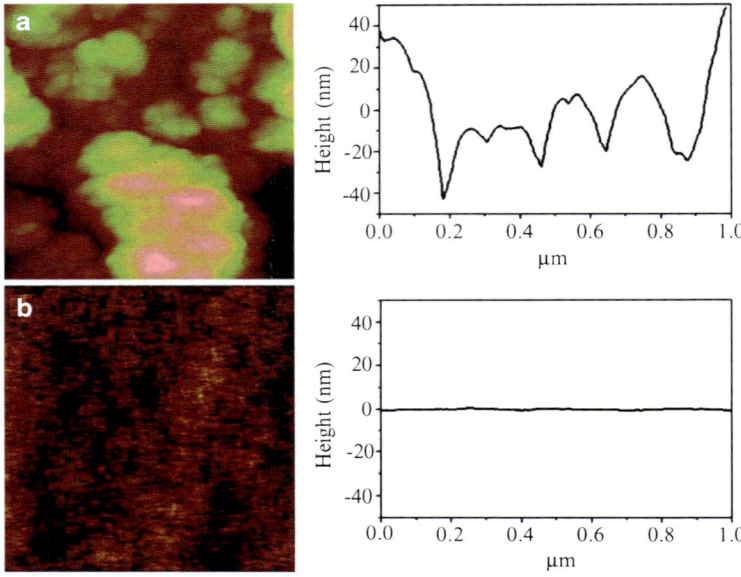

Fig. 3.4 Morphology of DES coating employing a novel electrochemically applied primer layer. 316L stainless steel was electrocoated using a tricopolymer consisting of three *N*-pyrrole derivatives: *N*-methylpyrrole, *N*-(2-carboxyethyl) pyrrole, and the butyl ester of *N*-(2-carboxyethyl)pyrrole. The material was then subjected to dip coating with an experimental paclitaxel-containing nonbiodegradable copolymer composed of methacrylate derivatives. The topography and cross section of the primer-coated (**a**) and subsequently dip-coated (**b**) surfaces was evaluated using atomic force microscopy. (Data are from Okner et al. [69]. Reprinted with permission, Copyright 2009 American Chemical Society)

3.6.2 Use of Electrochemistry to Generate SAM-Coated Stents

A completely different approach to preparing metal surfaces for use in implantable devices including stents is the application of self-assembled monolayers (SAMs). SAMs are single-layered molecular coatings that are deposited on metal or metal oxide surfaces by the adsorption of organic molecules from a solution [76–78]. In general, the molecules involved are long-chain (10–20 carbon) alkanes that are functionalized on either ends. Extensive van der Waals contacts

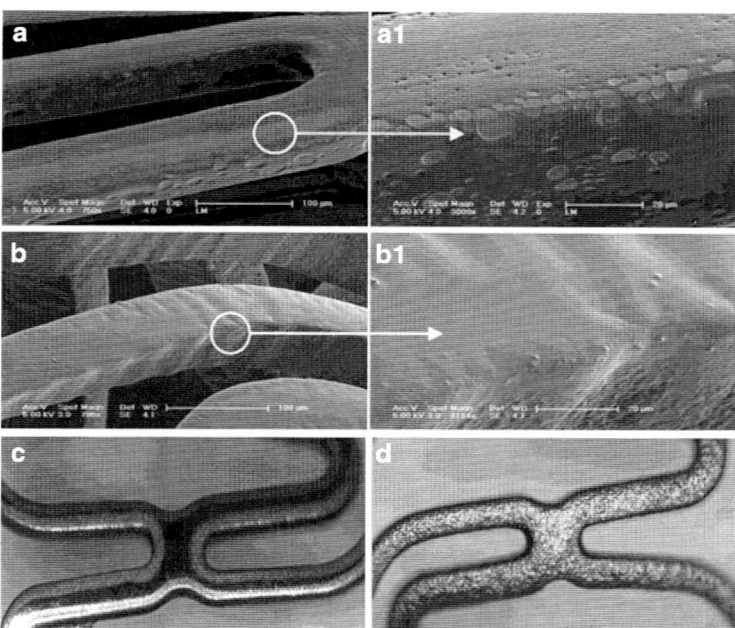

Fig. 3.5 Comparison of a paclitaxel-containing polymer coating on the surface of bare metal and primer-coated stents. 316L stainless steel stents were left as bare metal or electrocoated with the tricopolymer described in the legend to Fig. 3.5 prior to dip coating in paclitaxel-containing copolymer composed of methacrylate derivatives. Scanning electron microscopy of the paclitaxel-coated bare metal (*a*: 750× magnification; *a1*: 3,000× magnification) and primer-coated (*b*: 750× magnification; *b1*: 3,000× magnification) stents shows the increased smoothness and uniformity when using the primer coat. Light microscopy (*c*: paclitaxel-coated bare metal; *d*: paclitaxel-coated primer-coated; both 100× magnification) shows the increased uniformity of the drug-in-polymer coating when the primer coating is used. (Data are from Okner et al. [69]. Reprinted with permission, Copyright 2009 American Chemical Society)

between the alkyl chains allow them to form a stable, homogeneous, single molecule layer on the metallic surface (Fig. 3.6) [18] in much the same way that detergents, fatty acids, and phospholipids self-associate in aqueous solution to form micelles or bilayers. Several different kinds of functional groups, including carboxyl groups and thiols, allow the chains to be covalently or non-covalently attached to the metal through electrochemical (or other) means [79]. The use

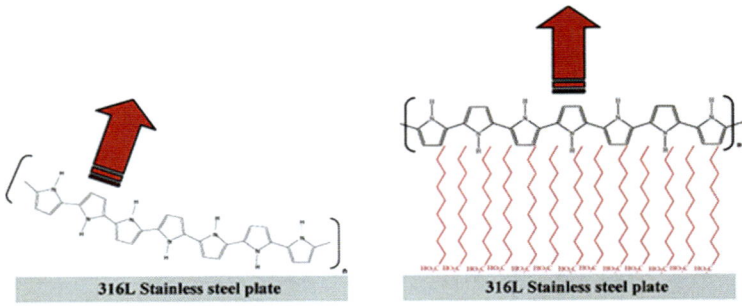

Fig. 3.6 Schematic diagram of adhesion-promoting self-assembling monolayers (SAMs). Electrodeposited polypyrrole adhered comparatively weakly to bare stainless steel (*left*). Formation of a SAM from decanoic acid using electrochemical methods leaves a surface that can still be coated by further electrodeposition methods, such as addition of polypyrrole shown on *right*. The SAM enhances the nucleation and polymerization of polypyrrole as a result of its hydrophobic nature. (Reproduced from Shustak et al. [87]. Reprinted with permission, Copyright 2006 American Chemical Society)

of electrochemistry affords a better control over the process and thus higher-quality, more well-defined surfaces [80]. The use of SAMs as a technique to coat and release therapeutic drugs directly from metal surfaces has been reported [76–78]. Surface modification of biomaterials using SAMs has important biomedical implications; a variety of biomolecules including proteins, peptides, DNA, and antibodies have been attached to SAMs [18]. Additionally, therapeutic SAMs (TSAMs) have been prepared by attaching a drug to Au and Ti surfaces through esterification reactions with hydroxyl-modified SAMs [18, 78].

Although stainless steel has not been as extensively studied as Au, Ti, or other metals, some studies dealing with the generation of SAMs on stainless steel have been reported [80]. Two groups [81, 82] have reported on the formation and characterization of *n*-alkanethiol SAMs on electronically reduced stainless steel surfaces. Their method involved applying a negative potential to remove the native oxide layer of the stainless steel followed by the addition of *n*-alkanethiols to the electrolyte solution to form the SAM.

As discussed above, drug-eluting organic polymers do not adhere well to bare metal stents, compromising their mechanical integrity.

Different types of primer layers have been investigated to improve the performance characteristics of the drug-eluting polymers (see above). The use of SAMs has also been reported for the creation of molecular adhesion promoters that improve the linkage between metals and organic coatings [83–86]. In 2006, Shustak and coworkers [87] reported the application of a decanoic acid SAM onto stainless steel prior to electropolymerization of pyrrole (Fig. 3.6); prior studies had reported formation of SAMs on stainless steel surfaces using other *n*-alkanoic acids including palmitic, stearic, and myristic acid. The monolayer affected the nucleation and growth of polypyrrole owing to its hydrophobic nature. The SAM was not too dense and thick to block electron transfer, thereby allowing electropolymerization to be used for the subsequent pyrrole layer. The carboxyl end groups of the SAM interact with the coat; the opposite end groups of the decanoic acid SAM can be modified to allow specific interactions with different coats [87].

3.6.3 Sol–Gel Coating of Stainless Steel Using Electrochemistry

More recently, electrochemistry has been utilized for the formation of thin sol–gel films formed by electrochemical deposition of organosilanes on stainless steel [88]. In this case, the deposition of aminopropyl-triethoxysilane (APTS) and its codeposition with propyl-trimthoxysilane (PrTMOS) and phenyltrimethoxysilane (PhMOS) was studied. This process was shown to provide a wide range of films with a variety of chemical and physical properties. The thickness of the coatings could be controlled by altering the sol concentration, applied potential, and electrodeposition time. The resulting films were analyzed using a variety of surface techniques to test correlations between deposition parameters and morphology, thickness, and elasticity of the films; their electrochemical characteristics were also tested including surface resistance and capacity. Importantly, the process was applied to the coating of a coronary stent. APTS was used for the coating and the stent was expanded on a balloon. Under these conditions, there was no evidence of cracking or delamination of the sol–gel coating [88]. The addition of a drug-polymer layer needs to be tested in order to determine the utility of sol–gel surfaces for the generation of DES.

3.7 Future Prospects

The development of stenting as a component of PCIs has been a watershed development in the practice of interventional cardiology. Furthermore, the development of DES has addressed the major deficiency in the original technology by dramatically lowering the rates of restenosis requiring re-intervention. However, the long-term safety of even the most recent DES remains to be fully appreciated. This is because large and long-term studies of complications such as late stage thrombosis are still ongoing. Moreover, since all stents are permanent implants, the clinical experience with these devices over the lifetime of the patients receiving them is still evolving.

In this context, there are clear avenues for improvement of the performance and safety of DES, related largely to the nature of the coatings applied. It is desirable to achieve maximally thin and uniform coating with enhanced biocompatibility and sustained hemocompatibility. Novel functionalization of the coating substances would allow for optimization of parameters such as anticoagulation as well as promotion of re-endothelialization concomitant with prevention of neointimal expansion. An ideal coating might also be biodegradable or bioabsorbable. The electrochemical approach is one that appears to have the potential to address most if not all of these considerations. A wide variety of coating chemistries can be manipulated electrochemically, and this approach affords exquisite control over the properties of the surfaces formed. Even DES variants in which a polymer is not used for holding the drug might benefit from certain electrochemically applied primer coats. Finally, it must also be stressed that the regulatory environment surrounding development of new medical devices in general and stents in particular is a justifiably challenging one, with an enormous investment required not only in preclinical work but also in the scale and duration of clinical trials. Thus, it is crucial that the combined efforts of materials chemists, biomedical engineers, and clinicians are focused on developing the best devices right from the design stage.

References

1. World Health Organization Fact Sheet N°310: http://www.who.int/mediacentre/factsheets/fs310/en/index.html.
2. Ross R. Atherosclerosis—an inflammatory disease. N Engl J Med. 1999; 340:115–26.

3. Cheng C, Tempel D, van Haperen R, van der Baan A, Grosveld F, Daemen MJ, et al. Atherosclerotic lesion size and vulnerability are determined by patterns of fluid shear stress. Circulation. 2006;113:2744–53.
4. Davignon J, Ganz P. Role of endothelial dysfunction in atherosclerosis. Circulation. 2004;109(23 Suppl 1):III27–32.
5. Le Brocq M, Leslie SJ, Milliken P, Megson IL. Endothelial dysfunction: from molecular mechanisms to measurement, clinical implications, and therapeutic opportunities. Antioxid Redox Signal. 2008;10:1631–74.
6. Higashi Y, Noma K, Yoshizumi M, Kihara Y. Endothelial function and oxidative stress in cardiovascular diseases. Circ J. 2009;73:411–8.
7. Allahverdian S, Pannu PS, Francis GA. Contribution of monocyte-derived macrophages and smooth muscle cells to arterial foam cell formation. Cardiovasc Res. 2012;95(2):165–72.
8. Mizuno Y, Jacob RF, Mason RP. Inflammation and the development of atherosclerosis. J Atheroscler Thromb. 2011;18:351–8.
9. Finn AV, Nakano M, Narula J, Kolodgie FD, Virmani R. Concept of vulnerable/unstable plaque. Arterioscler Thromb Vasc Biol. 2010;30:1282–92.
10. Falk E, Shah PK, Fuster V. Coronary plaque disruption. Circulation. 1995; 92:657–71.
11. Michaels AD, Chatterjee K. Angioplasty versus bypass surgery for coronary artery disease. Circulation. 2002;106:e187–90.
12. King SB, Schlumpf M. Ten-year completed follow-up of percutaneous transluminal coronary angioplasty: the early Zurich experience. J Am Coll Cardiol. 1993;22:353–60.
13. Ruygrok PN, de Jaegere PT, van Domburg RT, van den Brand MJ, Serruys PW, de Feyter PJ. Clinical outcome 10 years after attempted percutaneous transluminal coronary angioplasty in 856 patients. J Am Coll Cardiol. 1996; 27:1669–77.
14. Serruys PW, Jaegere PD, Kiemeneij F, Macaya C, Rutsch W, Heyndrickx G, et al. A comparison of balloon-expandable-stent implantation with balloon angioplasty in patients with coronary artery disease. N Engl J Med. 1994; 331:489–95.
15. Fischman DL, Leon MB, Baim DS, Schatz RA, Savage MP, Penn I, et al. A randomized comparison of coronary-stent placement and balloon angioplasty in the treatment of coronary artery disease. N Engl J Med. 1994;331: 496–501.
16. Serruys PW, Kutryk MJ, Ong AT. Coronary-artery stents. N Engl J Med. 2006;354:483–95.
17. Sigwart U, Puel J, Mirkovitch V, Joffre F, Kappenberger L. Intravascular stents to prevent occlusion and restenosis after transluminal angioplasty. N Engl J Med. 1987;316:701–6.
18. Mani G, Chandrasekar B, Feldman MD, Patel D, Agrawal CM. Interaction of endothelial cells with self-assembled monolayers for potential use in drug-eluting coronary stents. J Biomed Mater Res B Appl Biomater. 2009;90: 789–801.
19. Rothman MT, Davies SW. Intracoronary stents. Br Heart J. 1992;67:425–7.
20. Palmaz JC, Sibbitt RR, Reuter SR, Tio FO, Rice WJ. Expandable intraluminal graft: a preliminary study. Work in progress. Radiology. 1985;156:73–7.
21. Palmaz JC, Windeler SA, Garcia F, Tio FO, Sibbitt RR, Reuter SR. Atherosclerotic rabbit aortas: expandable intraluminal grafting. Radiology. 1986;160:723–6.

22. Schatz RA, Baim DS, Leon M, Ellis SG, Goldberg S, Hirshfeld JW, et al. Clinical experience with the Palmaz-Schatz coronary stent. Initial results of a multicenter study. Circulation. 1991;83:148–61.
23. Moussa I, Di Mario C, Reimers B, Akiyama T, Tobis J, Colombo A. Subacute stent thrombosis in the era of intravascular ultrasound-guided coronary stenting without anticoagulation: frequency, predictors and clinical outcome. J Am Coll Cardiol. 1997;29:6–12.
24. Topol EJ. Caveats about elective coronary stenting. N Engl J Med. 1994;331:539–41.
25. Hoffmann R, Mintz GS, Dussaillant GR. Patterns and mechanisms of in-stent restenosis. A serial intravascular ultrasound study. Circulation. 1996; 94: 1247–54.
26. Holmes J. State of the art in coronary intervention. Am J Cardiol. 2003; 91:50A–3.
27. Holmes Jr DR, Savage M, LaBlanche JM, Grip L, Serruys PW, Fitzgerald P, et al. Results of Prevention of REStenosis with Tranilast and its Outcomes (PRESTO) trial. Circulation. 2002;106:1243–50.
28. Sheiban I, Villata G, Bollati M, Sillano D, Lotrionte M, Biondi-Zoccai G. Next-generation drug-eluting stents in coronary artery disease: focus on everolimus-eluting stent (Xience V®). Vasc Health Risk Manag. 2008; 4:31–8.
29. Sheiban I, Leonardo F, Rosano GM, Pagnotta P, Marsico F, Montorfano M, et al. Predictors of long-term clinical outcome in patients undergoing multiple vessel stenting for coronary artery disease. Ital Heart J. 2000;1:480–6.
30. Sheiban I, Albiero R, Marsico F, Dharmadhikari A, Tzifos V, Pagnotta P, et al. Immediate and long-term results of "T" stenting for bifurcation coronary lesions. Am J Cardiol. 2000;85:1141–4.
31. Takebayashi H, Kobayashi Y, Dangas G, Fujii K, Mintz GS, Stone GW, et al. Restenosis due to underexpansion of sirolimus-eluting stent in a bifurcation lesion. Catheter Cardiovasc Interv. 2003;60:496–9.
32. Morice MC, Serruys PW, Sousa JE. A randomized comparison of a sirolimus-eluting stent with a standard stent for coronary revascularization. N Engl J Med. 2002;346:1773–80.
33. Ranade SV, Miller KM, Richard RE, Chan AK, Allen MJ, Helmus MN. Physical characterization of controlled release of paclitaxel from the TAXUS™ Express2™ drug-eluting stent. J Biomed Mater Res A. 2004;71: 625–34.
34. Morice MC, Serruys PW, Sousa JE, Fajadet J, Ban Hayashi E, Perin M, et al. A randomized comparison of a sirolimus-eluting stent with a standard stent for coronary revascularization. N Engl J Med. 2002;346:1773–80.
35. Moses JW, Leon MB, Popma JJ, Fitzgerald PJ, Holmes DR, O'Shaughnessy C, et al. SIRIUS Investigators. Sirolimus-eluting stents versus standard stents in patients with stenosis in a native coronary artery. N Engl J Med. 2003; 349:1315–23.
36. Grube E, Silber S, Hauptmann KE, Mueller R, Buellesfeld L, Gerckens U, et al. TAXUS I: six- and twelve-month results from a randomized, double-blind trial on a slow-release paclitaxel-eluting stent for de novo coronary lesions. Circulation. 2003;107:38–42.

37. Colombo A, Drzewiecki J, Banning A, Grube E, Hauptmann K, Silber S, et al. Randomized study to assess the effectiveness of slow- and moderate-release polymer-based paclitaxel-eluting stents for coronary artery lesions. Circulation. 2003;108:788–94.
38. Maisel WH. Unanswered questions—drug-eluting stents and the risk of late thrombosis. N Engl J Med. 2007;356:981–4.
39. Chitkara K, Pujara K. Drug-eluting stents in acute coronary syndrome: is there a risk of stent thrombosis with second-generation stents? Eur J Cardiovasc Med. 2010;1:20–4.
40. Levy Y, Tal N, Tzemach G, Weinberger J, Domb AJ, Mandler D. Drug-eluting stent with improved durability and controllability properties, obtained via electrocoated adhesive promotion layer. J Biomed Mater Res B Appl Biomater. 2009;91:819–30.
41. Belle EV, Susen S, Jude B, Bertrand ME. Drug eluting stents: trading restenosis for thrombosis. J Thrombosis Haemost. 2007;5:238–45.
42. Windecker S, Mierer B. Late coronary late thrombosis. Circulation. 2007;116:1952–65.
43. Austin D, Pell JP, Oldroyd KG. Drug-eluting stents: do the risks really overweigh the benefits? Heart. 2008;94:127–8.
44. Otsuka Y, Chronos NA, Apkarian RP, Robinson KA. Scanning electron microscopic analysis of defects in polymer coatings of three commercially available stents: comparison of BiodivYsio, Taxus, and Cypher Stents. J Invasive Cardiol. 2007;19:71–6.
45. Tesfamariam B. Local vascular toxicokinetics of stent-based drug delivery. Toxicol Lett. 2007;168:93–102.
46. McFadden E, Stabile E, Regar E, Cheneau E, Ong A, Kinnaird T, et al. Late thrombosis in drug-eluting coronary stents after discontinuation of antiplatelet therapy. Lancet. 2004;364:1466–7.
47. Iakovou I, Schmidt T, Bonizzoni E, Ge L, Sangiorgi G, Stankovic G, et al. Incidence, predictors, and outcome of thrombosis after successful implantation of drug-eluting stents. JAMA. 2005;293:2154–6.
48. Stone GW, Ellis SG, Cox DA, Hermiller J, O'Shaughnessy C, Mann JT, et al. TAXUS-IV Investigators. A polymer-based, paclitaxel-eluting stent in patients with coronary artery disease. N Engl J Med. 2004;350:221–31.
49. Rasmussen K, Maeng M, Kaltoft A, Thayssen P, Kelbaek H, Tilsted HH, et al. SORT OUT III study group. Efficacy and safety of zotarolimus-eluting and sirolimus-eluting coronary stents in routine clinical care (SORT OUT III): a randomised controlled superiority trial. Lancet. 2010;375:1090–9.
50. Abizaid A, Costa Jr JR. New drug-eluting stents: an overview on biodegradable and polymer-free next-generation stent systems. Circ Cardiovasc Interv. 2010;3:384–93.
51. Kereiakes DJ, Cox DA, Hermiller JB, Midei MG, Bachinsky WB, Nukta ED, et al. Guidant Multi-Link Vision Stent Registry Investigators. Usefulness of a cobalt chromium coronary stent alloy. Am J Cardiol. 2003;92:463–6.
52. Pelton AR, Schroeder V, Mitchell MR, Gong XY, Barney M, Robertson SW. Fatigue and durability of Nitinol stents. J Mech Behav Biomed Mater. 2008;1:153–64.

53. Burt HM, Hunter WL. Drug-eluting stents: a multidisciplinary success story. Adv Drug Deliv Rev. 2006;58:350–7.
54. Kukreja N, Onuma Y, Daemen J, Serruys PW. The future of drug-eluting stents. Pharmacol Res. 2008;57:171–80.
55. Haïdopoulos M, Turgeon S, Sarra-Bournet C, Laroche G, Mantovani D. Development of an optimized electrochemical process for subsequent coating of 316 stainless steel for stent applications. J Mater Sci Mater Med. 2006; 17:647–57.
56. Adams RO. A review of the stainless steel surface. J Vasc Sci Technol A. 1983;1:12–8.
57. Istephanous N, Bai Z, Gilbert JL, Rohly K, Belu A, Trausch I, et al. Oxide films on metallic biomaterials: myths, facts and opportunities. Materials Sci Forum. 2003;426–4:3157–63.
58. Köster R, Vieluf D, Kiehn M, Sommerauer M, Kähler J, Baldus S, et al. Nickel and molybdenum contact allergies in patients with coronary in-stent restenosis. Lancet. 2000;356:1895–7.
59. Lee YK, Hyung Park J, Tae Moon H, Yun Lee D, Han Yun J, Byun Y. The short-term effects on restenosis and thrombosis of echinomycin-eluting stents topcoated with a hydrophobic heparin-containing polymer. Biomaterials. 2007;28:1523–30.
60. Lewis AL, Vick TA, Collias AC, Hughes LG, Palmer RR, Leppard SW, et al. Phosphorylcholine-based polymer coatings for stent drug delivery. J Mater Sci Mater Med. 2001;12:865–70.
61. Lewis AL, Willis SL, Small SA, Hunt SR, O'byrne V, Stratford PW. Drug loading and elution from a phosphorylcholine polymer-coated coronary stent does not affect long-term stability of the coating in vivo. Biomed Mater Eng. 2004;14:355–70.
62. Acharya G, Park K. Mechanisms of controlled drug release from drug-eluting stents. Adv Drug Deliv Rev. 2006;58:387–401.
63. Basalus MW, Ankone MJ, van Houwelingen GK, de Man FH, von Birgelen C. Coating irregularities of durable polymer-based drug-eluting stents as assessed by scanning electron microscopy. EuroIntervention. 2009;5:157–65.
64. Okner R, Oron M, Tal N, Nyska A, Kumar N, Mandler D, et al. Electrocoating of stainless steel coronary stents for extended release of paclitaxel. J Biomed Mater Res A. 2009;88:427–36.
65. Lahann J, Klee D, Thelen H, Bienert H, Vorwerk D, Hoecker H. Improvement of haemocompatibility of metallic stents by polymer coating. J Mater Sci Mater Med. 1999;10:443–8.
66. Wiemer M, Butz T, Schmidt W, Schmitz KP, Horstkotte D, Langer C. Scanning electron microscopic analysis of different drug eluting stents after failed implantation: from nearly undamaged to major damaged polymers. Catheter Cardiovasc Interv. 2010;75:905–11.
67. Westedt U, Wittmar M, Hellwig M, Hanefeld P, Greiner A, Schaper AK, et al. Paclitaxel releasing films consisting of poly(vinyl alcohol)-graft-poly(lactide-co-glycolide) and their potential as biodegradable stent coatings. J Controlled Release. 2006;111:235–46.
68. Konno T, Watanabe J, Ishihara K. Enhanced solubility of paclitaxel using water-soluble and biocompatible 2-methacryloyloxyethyl phosphorylcholine polymers. J Biomed Mater Res A. 2003;65:210–5.

69. Okner R, Shaulov Y, Tal N, Favaro G, Domb AJ, Mandler D. Electropolymerized tricopolymer based on N-pyrrole derivatives as a primer coating for improving the performance of a drug-eluting stent. ACS Appl Mater Interfaces. 2009;1:758–67.
70. Okner R, Domb AJ, Mandler D. Electrochemical formation and characterization of copolymers based on N-pyrrole derivatives. Biomacromolecules. 2007;8:2928–35.
71. Shaulov Y, Okner R, Levi Y, Tal N, Gutkin V, Mandler D, et al. Poly(methyl methacrylate) grafting onto stainless steel surfaces: application to drug-eluting stents. ACS Appl Mater Interfaces. 2009;1:2519–28.
72. Verenitskaya TV, Efimov ON. Polypyrrole: a conducting polymer; its synthesis, properties and applications. Russ Chem Rev. 1997;66:443–57.
73. Zhang Z, Roy R, Dugré FJ, Tessier D, Dao LH. *In vitro* biocompatibility study of electrically conductive polypyrrole-coated polyester fabrics. J Biomed Mater Res. 2001;57:63–71.
74. Wang X, Gu X, Yuan C, Chen S, Zhang P, Zhang T, et al. Evaluation of biocompatibility of polypyrrole *in vitro* and *in vivo*. J Biomed Mater Res A. 2004;68:411–22.
75. Levy Y, Mandler D, Weinberger J, Domb AJ. Evaluation of drug-eluting stents' coating durability–clinical and regulatory implications. J Biomed Mater Res B Appl Biomater. 2009;91:441–51.
76. Mahapatro A, Johnson DM, Patel DN, Feldman MD, Ayon AA, Agrawal CM. Surface modification of functional self-assembled monolayers on 316L stainless steel via lipase catalysis. Langmuir. 2006;22:901–5.
77. Mahapatro A, Johnson D, Patel D, Feldman M, Ayon A, Agrawal C. Drug delivery from therapeutic self-assembled monolayers (T-SAMs) on 316L stainless steel. Curr Top Med Chem. 2008;8:281–9.
78. Mani G, Johnson DM, Marton D, Feldman MD, Patel D, Ayon AA, et al. Drug delivery from gold and titanium surfaces using self-assembled monolayers. Biomaterials. 2008;29:4561–73.
79. Ulman A. Formation and structure of self-assembled monolayers. Chem Rev. 1996;96:1533–54.
80. Shustak G, Domb AJ, Mandler D. Preparation and characterization of n-alkanoic acid self-assembled monolayers adsorbed on 316L stainless steel. Langmuir. 2004;20:7499–506.
81. Volmer-Uebing M, Stratmann M. A surface analytical and an electrochemical study of iron surfaces modified by thiols. Appl Surface Sci. 1992;55:19–35.
82. Ruan CM, Bayer T, Meth S, Sukenik CN. Creation and characterization of n-alkylthiol and n-alkylamine self-assembled monolayers on 316L stainless steel. Thin Solid Films. 2002;419:95–104.
83. Reinartz C, Fuerbeth W, Stratmann M. Adsorption and characterization of molecular adhesion promoter monolayers on iron surfaces under UHV conditions. Fresenius J Anal Chem. 1995;353:657–60.
84. Stratmann M. Chemically modified metal surfaces—a new class of composite materials. Adv Mater. 1990;2:191–5.
85. Bascom WD. Structure of silane adhesion promoter films on glass and metal surfaces. Macromolecules. 1972;5:792–8.
86. Sathyanarayana MN, Yaseen M. Role of promoters in improving adhesion of organic coatings to a substrate. Prog Org Coat. 1995;26:275–313.

87. Shustak G, Domb AJ, Mandler D. n-Alkanoic acid monolayers on 316L stainless steel promote the adhesion of electropolymerized polypyrrole films. Langmuir. 2006;22:5237–40.
88. Okner R, Favaro G, Radko A, Domb AJ, Mandler D. Electrochemical codeposition of sol–gel films on stainless steel: controlling the chemical and physical coating properties of biomedical implants. Phys Chem Chem Phys. 2010;12:15265–73.

Chapter 4
Screen Printed Electrodes Open New Vistas in Sensing: Application to Medical Diagnosis

Jonathan P. Metters and Craig E. Banks

4.1 Introduction

Although screen printing technology is a well-established technique for the fabrication of biosensors and chemical sensors [1–5], it continues to demonstrate great growth with further developments and honing of techniques being reported. One defining commercialisation of screen printed electrodes was the personal glucose biosensor used by those suffering with diabetes which is a billion dollar per annum global market [6–9]. With society in a constant state of growth and development, demands for medical-and industrial-orientated sensing devices continue to increase exponentially; this is particularly true in the former case where budget cuts within healthcare sectors due to the global recession are pushing the onus onto individual self-monitoring. In order to achieve this, inexpensive and disposable, yet highly accurate and rapid devices are highly desirable. Additionally the portability of such devices is of fundamental importance.

J.P. Metters • C.E. Banks (✉)
Division of Chemistry and Environmental Science, School of Chemistry and the Environment, Manchester Metropolitan University, Chester Street, Manchester, Lancashire M1 5GD, UK
e-mail: c.banks@mmu.ac.uk

Decentralised sensing is ever more prevalent and necessary, and thus traditional techniques utilising highly expensive, immovable analytical equipment such as Gas Chromatography—Mass Spectrometers are not feasible for sensing outside the realms of standard laboratories [10]. Portable, economical, and sensitive sensors are seen to be highly desirable in cases which include utilisation in hospitals where there is a suspected drug overdose, personal monitoring of diseases such as diabetes [6] as detailed earlier, the detection of potential pollutants or toxins within environmental samples [11] such as river water, the screening of drinking water at different sources [12], and also the rapid determination of naturally occurring biomolecules [2]; many other scenarios where an on-the-spot result is needed can be readily envisaged.

With increased development of scientific methods towards the monitoring of clinical-related sensing comes a requirement not only to address the issue of cost effectiveness, but also a constant need to satisfy the previously much sought after criterion of highly reproducible, sensitive methods of detection towards target analytes whilst maintaining the low cost production through scales of economy; such prerequisites all satisfied through screen printed sensors. The adaptability of screen printed electrodes is also of great benefit in a plethora of research; the ability to modify the electrodes with ease, through differing inks commercially available for the reference, counter and working electrodes allows for highly specific, finely calibrated electrodes to be produced for specific target analytes [1].

Screen printing involves printing a thixotropic fluid through a mesh screen which defines the shape and size of the desired electrode. The thixotropic fluid contains a variety of substances including graphite, carbon black, solvents, and polymeric binders [4, 13, 14]. Figure 4.1 depicts a schematic representation of the process of screen printing, where multiple layers and complex designs can be implemented. The inks utilised have a high viscosity (3–10 Pa.s at sheer rate of 230 s^{-1}) but when forced through the screen mesh by the squeegee blade, the ink undergoes sheer thinning allowing it to penetrate through the screen mesh which defines the final shape/design. Upon contact with the substrate, typically a ceramic or plastic material, the ink returns to its viscous state forming the intended shape/design with definition. Such final shapes/designs have thicknesses in the range of 20–100 μm and as such are thicker than that obtained by other printing methodologies, thus they are consequently termed

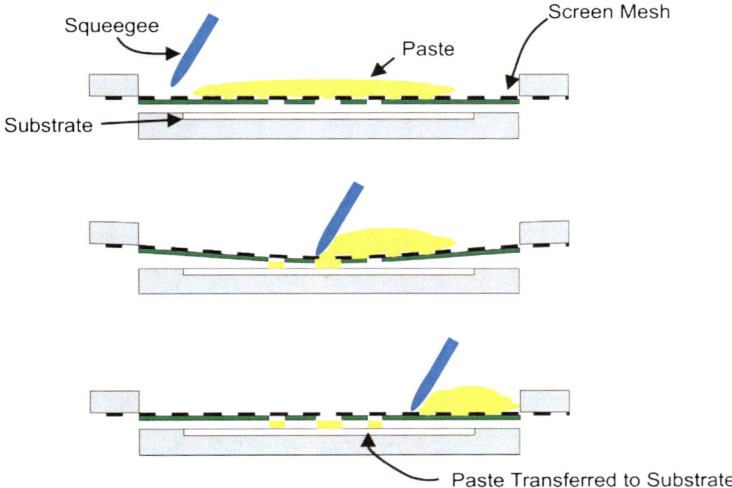

Fig. 4.1 Schematic representation of the screen printed manufacturing process (cross-sectional side view) outlining the basic processes involved for the production of the sensors. Reproduced from [5] with permission from the Royal Society of Chemistry

'thick film technology'. Note that the thickness can be readily controlled by the thickness of the stencil design and the mesh [15, 16]. Such designs can be printed onto ceramic substrates [17] or plastic substrates [17–19] depending on the intended application. In the latter case, the cost is generally lower and the carbon print is better adhered to the substrate than in the former case. Other fabrication techniques related to screen printing include the use of a cylinder press, rotary screen press, carousel textile press, and also a container printer [20].

Another approach worthy of mention, which has proven beneficial, is the use of pad printing [21]. Figure 4.2 shows the approach where the electrode pattern is etched into a stainless steel plate (typical depth ~25 μm) which defines the printed layer thickness. The ink is stored within a closed upturned cup which is magnetically held to the underlying plate, which minimises volatisation of the ink and helps to reduce print variability. The cup is mechanically transferred to the etch pattern and then returned to its initial position leaving a pool of ink within the etch reservoir. A flexible silicone rubber pad is immediately depressed onto the inked etch pattern, picks up the

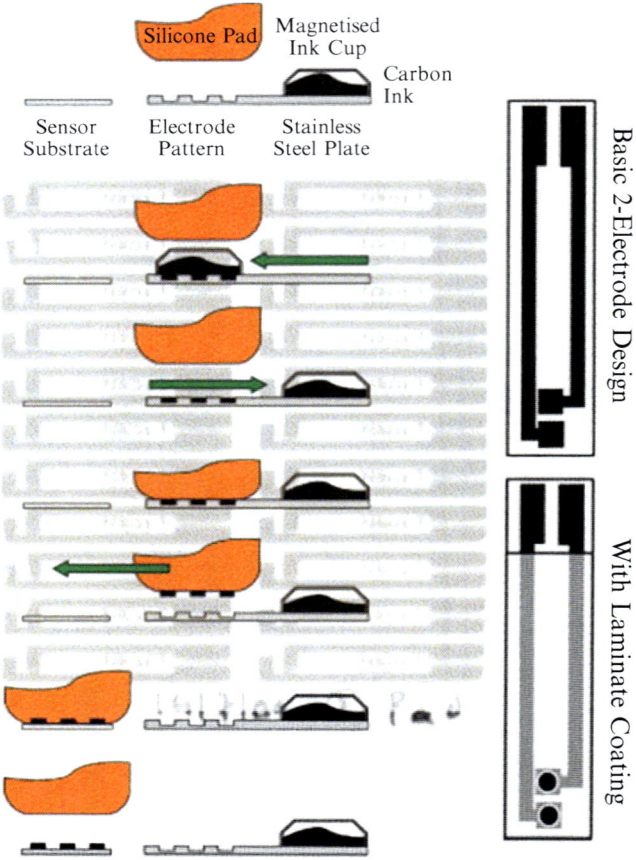

Fig. 4.2 Schematic outlining the pad printing process and the design of the electrode assemblies. Reproduced from [21] with permission from Elsevier

desired image and then transfers it to the base substrate which would form the basis of our proposed sensing assembly. The substrate can then be reprinted to form a thicker deposit or replaced to commence the printing of a new batch [21].

Conventional use of screen printed sensors focuses on the analysis of analytes exclusively within aqueous media, though many applications are increasingly found to be conducted in nonaqueous systems.

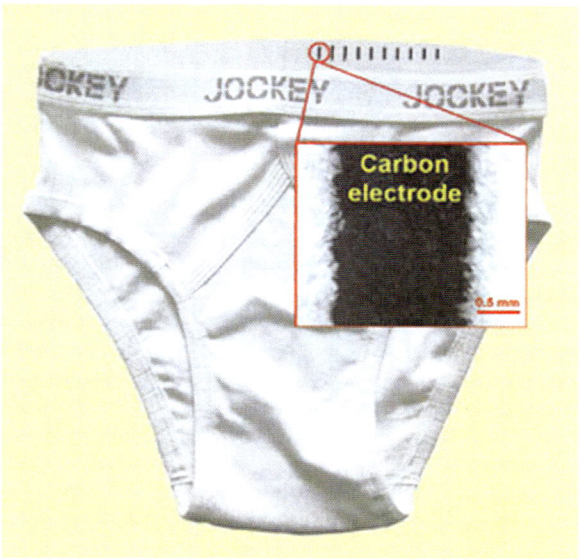

Fig. 4.3 The proposed on-body noninvasive textile-based sensing incorporated within men's briefs. Reproduced from [23] with permission from the Royal Society of Chemistry

In order to achieve this, solvent-resistant screen printed electrodes for use in biosensor applications have been reported [22].

Recently, Wang and colleagues have developed a method for printing biosensors directly onto clothing [23]. To form the sensors, Wang screen printed carbon electrode arrays directly onto the elastic bands of men's underwear as shown in Fig. 4.3. Sensitive detection of hydrogen peroxide and the enzyme NADH, both intrinsically linked with numerous biomedical processes was reported to be possible through the tight contact and direct exposure to the skin. Most interestingly, stresses associated with everyday use, where the screen printed electrode will undergo much strain and duress and are utilised for monitoring the wearer's health and surrounding environment [24–26], did not affect the performance of the sensor [23]. In fact the mechanical bending of carbon screen printed electrodes [24] modified with Nafion and glucose oxidase has been explored under extreme stress, ~180° pinch, and have found to display an enhancement in sensitivity which is attributed to a change in the bioactivity, viz glucose oxidase/Nafion layers rather than that of the carbon electrode surface [24]. Such additions are timely as the costs involved with hospital treatments; reliable, wearable, physiological monitoring systems would allow at-home physiological surveillance, which could potentially reduce the load on hospitals [23].

Due to the ever increasing attention deservedly paid to screen printing techniques and their vast potential applications not only within the field of electrochemistry, but equally within clinical applications, there is a substantial wealth of literature and as such there are authoritative reviews on utilising screen printed enzymatic electrode biosensors, see for example [5, 27–31]. This chapter aims to highlight and explore the many applications and novel developments related to screen printed electroanalytical sensors, with a keen focus being upon those having key medical applications.

4.2 Fundamental Understanding of Screen Printed Electroanalytical Sensors

Although the mechanical processes involved in screen printing techniques are almost uniform throughout, the composition of the inks used within the printing processes is seen to vary greatly. As mentioned earlier, inks consist of graphite particles, polymer binders, and other additives which are utilised for dispersion, printing, and adhesion tasks. Manufacturers, as would be expected, closely guard the formulation of their inks available for purchase, yet it has been shown that differences in ink composition e.g., type, size, or loading of graphite particles and the printing and curing conditions, can strongly affect the electron transfer reactivity and overall analytical performance of the resulting carbon sensors [13, 14, 32].

One key strength of screen printed electrochemical sensors is the attribute of providing excellent platforms for modification with a variety of nanoparticles and structurally related materials requiring no pretreatment such as electrode polishing or electrochemical pretreatment via electrodeposition, as is common when using solid non-disposable electrodes. While screen printed electrodes find widespread usage, the fundamental understanding of the electrochemical reactivity at these underlying electrodes has been seldom studied.

Choudhry et al. [34] have for the first time explored the role of the polymeric binder used in electrode fabrication via screen printing where it was demonstrated that a dramatic effect upon the electrode's morphology is evident as the amount of polymeric binder is dramatically increased; polymeric domains are readily formed which are clearly identifiable through SEM analysis. The effect on the voltammetric response from increasing the amount of binder was evaluated using an inner-sphere electron transfer redox probe. Analysis of the voltammetric peak-to-peak separations indicating that the observed heterogeneous electron transfer rate constant decreases as the polymer binder is increased, which is represented as % M_B/M_I, where M_B and M_I are the mass of the binder and the ink, respectively. The global coverage of edge plane sites of the electrode surface was found to decrease from 5.5 to 0.3% for the range of 0–80% (M_B/M_I) respectively, indicating that the electron transfer characteristics can be tailored from that of edge plane-like, to that of basal plane-like in nature [34]. Note that it is well documented that for graphite electrodes, an electrode consisting of basal plane domains (viz a basal plane pyrolytic graphite electrode) exhibits slow electron transfer while conversely, an electrode with a high proportion of edge plane sites (viz an edge plane pyrolytic graphite electrode) exhibits fast electron transfer, thus this understanding of screen printed electrodes is vital for their further development. This has significant impact in fundamental electrochemistry for example, when one is studying the electron transfer dynamics of say, metal nanoparticles, there is the need to have no contributions from the underlying electrode in order to study the electrochemical response of the metal nanoparticles which is usually achieved with a basal plane electrode fabricated from highly ordered pyrolytic graphite which exhibits relatively slow electron transfer kinetics but unfortunately has the drawbacks of being expensive and requiring surface renewal before each analysis/measurement; the use of a basal plane-like electrode which are disposable and cost effective therefore has clear advantages.

It has been shown that due to non-linear diffusion over an electrode surface the individual contributions of edge and basal plane graphite do not scale with relative areas [35]. It has been conveniently demonstrated for electroactive species with diffusion coefficients of ~1×10^{-6} cm^2 s^{-1} and greater, that there is no contribution from basal plane [35]. Additionally the demonstration that

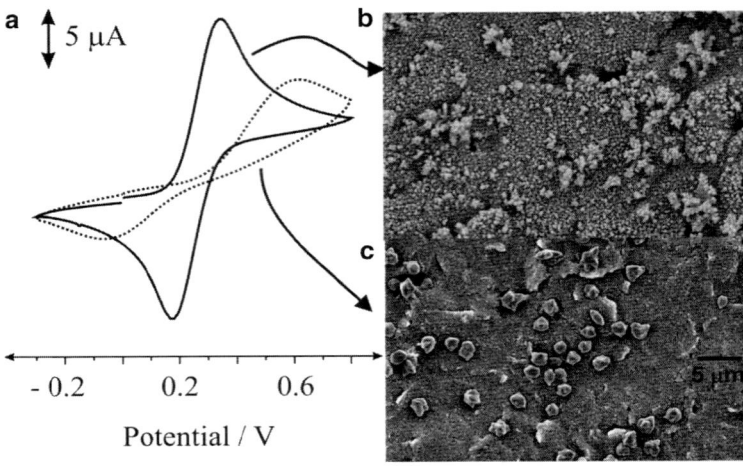

Fig. 4.4 Cyclic voltammetric profiles (**a**) obtained in 1 mM potassium ferrocyanide in 1 M KCl using the standard electrochemical platform (*solid line*) with that of a bespoke electrochemical platform (*dashed line*). Scans recorded at 100 mV s^{-1} vs. SCE. SEM images of the copper plated standard (**b**) and bespoke (**c**) screen printed electrochemical platforms. Reproduced from [37] with permission from Elsevier

the polymeric formulation used within the screen printed ink is electrochemically inert [34] and does not contribute to the cyclic voltammetric response allows for a convenient methodology for modifying the electrochemical reactivity of screen printed electrochemical platforms. It is thus understood that the electrochemical activity of screen printed electrodes depend on the global coverage of edge plane-like sites/defects. This work clearly contributes to the understanding that the binder/ink has a detrimental effect on the voltammetric performance and also confirming that edge plane-like sites are the origin of electron transfer.

Harnessing the ability to tailor the electrode surface's electron transfer properties has clear analytical implications. As depicted by Fig. 4.4, the voltammetric performance of a standard unmodified screen printed electrode is compared with that of a polymeric-modified screen printed electrode with the former exhibiting fast electron transfer (small peak-to-peak separation) while the latter has slow electron transfer (large peak-to-peak separation); copper plating these two screen printed electrodes has a clear distinction in the copper metal morphology. Choudhry et al. have shown that such properties can be beneficially utilised for the electrolytic modification of various electrocatalytic metals where the bespoke electrode surface acts as micron-sized template [34]. Note that clear cost reductions are enabled through such modification as less metal is utilised; a vital trait, especially where precious materials are used, but also has the beneficial view point of creating a *metal ensemble*. The term ensemble is used to indicate that the spacing between each metal deposit, viz metal domain is not uniform, as in the case of arrays [36]. When an ensemble is used over that of an electrolytically modified macroelectrode, a dramatic change in the mass transport prevails facilitating low detection limits and improvements in the analytical performance even though only a fraction of the electrode surface is covered.

Further beneficial modification of bespoke screen printed electrodes seen to be electrolytically modified with palladium which is electrocatalytic towards the electrochemical oxidation of hydrazine where the underlying (unmodified) electrode exhibits slow electron transfer has been elegantly reported by Choudhry et al. [37]. An average palladium particle size of 2.7 μm was found possible where the screen printed electrode substrates act as a template for the deposition of the target metal. This equates to a global coverage of only 0.27 (where 1.0 indicates a fully covered electrode surface). The electrochemical performance towards hydrazine was explored where a limit of detection of ~9 μM was found to be possible which compared well with nanoparticle-modified electrodes, even though only 27% of the surface was modified! Clearly the mass transfer from employing a microdomain electrode, over that of a macroelectrode is highly beneficial and acts akin to a nanoparticle-modified electrode [37].

In the case of the nucleation of metal deposits on screen printed electrodes, the deposition and stripping mechanisms in terms of nucleation processes are important when quantifying metals at screen printed electrodes; the authoritative work of Honeychurch et al. [38] and Brainina et al. [39] is highly recommended. Honeychurch et al. have shown that the redox behaviour of lead is not as straight forward as one would first think, as during the stripping step two or three peaks could be observed, which was attributed to the heterogeneity of the screen printed electrode surface where deposition occurs in a competitive nature with the target metal depositing onto favourable active sites as a monolayer. As deposition proceeds, these sites are depleted and the process becomes competitive between metal deposition on unoccupied bare electrode sites or on already deposited metal [38].

Recently Choudhry et al. [40] have shown that when an electrochemical modification designed for producing nickel nanoparticles on boron-doped diamond is explored using a standard screen printed electrodes, the observed resulting nickel morphology is actually microrod like. The production of such structures requires complicated fabrication strategies (non-electrochemical) and this is the first example of producing such structures electrochemically; the change in deposit morphology from using a boron-doped diamond electrode to a screen printed electrode is due to the differing nucleation dynamics, allowing new and exciting structures to be derived. These nickel microrods were shown to be analytically useful towards the sensing of alcohols and have potential in a plethora of areas.

Other fundamental work on understanding screen printed sensors has been to explore surface oxygen functionalities residing on the electrode surface. Zen et al. [41] have reported on a 'pre-anodized' screen printed electrode for the detection of a range of target analytes. This has the most beneficial effect on the simultaneous sensing of ascorbic acid, dopamine and uric acid where on a standard screen printed electrode only broad and overlapping voltammetric peaks are observed while for the case of the pre-anodized screen printed electrodes, three distinguishable signals are readily observed which, using Raman and XPS, were found to be due to an increment of carbon-oxygen species residing on the graphite surface and/or surface reorientation through the generation of edge plane sites [42] which we interpret as likely through the removal of the ink from edge plane like sites/defects on the graphite. This work has been extended for application in the direct electron transfer to glucose oxidase [43] and for the sensing of lincomycin [41].

4.3 Direct Oxidation/Reduction of Analytes Using Screen Printed Sensors

Recent work by Metters and coworkers has highlighted the potential of screen printed sensors for the direct oxidation of key analytes of interest [44, 45]. The incorporation of noble metals such as gold and platinum within the screen printed sensors were found to offer highly beneficial properties over other, previously reported electrode materials, offering not only direct oxidation of the analytes of interest, but also much more modest costs relating to the electrode material.

As is the case with many heavy metals, there is a keen interest in the monitoring of chromium (III) and (VI) [46]. Although chromium (III) possesses no significant threat, being relatively harmless, it maintains an important biological role [47]. In contrast, chromium (VI) poses a great environmental threat, being around 100–1,000 times more toxic than chromium (III) [48]. This increased hazardous status is attributed to the high oxidation potential and World Health Organisation (WHO) recommends chromium (VI) to be limited to 0.05 mg L^{-1} (0.17 µM) within groundwater [49]. Due to the previously discussed imposed restrictions relating to chromium (VI) levels within water, areas of industry such as plating industries, cooling towers, timber treatment, leather tanning, wood preservation, and steel manufacturing require sensitive and reliable techniques to monitor anthropogenic chromium pollution in ground water [50, 51].

Currently available techniques for the detection of chromium (VI) fall into two distinct categories: direct and indirect detection. Chromatography [52, 53], extraction [54, 55], and co-precipitation [56, 57] are techniques which all require prior separation, which increase detection costs, time, and accessibility. Each of the indirect methods also suffers from chromium (III) interference, often found in relatively high concentrations within natural samples [58]. Direct detection of chromium (VI) can be achieved through spectrophotometric techniques boasting limits of detection of ~0.85 µM though no studies have reported the possible effects of inferences upon the detection of chromium (VI) [59–61]. In addition to spectrophotometric techniques, electrochemical methods facilitate the direct and rapid detection of chromium (VI) even within samples considered to contain excessive concentrations of chromium (III), thus highlighting the robust nature of the electrochemical techniques towards chromium (VI) detection. Table 4.1 highlights the various different electrochemical strategies utilised for the detection of chromium (VI). Limits of detection for chromium (VI) far below WHO guidelines have been reported through the use of electrochemical methods, such as 0.01 and 0.032 µM at sol-gel-modified glassy carbon [62] and mercury thin-film [63] electrodes respectively, however, as is exposed within Table 4.1 many of the electrochemical techniques require an

Table 4.1 A summary of electrochemical reports for towards the detection of chromium (VI). Reproduced from [44] with permission from the Royal Society of Chemistry

Electrode	Linear range (μM)	Limit of detection (μM)	Comments	Ref	
Gold macro	20–2000	4.5	Detection in the presence of known interferon's	43	Above recommended limit
Gold nanoparticle-modified screen printed	θ	0.4	Electrochemical deposition pretreatment was required for the formation on the nanoparticle-modified surface	21	
Silver nanoparticle-modified screen printed	θ	0.85	Electrochemical deposition pretreatment was required for the formation on the nanoparticle-modified surface	21	
Gold nanoparticle-modified indium tin oxide	5–100	2	Gold nanoparticles induced through electrodeposition	45	

World Health Organisation Recommendation: 0.17 μM

Electrode	Range	LOD	Notes	Ref
Gold nanoparticle-modified screen printed	0.03–17	0.015	Nanoparticles introduced through electrodeposition	46
Graphite screen printed	0.34–3.4	0.065		20
Sol-gel-modified glassy carbon	0.04–1.35	0.01	Pretreatment to allow formation of electrodeposited sol-gel film followed by 12 h curing in an oven	17
Mercury thin-film electrode	0.01–0.6	0.032	Electrodeposition of a mercury film performed prior to analysis	18
Poly(4-vinvlpyridine)-coated platinum	0.1–10	Θ	Pretreatment required for the formation of the thin film	47
				Below recommended limit

Θ = Not stated

additional pretreatment phase prior to analysis, such as potential cycling, film, and nanoparticle formation.

Screen printed electrodes offer further advantages for the detection of such environmentally important analytes due to their scales of economy, ease of fabrication, facile modification, and disposable nature [5]. Metters and coworkers determined admirable limits of detection (3σ) of 4.4 µM towards chromium (VI) when utilising the gold screen printed macro electrode with linearity observed over the range 10–1,600 µM (as seen in Table 4.1). In addition the analytical applicability of the screen printed sensors is also tested through chromium (VI) detection in environmental samples [44].

Notably it is found that while the gold screen printed electrodes have electrode kinetics typically one order of magnitude lower than polycrystalline gold macroelectrodes, as is measured via a standard redox probe, in terms of analytical sensing, these gold screen printed macro electrodes mimic polycrystalline gold in terms of their analytical performance towards the sensing of chromium (III) and (VI), whilst boasting additional advantages over the macro electrode due to their disposable one-shot nature and the ease of mass production.

Interestingly, and arguably most importantly, an additional advantage of these gold screen printed macro electrodes compared to polycrystalline gold is the alleviation of the requirement to potential cycle the latter to form the required gold oxide which aids in the simplification of the analytical protocol [44, 46, 64].

In similar fashion Metters et al. fabricated further novel, noble-based screen printed sensors; platinum screen printed macro electrodes [45]. Very few literature reports concerning the fabrication of platinum screen printed macroelectrodes exist [65–67]; Metters and coworkers report the fabrication and characterisation of platinum screen printed macroelectrodes and critically evaluate their response towards the important target analytes hydrazine and hydrogen peroxide. Hydrazine, an increasingly important industrial molecule utilised within rocket fuel and missile systems and corrosive inhibitors, in addition to the ever expanding fuel cell development [68, 69], highlights the need for facile detection of hydrazine, a matter which is further compounded with the knowledge that it is a neurotoxin and molecule which is found to have carcinogenic and mutagenic effects [45, 69, 70].

Previous reports have detailed the sensing of hydrazine including methods based on flow injection analysis (FIA) [71, 72], ion chromatography [73], chemiluminescence (CL), and various types of spectroscopy [74–76]. However, the processes involved in many of these methods are extremely complex, and the linear ranges are relatively narrow and have low precision. Fortunately, electrochemical techniques offer the opportunity for portable, cheap, and rapid methodologies. Platinum is one such electrode which has been successfully utilised for the sensing of hydrazine, offering faster electrode kinetics over that of the traditional, more sluggish carbon-based electrodes. Drawbacks to the use of a traditional platinum electrode evidently centre on the expense and uneconomical nature of such materials. Similarly hydrogen peroxide is of importance within acting as a common intermediate in both environmental and biological systems and is a product of the oxidation of glucose by glucose oxidase [77]. As such the investigation of its detection is of interest and has had substantial attention, with the application of various techniques such as chemiluminescence [78], oxidimetry [79], and electrochemistry [80–82].

Metters et al. determined that highly novel platinum-based sensors allowed commendable detection limits for the two analytes of interest; in the case of hydrazine a linear range of 50–500 µM was reported with a limit of detection of 0.15 µM achievable, whilst the electroanalytical reduction of hydrogen peroxide is also shown to be feasible with a linear range over 100–1000 µM and a limit of detection of 0.14 µM which is competitive to other reported analytical protocols. The novelty of these platinum screen printed electrodes is highlighted in that the platinum on the screen printed electrode surface resides as an oxide which is favourable for the electrochemical oxidation of hydrazine, alleviating the need for extensive potential cycling when platinum macroelectrodes are used [45, 83, 84].

Further to the studies reported by Kadara et al. using platinum screen printed macro electrodes, similar screen printed sensors utilising a platinum working surface have been fabricated by Dominguez-Renedo and colleagues with the screen printed sensors being applied towards the detection of formaldehyde. The platinum-based sensors were found to exhibit low levels of detection ~60 µM for formaldehyde. Unlike the afore mentioned study by Kadara [45], the platinum-based sensors fabricated by Dominguez-Renedo utilised an ink which consisted of platinised carbon, which would thus result in a working electrode surface from which any response could be attributed to either of the two ink components, in comparison to the uniform platinum working surface reported by Kadara [45].

4.4 Mediator-Modified Screen Printed Sensors

In numerous instances the target molecule may be electrochemically challenging or verging on impossible to obtain qualitative or quantitative voltammetric responses. In such cases, a common approach is to employ an 'electrocatalytic' mediator and methodologies to immobilise the chosen mediator onto the screen printed sensor include drop casting, physical attachment (the use of polymer coatings), covalent bonds sorption (physisorption, and chemisorption), or mixing into a carbon paste, are all viable approaches [85]. The last approach is commonly applied and has been used for many decades such as in the pioneering work by Adams et al. [86], but has the drawback of non-stability and is not easily mass produced with reproducibility which in some instances is not optimal.

The mixing of mediators into screen printed ink which are then readily mass produced has been a successful approach for a range of mediators such as Meldola's blue [87, 88], Prussian blue [89], crown-ether [90], and nickel hexacyanoferrate [91, 92], to name just a few. Undoubtedly one of the most widely utilised mediators is cobalt phthalocyanine [93] (CoPC) which has been used in, for example the detection of dimethyldisulfide [94], ethanethiol [94], glutathione [95], hydrogen peroxide [96–98], hydrogen sulphide [94], 2-mercaptoethanesulfonic acid [95], methanethiol [94], propanethiol [99], thioglycolic [100], and finally citric acid [101]. However Kozub and Compton [102] have cast doubt over its claimed electrocatalytic properties, stating that CoPC undergoes three distinct types of redox processes in nonaqueous solution and that the thermodynamics of these couples are solvent sensitive. When CoPC is immobilised on graphite it is believed to take the form of water insoluble microcrystallites of CoPC which when oxidised from $Co^{II}PC$ to $Co^{III}PC$ likely undergoes solubilisation with loss of surface modification. The role of microcrystallites, as opposed to molecularly absorbed CoPC, has been neglected in the study of CoPC-mediated oxidations, as has the scope for loss of modification from the surface [102].

One such mediator-modified sensor based upon a screen printed reported a novel enzyme entrapment approach based on an electropolymerization process utilising multi-walled carbon nanotubes (MWCNT), β-cyclodextrin, and glucose oxidase [103]. The authors propose a new way to entrap the enzyme based on cyclodextrin

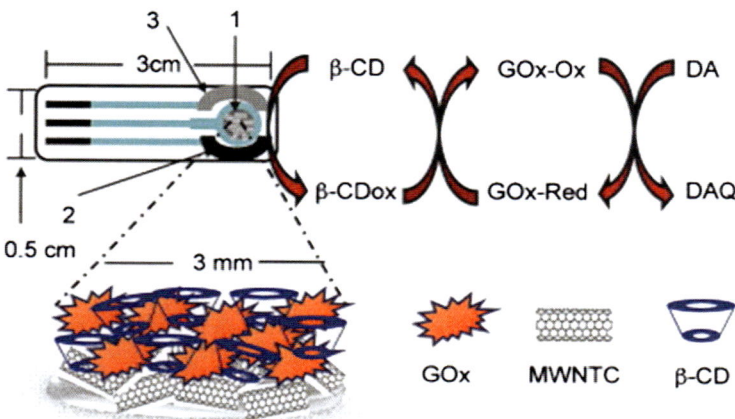

Fig. 4.5 Schematic representation of the biosensor and its different parts, where *1* is the working electrode, *2* is the counter electrode, and *3* is the reference electrode. Also the scheme depicts the dopamine (DA) oxidation at the working electrode surface. The reactions describe the redox process of dopamine to dopamine quinone by GO_x and using β-cyclodextrin as redox mediator co-immobilised with the enzyme. Screen printed electrode/MWCNT is used as transducer. Reproduced from [103] with permission from Elsevier

electropolymerization onto screen printed electrode modified with MWCNT. β-cyclodextrin is used as an electron transfer mediator between enzyme and the electrode surface (Fig. 4.5). Dopamine quantification is presented using a screen printed electrode modified by electropolymerization of cyclodextrin with glucose oxidase. It was shown that although dopamine determination can be achieved with all of them, the electrodes modified with MWCNT presented better analytical features that those built without MWCNT, the best being the one including all components [103].

This biosensor was found to display good reproducibility, repeatability, and durability over time. The limit of detection towards dopamine was reported to be 0.48 μM with a linear range of 10–50 μM which is reported to be comparable or even better than many other electrodes reported in the literature. Moreover, it was also shown that using this electrode, dopamine quantification is possible in the presence of interfering agents such as ascorbic and uric acid. These findings demonstrate that the approach employed is feasible for enzyme entrapment and may find applications in other biosensing systems, where better sensitivity, stability, and fast response are required [103].

Notable work by Ping and coworkers report an ionic liquid bulk-modified screen printed electrode [104]. Ionic liquids have been extensively utilised in analytical chemistry due to their physicochemical properties and excellent biocompatibility and have been used in carbon composites but never in screen printed electrodes. This elegant work by Ping et al. is a platform for incorporating a whole variety of ionic liquids where the ionic liquid can provide benefits in terms of sensing. This has been extended by Ren et al. [105] towards a DNA sensor allowing detection limits of the order 10^{-17} M to be possible.

Finally, carbon nanotube screen printed electrodes have been reported for the detection of p-aminophenol in alkaline phosphatise-based assays [106], which also has the benefit of increasing the electrochemically active area by up to ~50% [107]. This has the benefit that other carbon nanotubes such as single-walled carbon nanotube and multi-walled carbon nanotube derivatives (bamboo, herringbone, and so on) can be readily incorporated allowing mass produced and reproducible nanotube-modified screen printed electrodes to be realised.

One further instance where the underlying electrode substrate is consequently modified with a compound to allow for sensitive analyte detection is modification using alcohol oxidase, for example, towards the amperometric assay of formaldehyde entrapped within a gel [108] for which it was reported that functionalised platinum screen printed electrodes modified with alcohol oxidase immobilised by cross-linking with glutaraldehyde (GA) and bovine serum albumin (BSA) which has been reported to improve the analytical sensitivity and selectivity towards formaldehyde.

Exciting work by Ding et al. [109] describes the utilisation of a screen printed electrode consisting of a colloidal gold-modified graphite ink working electrode, modified using nafion for the monitoring of 5-HT levels within laboratory rats. The study demonstrated its applicability in enabling the monitoring of depression within the rats observed during treatment with commercially available antidepressants. It is suggested that the nafion-modified screen printed electrode utilised within this study would be suitable for 5-HT detection and monitoring in biological samples, with potential to be used for the screening of antidepressant presence in addition to clinical diagnosis [109].

4.5 Metal Oxide-Modified Screen Printed Sensors

As mentioned in the above section mediator bulk-modified screen printed electrodes have been the backbone of sensors towards portable and decentralised testing which are also easily mass produced and consequently have scales of economy. Decoration or the use of metallic screen printed inks for the modification of the electrode surface with electrocatalytic metals is a common approach such as producing film-modified screen printed electrodes [110]. Another commonly reported approach is to incorporate metallic electrocatalysts into a carbon paste electrode [111–113]. However, this methodology suffers from highlighted drawbacks (see earlier). Wu et al. [114] have utilised ruthenium oxide which is commonly used in resistive pastes for screen printing and demonstrated this towards the sensing of ascorbic acid which provided an analytical output of 0–4 mM with little retardation being observed from common interferents: uric acid and hydrogen peroxide [114]. This work has been extended for example for the determination of hydrogen peroxide and with enzymatic modification for sensing hypoxanthine and glucose, whilst also being applied to food analysis [115]. Similarly the bulk modification of screen printed electrodes with copper oxide for carbohydrate sensing [116], nickel oxide for hydroxide detection [117], manganese oxide for nitrite, oxygen, and ascorbic acid [118], and finally bismuth oxide have been reported [119].

Of this work, the use of bismuth oxide is particularly interesting as it is well documented that modifying an electrode surface with bismuth aids to improve the electroanalytical performance and can act akin to a mercury-modified electrode yet having negligible toxicity [120–123]. Since this pioneering work, by Wang and coworkers, the use of bismuth-modified electrodes has been greatly expanded. Recent prominent examples include lead in human blood [124], zinc and cadmium via sono-electroanalysis [125], sensing of *Escherichia coli* [126], amino-salicylate drugs [127], indium [128], and metallotheionein [129]. In these examples bismuth-modified electrodes are usually prepared via *ex situ* or *in situ* electrodeposition. This approach involves careful preparation of the electrode surface between samples and to eliminate this preparative step, the use of disposable screen printed electrodes as underlying electrode substrates has been reported [130]. Towards simplifying the electrochemical methodology further, bismuth-powder-modified carbon paste electrodes [131] have been reported as well as bismuth nanopowder-modified electrodes where the bismuth nanopowder is immobilised with Nafion [131, 132].

Quintana and colleagues [133] have recently reported on a comparative study of bismuth-modified screen printed sensors for the detection of lead. The comprehensive study observed not only the effect of *in situ*, *ex situ* and 'Bi_2O_3 bulk' bismuth modification of the screen printed sensor, but also the influence of different analytical parameters such as the electrolyte solution composition (acetate buffer, chloridric acid, nitric acid, perchloric acid) and the ionic strength ensure that the most effective preparatory steps were taken prior to analysis. The data is seen to show that lead detection by means of *in situ* bismuth-modified screen printed electrode, under the applied conditions, depends mainly on the ionic strength and type of electrolyte solution used whilst determining that the most appropriate electrolyte for use (sample permitting) during analysis was an acetate buffer. In the subsequent test case, using aqueous samples, it was shown that the screen printed electrode produced using electrochemically pretreated substrate and *in situ* bismuth modification allows the detection of lead ions at levels as low as 0.15 µg L^{-1}. When the *in situ* bismuth-modified screen printed electrode was challenged with samples of tap and bottled water samples spiked with known concentration of lead ions, the high reproducibility of the results could be inferred by the relative standard deviation (from 3.2 to 7.5% RSD) as well as by the satisfactory values of the resulting recovery percentages (81 and 98%) [133].

Bulk-modified screen printed electrodes, where micron-sized particles of the electrocatalytic metal are incorporated into the electrode, are generally found to act analytically similarly to that of a nanoparticle-modified carbon electrode. The pertinent question here is, *why is this the case?* The reason can be found by considering the diffusion zones at each electrode surface viz either the nano- or micron-particles. Figure 4.6 depicts a schematic representation of the diffusional zones at microelectrode and nanoparticle ensembles. As depicted by figure 4.6, the distance between neighbouring particles is

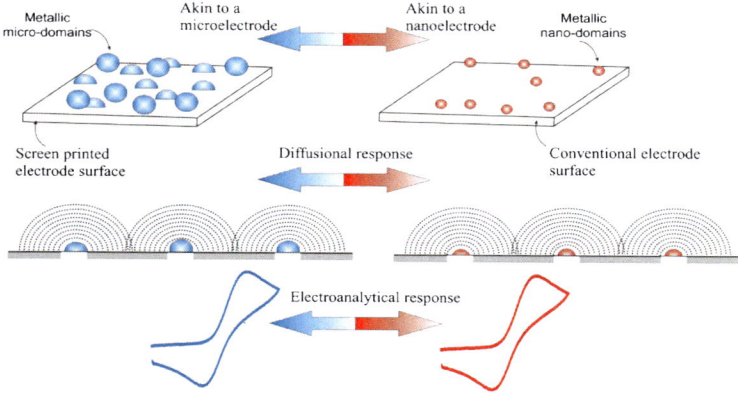

Fig. 4.6 Schematic representation of the diffusional zones at micro- and nanoelectrode ensembles. Reproduced from [5] with permission from the Royal Society of Chemistry

critical in influencing the performance of the ensemble, as in the case of microelectrode arrays. Again, diffusion zones will build up at each electrode, in this case the micro- or nanoparticles, and due to no regular spacing, diffusional overlap occurs to a substantial degree at modest scan rates, which thus, as would be expected results in essentially the same voltammetric profiles as those observed at the microelectrode arrays. The degree of overlap is important, and if the overlap is not significant the ensemble has faster mass transport and hence improves electroanalytical performance over that of a heavily overlapping diffusional regime. The voltammetric response observed is identical due to the fact that the diffusional profiles are similar indicating that a nanoparticle-modified electrode for use in analysis does not always confer enhancements in electroanalysis. It is clear that in going from a micro- to nanoparticle size ensemble is not always beneficial since close neighbouring nanoparticles will heavily overlap acting akin to a microparticle of the same geometric area [134]. The advantageous approach of the methodology, as described above, is that any metallic oxide can be readily incorporated, allowing a true platform technology.

4.6 Improving Mass Transport Through Advantageous Designs

Screen printing technology has been used extensively to develop unique sensors, intelligently designed to improve mass transport increasing sensitivity for applications such as for biosensors for important biomedical markers [135]. Microelectrodes can be formed by a variety of methods including laser micromachining [136], sonoelectrochemistry [137, 138], and photolithography techniques [139], and by direct cutting as shown by Chang and Zen [140] and Authier et al. [141]. The method of Authier et al. and a further similar method, reported by Zen et al. [140] is considered to be the simplest to use and a similar method was used in the present study to produce microband electrodes. Microband electrodes have several advantages over conventional sized electrodes (mm^2) including low ohmic drop, steady-state currents, independence of signal on stir rate, and increased signal-to-noise ratios, leading to lower detection limits [142–144].

Hart and coworkers have elegantly demonstrated for the first time the fabrication of screen printed carbon microband electrodes utilised to readily detect H_2O_2 and that the same ink used, with the addition of lactate oxidase, can be used to construct microband biosensors to measure lactate [144]. These inks were screen printed on one side of the 0.5 mm thick PVC substrate and a defined area of 9 mm^2 was created using dielectric tape; an Ag/AgCl reference electrode was printed around the working electrode (conventional sized Fig. 4.7a). Microbands were formed by cutting through the working electrode and insulator layer using scissors, thus exposing the edge of the electrodes (Fig. 4.7b). The geometrical area of the edge was calculated to be 5.28×10^{-4} cm^2 by multiplying the length (l) by the width (w). The value of w (17.6 μm) was measured using a TESA digital micrometre obtained from Radio Spares, Switzerland. The microband devices were fabricated by a simple cutting procedure using conventional sized carbon-based screen printed electrodes containing the electrocatalyst CoPC. A scan rate study revealed that the mass transport mechanism was a mixture of radial and planar diffusion. It was determined that amperometric studies under quiescent and hydrodynamic conditions indicated that radial diffusion predominated, whilst chronoamperometric study indicated that steady-state currents were obtained with these devices for a variety of H_2O_2 concentrations and

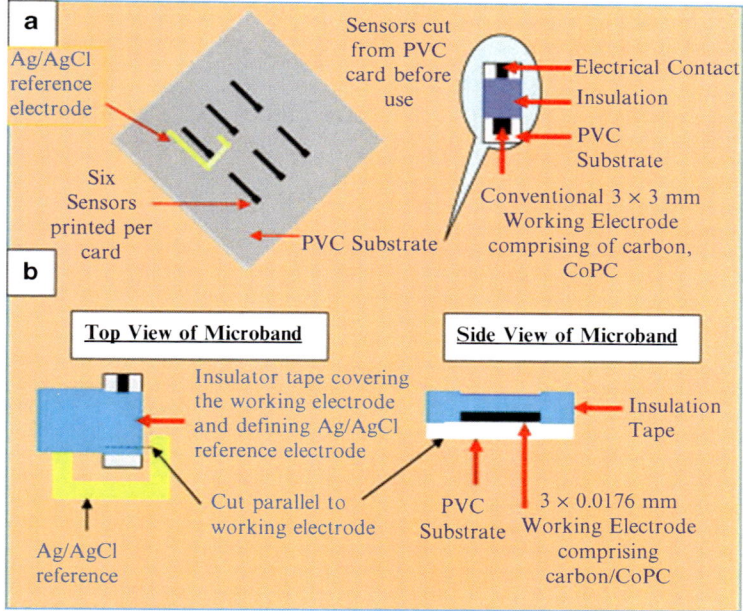

Fig. 4.7 Schemic diagram of SPCEs used and method to produce (**a**) conventional and (**b**) microband electrodes. Reproduced from [144] with permission from Elsevier

that the currents were proportional to the analyte concentration. The incorporation of lactate oxidase within the inks used during the printing process allowed for the fabrication of lactate microband biosensors. Interestingly, it was noted that the presence of lactate oxidase, did not compromise the viability of the sensor towards the detection of H_2O_2 [144]. Hart and coworkers reported commendable detection limits of 289 µM with a linear range of 1–6 mM for lactate using the lactate oxidase modified, providing a potential platform for monitoring cell metabolism *in vitro* by measuring lactate electrochemically via a microband biosensor [144].

In similar fashion the measurement of lead in acetate leachates from ceramic glazed plates utilising microband screen printed electrodes has been reported by Hart and Honeychurch [145]. The microband sensors enabled efficient detection of the lead; in ceramics fortified with 2.10 µg of lead (equivalent to 100 ng/mL in the

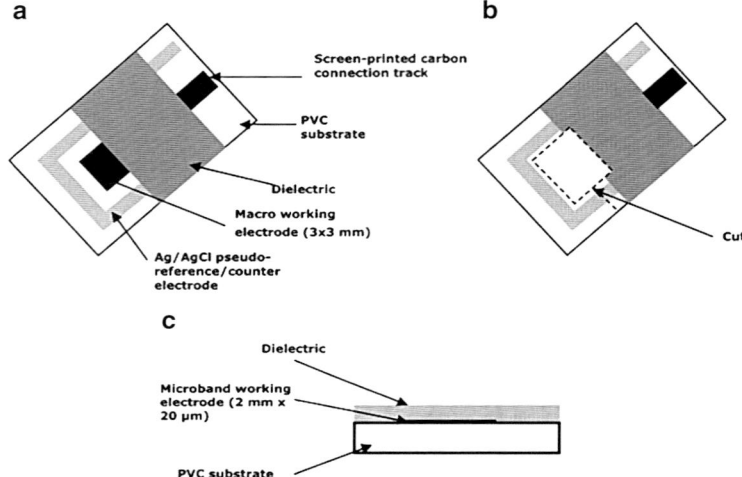

Fig. 4.8 Diagrammatic representation of the method for the manufacture of microband senors. (**a**) Plain view of 3×3 mm screen printed electrode, (**b**) and the modification to fabricate the microband sensor (**c**) end on view of the microband sensor. Reproduced from [145] with permission from Elsevier

leachate), a mean recovery of 82.08% (% CV = 4.07%) was obtained. The microbands used consisted of 2 mm in length with a width of 20 μM achieved as depicted in Fig. 4.8, through the end of a generic mass produced screen printed sensor's working electrode being cut off perpendicular to the screen printed carbon track to expose a 2 mm wide section of the screen printed carbon connection strip (Fig. 4.8b), sandwiched between the dielectric (insulting tape) and the PVC substrate, hence forming the desired microband electrode.

Other electrode designs have been reported such as a ring electrode where by fabrication process involves a hole of 1 mm in diameter drilled through both polymer and electrode pad layers with the ink layer within the drill cavity forming the basis of the working electrode. The basic layout of the electrode is depicted in Fig. 4.9a. The efficacy of pursuing this approach a means of preparing microelectrodes was assessed again within the study, through comparison of the response to ferrocene carboxylic acid at slow scan rates of ~5 mVs. The voltammetric profile of the ring is shown in Fig. 4.9b [21].

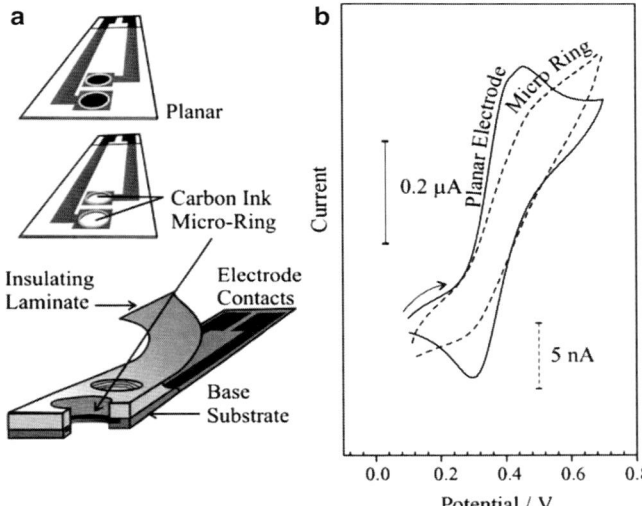

Fig. 4.9 (**a**) Schematic of the ring electrode. (**b**) Cyclic voltammogram detailing the response to 1 mM ferrocence carboxylic acid (pH 7). Reproduced from [21] with permission from Elsevier

4.7 Enzyme-Modified Screen Printed Electrodes and Immunosensors

Such a mention of screen printed electrodes is not complete without recourse to mention the glucose sensor and related areas. An authoritative review by Heller and Feldman is a must read [7] which provides a truly thorough overview of glucose sensors in general. The concept of the use of screen printed electrodes for the sensing of glucose is well established and involves, in simplification, a conductive carbon working electrode, usually made via screen printing, modified with the enzyme glucose oxidase where glucose is oxidised by the enzyme and the electrons involved are usually relayed to the electrode via a suitable mediator resulting in the observed electric current proportional to the level of glucose in the sample matrix. For example, glucose oxidase can be immobilised by cross-linking saturated glutaraldehyde (GA) and tetrathiafulvalene (TFF) with the latter as a mediator which can be described by [146]:

$$\text{Glucose} + 2\text{TTF}^+ \rightarrow \text{gluconic acid} + 2\text{TTF} + 2\text{H}^+ \quad (4.1)$$

$$2\text{TTF} \rightarrow 2\text{TTF}^+ + 2e^- \quad (4.2)$$

The last oxidation reaction takes place on the electrode surface. A substantial proportion of the disposable biosensors used in the sensing of glucose are based on electrochemical determination of enzymatic-generated hydrogen peroxide:

$$\text{Glucose} + O_2 + GO_X \rightarrow \text{gluconic acid} + H_2O_2 \quad (4.3)$$

The electrochemical oxidation or reduction of H_2O_2 generally requires high over potentials at bare electrodes, which exhibits a very poor sensitivity such that other electro-active analytes consequently electrochemically oxidised. Accordingly, most of the glucose sensors utilise mediators that enable the sensing of hydrogen peroxide at low potentials, thereby avoiding any kind of electrochemical interferent. There are many methods for the immobilisation of GO_X on carbon-based screen printed electrodes using different mediators, which may be found in the literature.

Figure 4.10 depicts an expanded view of a typical electrochemical blood-glucose strip showing the various components making up the electrochemical sensor. Note that there are many configurations on the above themes for sensing glucose but however the glucose is measured, it is clear that the underlying electrode has allowed implementation into the market such that very low glucose screen printed electrodes can be fabricated on the million-sensor scale [7]. The ease of fabrication and even more importantly, modification is one of the predominantly attractive features of screen printing. The ease of modification and incorporation of mediators such as enzymes as explored in relation to the sensing of glucose has been seen to expand the possibilities of clinical applications utilising screen printed sensors.

Amphetamine and its analogues are popular recreational drugs of abuse due to the fact that they are potent stimulants of the central nervous system. Butler et al. [147] and Luangaram et al. [148] describe amperometric immunosensors for their determination in urine and saliva, based on screen printed carbon electrodes and amperometic techniques; such approaches could be useful in medical and forensic situations.

Similarly, the development of an amperometric immunosensor for the diagnosis of Chagas' disease using a specific glycoprotein of the trypomastigote surface, which belongs to the Tc85-11 protein family of *Trypanosoma cruzi* (*T. cruzi*), has been reported [149]. In this process, atomically flat gold surfaces on a silicon substrate and gold

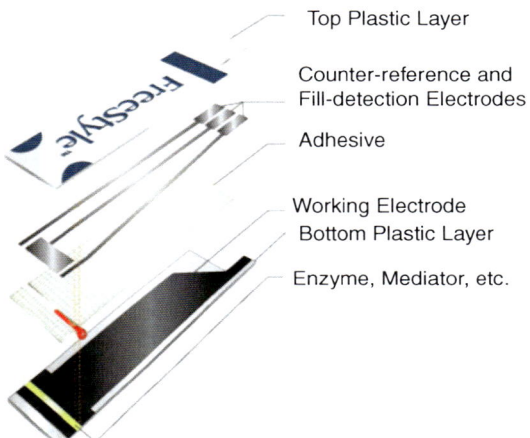

Fig. 4.10 Expanded view of an electrochemical blood-glucose monitoring strip used for diabetes management. The capillary chamber is shown partially filled with a droplet of blood. More than a billion of the strips shown are produced annually. Reprinted with permission from [7]. Copyright (2008) American Chemical Society

screen printed electrodes were functionalized with cystamine and later activated with GA, which was used to form covalent bonds with the purified recombinant antigen (Tc85-11). The antigen reacts with the antibody from the serum, and the affinity reaction was monitored directly using atomic force microscopy or amperometry through a secondary antibody tagged to HRP. In the amperometric immunosensor, peroxidase catalyses the L2 formation in the presence of hydrogen peroxide and potassium iodide, and the reduction current intensity was measured at a given potential with screen printed electrodes. The immunosensor was applied to sera of chagasic patients and patients having different systemic diseases [149].

The presence of mycotoxins, which are produced by fungi in a high amount of food, in levels higher than the accepted ones, represent a threat for the food harmlessness, as well as an important risk in alimentary health; milk is usually contaminated with small amounts of aflatoxin.

M1 (AFM1) as a consequence of the metabolism by the cow of aflatoxin B1 (AFB1), a mycotoxin that is commonly produced by the fungal strains *Aspergillus flavus* and *Aspergillus parasiticus* and found in certain animal foodstuff. Toxicological concern about AFM1 arises principally from its close structural similarity to AFB1, the latter having been shown to be one of the most potent known carcinogens. European Community limits the concentration of AFB1 in foodstuffs for dairy cows to a maximum of 10 μg/kg. This limit was chosen taking into account the quantities of feed consumed and the fact that 1–4% of the ingested AFB1 appears as AFM1 in the milk.

Screen printed carbon electrodes bearing a surface-adsorbed antibody against AFB1 were also used in a competitive immunoassay, based on competition of free analyte with a biotinayleted AFB1 conjugate [150]. Subsequent addition of streptavidin-ALP conjugate, followed by a 1-naphtyl phosphate substrate resulted in the production of the electrochemically active product, 1-naphtol; this was oxidised using linear sweep voltammetry (LSV) and constituted the measurement step. These immunosensors were fabricated in an array configuration, which represent the initial studies towards the development of an automated instrument for multi-analyte determinations. Alarcon et al. [151, 152] describe a direct, competitive ELISA for the quantitative determination of another mycotoxin, ochratoxin A (OTA), using polyclonal antibodies and carbon-based screen printed electrodes. The immunosensor appears to be suitable for OTA contamination screening in food samples.

Exciting work by Rusling et al. [153] demonstrates the potential applications of screen printed sensors for the detection of two cancer biomarker proteins, whereby a microfluidic electrochemical immunoassay was fabricated using a moulded polydimethylsiloxane channel and routine machined parts interfaced with a pump and sample injector. Using off-line capture of analytes by heavily enzyme-labelled 1 μm superparamagnetic particle (MP)-antibody bioconjugates and capture antibodies attached to an 8-electrode screen printed sensor, simultaneous detection of cancer biomarker proteins prostate specific antigen (PSA) and interleukin-6 (IL-6) in serum was achieved at sub-pg mL^{-1} levels. MPs were conjugated with ~90,000 antibodies and ~200,000 horseradish peroxidase (HRP) labels to provide efficient off-line capture and high sensitivity. The sensors utilised for this measurement were screen printed electrodes which featured a layer of 5 nm glutathione decorated gold nanoparticles to attach antibodies that capture MP-analyte bioconjugates. It was determined that detection limits of 0.23 pg mL^{-1} for PSA and 0.30 pg mL^{-1} for IL-6 were possible in diluted serum mixtures. PSA and IL-6 biomarkers were also measured in serum of prostate cancer patients in total assay time 1.15 h and the screen printed sensor array results gave excellent correlation with standard enzyme-linked immunosorbent assays (ELISA). Undoubtedly, the use of screen printed electrodes within this study, not only demonstrates the wide potential applications of screen printed sensors, but in addition the ever expanding horizon of medically based sensing devices utilising the functionality of screen printed electrodes, in this instance offering great promise for accurate, sensitive multiplexed detection of diagnostic cancer biomarkers [153].

4.8 Conclusions

In this chapter we have overviewed the recent developments in screen printing from their fundamental understanding through to highly novel and innovative designs which in part improved analytical performance towards target analytes. The applications of electrochemistry are truly wide and in this volume we have seen that medical applications are on the rise. Thus, from understanding how screen printed electrodes can be fabricated and modified for the end users needs, new medical electrochemical sensors based on screen printing will undoubtedly emerge.

References

1. Hart JP, Wring SA (1997) Recent developments in the design and application of screen-printed electrochemical sensors for biomedical, environmental and industrial analyses. Trends Anal Chem 16:89.
2. Hart JP, Wring SA (1994) Screen-printed voltammetric and amperometric electrochemical sensors for decentralized testing, Configurations used in the design of screen-printed enzymatic biosensors. A review. Electroanalysis 6:617.
3. Sirvent MA, Merkoci A, Alegret S (2000) Configurations used in the design of screen-printed enzymatic biosensors. A review. Sens Actuators B 69:153.
4. Morrin A, Killard AJ, Smyth MR (2003) Electrochemical characterization of commercial and home-made screen-printed carbon electrodes. Anal Lett 36:2021.
5. Metters JP, Kadara RO, Banks CE (2011) New directions in screen printed electroanalytical sensors: an overview of recent developments. Analyst 136:1067.

6. Honeychurch KC, Hart JP (2003) Screen-printed electrochemical sensors for monitoring metal pollutants. Trends Anal Chem 22:456.
7. Heller A, Feldman B (2008) Electrochemical glucose sensors and their applications in diabetes management. Chem Rev 108:2482.
8. Newman JD, Turner APF (2005) Home blood glucose biosensors: a commercial perspective. Biosens Bioelectron 20:2435.
9. Wilson R, Turner APF (1992) Glucose oxidase: an ideal enzyme. Biosens Bioelectron 7:165.
10. Brainina KZ (2001) Electroanalysis: from laboratory to field versions. Anal Chem 56:303.
11. Gurban AM, Rotariu L, Tudorache M, Bala C, Noguer T (2009) Sensors for environment, health and security, 2. Springer, Dordrecht, p 401.
12. Simm AO, Banks CE, Compton RG (2004) Sonoelectroanalytical detection of ultra-trace arsenic. Electroanalysis 17:335.
13. Kadara RO, Jenkinson N, Banks CE (2009) Characterisation of commercially available electrochemical sensing platforms. Sens Actuators B 138:556.
14. Wang J, Tian B, Nascimento VB, Angnes L (1998) Performance of screen-printed carbon electrodes fabricated from different carbon inks. Electrochim Acta 43:3459.
15. Kipphan H (2001) Handbook of print media. Springer, Heidelberg
16. Barlow FD, Elshabini A (2007) Ceramic interconnect technology handbook. CRC Press, Florida.
17. www.dropsens.com; Accessed 2012.
18. www.kanichi-research.com; Accessed 2012.
19. http://www.zensor.com.tw/, Accessed 2012.
20. http://www.mascoprint.co.uk/resources/screenprint/index.htm, Accessed 2012.
21. Mooring L, Karousos NG, Livingstone C, Davis J, Wildgoose GG, Wilkins SJ et al (2004) Evaluation of a novel pad printing technique for the fabrication of disposable electrode assemblies. Sens Actuators B Chem 107:491.
22. Kroger S, Turner APF (1997) Solvent-resistant carbon electrodes screen printed onto plastic for use in biosensors. Anal Chim Acta 347:9.
23. Yang Y-L, Chuang M-C, Lou S-L, Wang J (2010) Thick-film textile-based amperometric sensors and biosensors. Analyst 135:1230.
24. Chuang MC, Yang YL, Tseng TF, Lou SL, Wang J (2009) Flexible thick-film glucose biosensor: influence of mechanical bending on the performance. Talanta 81:15.
25. Coyle S, Wu Y, Lau KT, De Rossi D, Wallace G, Diamond D (2007) Smart nanotextiles: a review of materials and applications. MRS Bull 32:434.
26. Diamond D, Coyle S, Scarmagnani S, Hayes J (2008) Wireless sensor networks and chemo-/biosensing. Chem Rev 108:652.
27. Tudorache M, Bala C (2007) Biosensors based on screen-printing technology, and their applications in environmental and food analysis. Anal Bioanal Chem 388:565.
28. Renedo OD, Alonso-Lomillo MA, Arcos Martinez MJ (2007) Recent developments in the field of screen-printed electrodes and their related applications. Talanta 73:202.

29. Hart JP, Crew A, Crouch E, Honeychurch KC, Pemberton RM (2004) Some recent designsand developments of screen-printed carbon electrochemical sensors/biosensors for biomedical, environmental, and Industrial analyses. Anal Lett 37:789.
30. Newman JD, Turner APF (1992) Biosensors: principles and practice. Essays Biochem 27:147.
31. Albareda-Servent M, Merkoci A, Alegret S (2000) Configurations used in the design of screen-printed enzymatic biosensors. A review. Sens Actuators B 69:153.
32. Fanjul-Bolado P, Hernandez-Santos D, Lamas-Ardisana PJ, Martin-Pernia A, Costa-Garcia A (2008) Electrochemical characterization of screen-printed and conventional carbon paste electrodes. Electrochim Acta 53:3635.
33. Kadara RO, Jenkinson N, Banks CE (2009) Screen printed recessed microelectrode arrays. Sens Actuators B 142:342.
34. Choudhry NA, Kampouris DK, Kadara RO, Banks CE (2009) Disposable highly ordered pyrolytic graphite-like electrodes: tailoring the electrochemical reactivity of screen printed electrodes. Electrochem Commun 12:6.
35. Davies TJ, Hyde ME, Compton RG (2005) Nanotrench arrays reveal insight into graphite electrochemistry. Angew Chem 117:5251.
36. Choudhry NA, Kadara RO, Banks CE (2010) "Cosmetic electrochemistry": the facile production of graphite microelectrode ensembles. Phys Chem Chem Phys 12:2285.
37. Choudhry NA, Kadara RO, Jenkinson N, Banks CE (2010) Screen printed electrodes provide micro-domain sites for fabricating disposable electrocatalytic ensembles. Electrochem Commun 12:406.
38. Honeychurch KC, Hart JP, Cowell DC (2000) Voltammetric behavior and trace determination of lead at a mercury-free screen-printed carbon electrode. Electroanalysis 12:171.
39. Brainina KZ, Zakharchuk NF, Synkova DP, Yudelevich IG (1972) Discharge-ionization of metals on an indifferent electrode. Electroanal Chem 35:165.
40. Choudhry NA, Banks CE (2011) Electrolytically fabricated nickel microrods on screen printed graphite electrodes: electro-catalytic oxidation of alcohols. Anal Methods 3:74.
41. Chiu MH, Yang HH, Liu CH, Zen JM (2009) Determination of lincomycin in urine and some foodstuffs by flow injection analysis coupled with liquid chromatography and electrochemical detection with a preanodized screen-printed carbon electrode. J Chromatogr 877:991.
42. Prasad KS, Muthuraman G, Zen JM (2008) The role of oxygen functionalities and edge plane sites on screen-printed carbon electrodes for simultaneous determination of dopamine, uric acid and ascorbic acid. Electrochem Commun 10:559.
43. Yang TH, Hung CL, Ke JH, Zen JM (2008) An electrochemically preanodized screen-printed carbon electrode for achieving direct electron transfer to glucose oxidase. Electrochem Commun 10:1094.
44. Metters JP, Kadara RO, Banks CE (2012) Electroanalytical sensing of chromium (II) and (VI) utilising gold screen printed macro electrodes. Analyst 137:896.
45. Metters JP, Tan F, Kadara RO, Banks CE (2012) Platinum screen printed electrodes for the electroanalytical sensing of hydrazine and hydrogen peroxide. Anal Methods 4:1272.

46. Welch CM, Hyde ME, Nekrassova O, Compton RG (2004) The oxidation of trivalent chromium at polycrystalline gold electrodes. Phys Chem Chem Phys 6:3153.
47. Vincent JB (2003) Trace elements in viral hepatitis. J Trace Elem Med Biol 16:227.
48. Cespon-Romero RM, Yebra-Biurrun MC, Bermejo-Barrera MP (1996) Preconcentration and speciation of chromium by the determination of total chromium and chromium(III) in natural waters by flame atomic absorption spectrometry with a chelating ion-exchange flow injection system. Anal Chim Acta 327:37.
49. WHO (1993) Guidance for drinking water quality, 2nd edn, vol 1. Recommendations.WHO, Geneva.
50. Golub D, Oren Y (1989) Removal of chromium from aqueous solution by electrochemical treatment on fibrons carbon and graphite electrodes. J Appl Electrochem 19:311.
51. Kieber RJ, Willey JD, Zvalaren SD (2002) Chromium speciation in rainwater: temporal variability and atmospheric deposition. Environ Sci Technol 36:5321.
52. Larochelle JH, Johnson DC (1978) Chromatographic determination of chromium(VI) with coulometric detection based on the electrocatalysis by adsorbed iodine of the reduction at platinum electrodes in acidic solutions. Anal Chem 50:240.
53. Bond AM, Wallace GH (1982) Simultaneous determination of copper, nickel, cobalt, chromium(VI), and chromium(III) by liquid chromatography with electrochemical detection. Anal Chem 54:1706.
54. de Joung J, Brinkman AUT (1978) Determination of chromium(III) and chromium(VI) in sea water by atomic absorption spectrometry. Anal Chim Acta 98:243.
55. Manning DC, Slavin W (1983) The determination of trace elements in natural waters using the stabilized temperature platform furnace. Appl Spectrosc 37:1.
56. Pik AJ, Eckert JM, Williams KL (1981) The determination of dissolved chromium(III) and chromium(VI) and particulate chromium in waters at mg l-1 levels by thin-film X-ray fluorescence spectrometry. Anal Chim Acta 124:351.
57. Cranston RE, Murray JW (1978) The determination of chromium species in natural waters. Anal Chim Acta 99:275.
58. Turyan I, Mandler D (1997) Selective determination of Cr(VI) by a self-assembled monolayer-based electrode. Anal Chem 69:894.
59. Kalembkiewiez J, Soco E (2002) Separation and determination of chromium (III) and chromium (VI) in environmental samples. Wiadomosci Chameczne 56:855.
60. Li Q, Morris KJ, Dasgupta PK, RaimundoJr IM, Temkin H (2003) Determination of chromium, cadmium and lead in food-packaging materials by axial inductively coupled plasma time-of-flight mass spectrometry. Anal Chim Acta 479:151.
61. Korolczuk M, Grabareczyk M (1999) Voltammetric determination of traces of Cr(VI) in the flow system in the presence of bipyridyne. Talanta 49:703.

62. Carrington NA, Yong L, Xue Z-L (2006) Electrochemical deposition of sol–gel films for enhanced chromium(VI) determination in aqueous solutions. Anal Chim Acta 572:17.
63. Brett CMA, Fillipe OMS, Neves CS (2003) Determination of chromium(VI) by batch injection analysis and adsorptive stripping voltammetry. Anal Lett 36:955.
64. Welch CM, Nekrassova O, Compton RG (2005) Reduction of hexavalent chromium at solid electrodes in acidic media: reaction mechanism and analytical applications. Talanta 65:74.
65. del Torno-de RL, Alonso-lomillo MA, Dominguez-Renedo O, Merino-Sanchez C, Merino-Amayuelas MP, Arcos-Martinez MJ (2011) Fabrication and characterization of disposable sensors and biosensors for detection of formaldehyde. Talanta 86:324.
66. de Mattos IL, Gorton L, Ruzgas T (2003) Sensor and biosensor based on prussian blue modified gold and platinum screen printed electrodes. Biosens Bioelectron 18:193.
67. Dutronc P, Carbonne B, Menil F, Lucat C (1992) Influence of the nature of the screen-printed electrode metal on the transport and detection properties of thick-film semiconductor gas sensors. Sens Actuators B 6:279.
68. Zelnick SD, Mattie DR, Stepaniak PC (2003) Occupational exposure to hydrazines: treatment of acute central nervous system toxicity. Aviat Space Environ Med 74:1285.
69. Mo J-W, Ogoreve B, Zhang X, Pihlar B (2000) Cobalt and copper hexacyanoferrate modified carbon fiber microelectrode as an all-solid potentiometric microsensor for hydrazine. Electroanal 12:48.
70. Garrod S, Bollard ME, Nicholls AW, Connor SC, Connelly J, Nicholson JK et al (2005) Integrated metabonomic analysis of the multiorgan effects of hydrazine toxicity in the rat. Chem Res Toxicol 18:115.
71. Wang J, Lu Z (1989) Electrocatalysis and determination of hydrazine compounds at glassy carbon electrodes coated with mixed-valent ruthenium(III, II) cyanide films. Electroanal 1:517.
72. Ebadi M (2003) Electrocatalytic oxidation and flow amperometric detection of hydrazine on a dinuclear ruthenium phthalocyanine-modified electrode. Can J Chem 81:161.
73. Gilbert R, Rioux R (1984) Ion chromatographic determination of morpholine and cyclohexylamine in aqueous solutions containing ammonia and hydrazine. Anal Chem 56:106.
74. Safavi A, Ensafi AA (1995) Kinetic spectrophotometric determination of hydrazine. Anal Chim Acta 300:307.
75. Golabi SM, Zare HR (1999) Electrocatalytic oxidation of hydrazine at glassy carbon electrode modified with electrodeposited film derived from caffeic acid. Electroanal 11:1293.
76. Guerra SV, Kubota LT, Xavier CL, Nakagaki S (1999) Experimental optimization of selective hydrazine detection in flow injection analysis using a carbon paste electrode modified with copper porphyrin occluded into zeolite cavity. Anal Sci 15:1231.
77. White JW, Subers MH, Schepartz AI (1963) The identification of inhibine, the antibacterial factor in honey, as hydrogen peroxide and its origin in a honey glucose-oxidase system. BiochimBiophysActa 73:57.

78. Kiba N, Tokizawa T, Kato S, Tachibana M, Tanai K, Koizumi H et al (2003) Flow-through micro sensor using immobilized peroxidase with chemiluminometric FIA system for determining hydrogen peroxide. Anal Sci 19:823.
79. Hurdis EC, Romeyn H (1954) Accuracy of determination of hydrogen peroxide by cerate oxidimetry. Anal Chem 26:320.
80. Miao X-M, Yuan R, Chai Y-Q, Shi Y-T, Yuan Y-Y (2008) Direct electrocatalytic reduction of hydrogen peroxide based on Nafion and copper oxide nanoparticles modified Pt electrode. J Electroanal Chem 612:157.
81. Welch CM, Banks CE, Simm AO, Compton RG (2005) Direct electrocatalytic reduction of hydrogen peroxide based on Nafion and copper oxide nanoparticles modified Pt electrode. Anal Bioanal Chem 382:12.
82. Batchelor-McAuley C, Du Y, Wildgoose GG, Compton RG (2008) The use of copper(II) oxide nanorod bundles for the non-enzymatic voltammetric sensing of carbohydrates and hydrogen peroxide. Sens Actuators B 135:230.
83. Aldous L, Compton RG (2011) The mechanism of hydrazine electro-oxidation revealed by platinum microelectrodes: role of residual oxides. PhysChemChemPhys 13:5279.
84. Aldous L, Compton RG (2011) Towards mixed fuels: the electrochemistry of hydrazine in the presence of methanol and formic acid. Chemphyschem 12:1280.
85. Zen J, Kumar AS, Tsai D (2003) Recent updates of chemically modified electrodes in analytical chemistry. Electroanalysis 15:1073.
86. Rice ME, Galus Z, Adams RN (1983) Graphite paste electrodes: effects of paste composition and surface states on electron-transfer rates. Electroanal Chem 143:89.
87. Sprules SD, Hartley IC, Wedge R, Hart JP, Pittson R (1996) A disposable reagentless screen-printed amperometric biosensor for the measurement of alcohol in beverages. Anal Chim Acta 329:215.
88. Carsol M-A, Volpe G, Mascini M (1997) Amperometric detection of uric acid and hypoxanthine with xanthine oxidase immobilized and carbon based screen-printed electrode. Application for fish freshness determination. Talanta 44:2151.
89. Ricci F, Palleschi G (2005) Sensor and biosensor preparation, optimisation and applications of Prussian blue modified electrodes. BiosensBioelectron 21:389.
90. Parat C, Betelu S, Authier L, Potin-Gautier M (2006) Determination of labile trace metals with screen-printed electrode modified by a crown-ether based membrane. Anal Chim Acta 573–574:14.
91. Lin J, Zhou DM, Hocevar SB, McAdams ET, Ogorevc B, Zhang XJ (2005) Nickel hexacyanoferrate modified screen-printed carbon electrode for sensitive detection of ascorbic acid and hydrogen peroxide. Front Biosci 10:483.
92. Jazek J, Dilleen JW, Haggett BGD, Fogg AG, Birch BJ (2007) Hexacyanoferrate(III) as a mediator in the determination of total iron in potable waters as iron(II)-1,10-phenanthroline at a single-use screen-printed carbon sensor device. Talanta 71:202.

93. Wring SA, Hart JP, Bracey L, Birch BJ (1990) Development of screen-printed carbon electrodes, chemically modified with cobalt phthalocyanine, for electrochemical sensor applications. Anal Chim Acta 231:203.
94. Napier A, Hart JP (1996) Voltammetric and amperometric studies of selected thiols and dimethyldisulfide using a screen-printed carbon electrode modified with cobalt phthalocyanine: studies towards a gas sensor. Electroanalysis 8:1006.
95. Sehlotho N, Griveau S, Ruille N, Boujtita M, Nyokong T, Bedioui F (2008) Electro-catalyzed oxidation of reduced glutathione and 2-mercaptoethanol by cobalt phthalocyanine-containing screen printed graphite electrodes. Mater Sci Eng 28:606.
96. Gilmartin MAT, Ewen RJ, Hart JP, Honeybourne CL (1995) Voltammetric and photoelectron spectral elucidation of the electrocatalytic oxidation of hydrogen peroxide at screen-printed carbon electrodes chemically modified with cobalt phthalocyanine. Electroanalysis 7:547.
97. Gilmartin MAT, Ewen RJ, Hart JP (1996) Efficacy of organometallic-containing screen-printed carbon strips as catalysts for the oxidation of hydrogen peroxide: a voltammetric and X-ray photoelectron spectroscopic investigation. J Electroanal Chem 401:127.
98. Rawson FJ, Purcell WM, Xu J, Cowell DC, Fielden PR, Biddle N et al (2007) Fabrication and characterisation of novel screen-printed tubular microband electrodes, and their application to the measurement of hydrogen peroxide. Electrochim Acta 52:7248.
99. Abass AK, Hart JP (1997) Electrocatalytic, diffusional and analytical characteristics of a cobalt phthalocyanine modified, screen-printed, amperometric gas sensor for propanethiol. Sensor Actuator 41:169.
100. Hart JP, Hartley IC (1994) Voltammetric and amperometric studies of thiocholine at a screen-printed carbon electrode chemically modified with cobalt phthalocyanine: studies towards a pesticide sensor. Analyst 119:259.
101. Honeychurch KC, Gilbert L, Hart JP (2010) Electrocatalytic behaviour of citric acid at a cobalt phthalocyanine-modified screen-printed carbon electrode and its application in pharmaceutical and food analysis. Anal Bioanal Chem 396:3103.
102. Kozub BR, Compton RG (2010) Voltammetric studies of the redox mediator, cobalt phthalocyanine, with regard to its claimed electrocatalytic properties. Sens Actuators B 147:350.
103. Alarcon-Angles G, Guix M, Silva WC, Ramirez-Silva MT, Palomar-Pardave M, Romero-Romo M et al (2010) Enzyme entrapment by b-cyclodextrin electropolymerization onto a carbon nanotubes-modified screen-printed electrode. BiosensBioelectron 26:1768.
104. Ping J, Wu J, Ying Y (2010) Development of an ionic liquid modified screen-printed graphite electrode and its sensing in determination of dopamine. ElectrochemCommun 12:1738.
105. Ren R, Leng C, Zhang S (2010) A chronocoulometric DNA sensor based on screen-printed electrode doped with ionic liquid and polyaniline nanotubes. BiosensBioelectron 25:2089.

106. Lamas-Adrisana PJ, Queipo P, Fanjul-Bolado P, Costa-Garcia A (2008) Multiwalled carbon nanotube modified screen-printed electrodes for the detection of p-aminophenol: optimisation and application in alkaline phosphatase-based assays. Anal Chim Acta 615:30.
107. Fanjul-Bolado P, Queipo P, Lamas-Ardisana PJ, Costa-Garcia A (2007) Manufacture and evaluation of carbon nanotube modified screen-printed electrodes as electrochemical tools. Talanta 74:427.
108. Khlupova M, Kuznetsov B, Demkiv O, Gonchar M, Csoregi E, Shleev S (2007) Intact and permeabilized cells of the yeast Hansenula polymorpha as bioselective elements for amperometric assay of formaldehyde. Talanta 71:934.
109. Liu M, Hu J, Ma T, Wang S, Ding H (2011) Application of a disposable screen-printed electrode to depression diagnosis for laboratory rats based on blood serotonin detection. Anal Sci 27:839.
110. Stozhko NY, Malakhova NA, Fyodorov MV, Brainina KZ (2008) Modified carbon-containing electrodes in stripping voltammetry of metals. Part II. Composite and microelectrodes. Solid State J Electrochem 12:1219.
111. Wang J, Naser N, Angnes L, Wu H, Chen L (1992) Metal-dispersed carbon paste electrodes. Anal Chem 64:1285.
112. Yoon JH, Muthuraman G, Yang JE, Shim YB, Won MS (2007) Pt-Nanoparticle Incorporated carbon paste electrode for the determination of Cu(II) ion by anodic stripping voltammetry. Electroanalysis 19:1160.
113. Ojani R, Raoof JB, Fathi S (2009) Ferricyanide immobilized within organically modified MCM-41; application for electrocatalytic reduction of hydrogen peroxide. Solid State J Electrochem 13:927.
114. Wu J, Suls J, Sansen W (2000) Amperometric determination of ascorbic acid on screen-printing ruthenium dioxide electrode. ElectrochemCommun 2:90.
115. Kotzian P, Brazdilova P, Kalcher K, Vytras K (2005) Determination of hydrogen peroxide, glucose and hypoxanthine using (bio)sensors based on ruthenium dioxide modified screen printed electrodes. Anal Lett 38:1099.
116. Choudhry NA, Kampouris DK, Kadara RO, Jenkinson N, Banks CE (2009) Next generation screen printed electrochemical platforms: non-enzymatic sensing of carbohydrates using copper(II) oxide screen printed electrodes. Anal Methods 1:183.
117. Hallam PM, Kampouris DK, Kadara RO, Jenkinson N, Banks CE (2010) Nickel oxide screen printed electrodes for the sensing of hydroxide ions in aqueous solutions. Anal Methods 2:1152.
118. Beyene NW, Kotzian P, Schachl K, Alemuc H, Turkušic E, C˘opra A et al (2004) (Bio)sensors based on manganese dioxide-modified carbon substrates: retrospections, further improvements and applications. Talanta 64:1151.
119. Kadara RO, Tothill IE (2008) Development of disposable bulk-modified screen-printed electrode based on bismuth oxide for stripping chronopotentiometric analysis of lead (II) and cadmium (II) in soil and water samples. Anal Chim Acta 623:76
120. Kokkinos C, Economou A (2008) Stripping analysis at bismuth-based electrodes (a review). Curr Anal Chem 4:183.

121. Wang J (2005) Stripping analysis at bismuth electrodes: a review. Electroanalysis 17:1341.
122. Economou A (2005) Bismuth-film electrodes: recent developments and potentialities for electroanalysis. Trends Anal Chem 24:334.
123. Khairy M, Kadara RO, Kampouris DK, Banks CE (2010) Disposable bismuth oxide screen printed electrodes for the sensing of zinc in seawater. Electroanalysis 22:1455.
124. Kruusma J, Banks CE, Compton RG (2004) Mercury-free sono-electroanalytical detection of lead in human blood by use of bismuth-film-modified boron-doped diamond electrodes. Anal Bioanal Chem 379:700.
125. Banks CE, Kruusma J, Hyde ME, Salimi A, Compton RG (2004) Sonoelectroanalysis: investigation of bismuth-film-modified glassy carbon electrodes. Anal Bioanal Chem 379:227.
126. Zhang W, Tang H, Geng P, Wang QH, Jin LT (2007) Amperometric method for rapid detection of Escherichia coli by flow injection analysis using a bismuth nano-film modified glassy carbon electrode. ElectrochemCommun 9:833.
127. Nigovic B, Simunic B, Hocevar S (2009) Voltammetric measurements of aminosalicylate drugs using bismuth film electrode. Electrochim Acta 54:5678.
128. Charalambous A, Economou A (2005) A study on the utility of bismuth-film electrodes for the determination of In(III) in the presence of Pb(II) and Cd(II) by square wave anodic stripping voltammetry. Electrochim Acta 547:53.
129. Yang ML, Zhang ZJ, Hu ZB, Li JH (2006) Differential pulse anodic stripping voltammetry detection of metallothionein at bismuth film electrodes. Talanta 69:1162.
130. Zaouak O, Authier L, Cugnet C, Castetbon A, Potin-Gautier M (2009) Bismuth-coated screen-printed microband electrodes for on-field labile cadmium determination. Electroanalysis 21:689.
131. Hocevar SB, Svancara I, Vytras K, Ogorevc B (2005) Novel electrode for electrochemical stripping analysis based on carbon paste modified with bismuth powder. Electrochim Acta 51:706.
132. Lee GJ, Lee HM, Uhm YR, Lee MK, Rhee CK (2008) Square-wave voltammetric determination of thallium using surface modified thick-film graphite electrode with Bi nanopowder. ElectrochemCommun 10:1920.
133. Quintana JC, Arduini F, Amine A, Punzo F, Destri GL, Bianchini C et al (2011) Part I: a comparative study of bismuth-modified screen-printed electrodes for lead detection. Anal Chim Acta 707:171.
134. Kampouris DK, Banks CE (2010) Electrode design of microelectrode arrays can significantly affect the electro-analytical performance towards chromium (VI) sensing. Environmentalist 104:14.
135. Hart JP, Crew A, Crouch E, Honeychurch KC, Pemberton RM (2007) In: Alegret S, Merkoci A (Eds) Comprehensive Analytical Chemistry, 49, 497.
136. Ball C, Scott DL, Lump JK, Daunert S, Wang J, Bachas LG (2000) Electrochemistry in nanovials fabricated by combining screen printing and laser micromachining. Anal Chem 72:497.
137. Law KA, Higson SPJ (2005) Sonochemically fabricated acetylcholinesterase micro-electrode arrays within a flow injection analyser for the determination of organophosphate pesticides. BiosensBioelectron 20:1914.

138. Pritchard J, Law K, Vakurvov A, Millner P, Higson SPJ (2004) Sonochemically fabricated enzyme microelectrode arrays for the environmental monitoring of pesticides. BiosensBioelectron 20:765.
139. Burmeister J, Palmer M, Gerhardt GA (2005) L-lactate measures in brain tissue with ceramic-based multisite microelectrodes. BiosensBioelectron 20:1772.
140. Chang J-L, Zen J-M (2006) Fabrication of disposable ultramicroelectrodes: characterization and applications. ElectrochemCommun 8:571.
141. Authier L, Groissiord C, Brossier P (2001) Gold nanoparticle-based quantitative electrochemical detection of amplified human cytomegalovirus DNA using disposable microband electrodes. Anal Chem 73:4450
142. Xie X, Stueben D, Berner Z (2005) The application of microelectrodes for the measurements of trace metals in water. Anal Lett 38:2281.
143. Bond AM (1994) Past, present and future contributions of microelectrodes to analytical studies employing voltammetric detection, a review. Analyst 119:R1.
144. Rawson FJ, Purcell WM, Xu J, Pemberton RM, Fielden PR, Biddle N et al (2009) A microband lactate biosensor fabricated using a water-based screen-printed carbon ink. Talanta 77:1149.
145. Honeychurch KC, Al-Berezanchi S, Hart JP (2011) The voltammetric behaviour of lead at a microband screen-printed carbon electrode and its determination in acetate leachates from glazed ceramic plates. Talanta 84:717.
146. Newman JD, Turner APF, Marrazza G (1992) Ink-jet printing for the fabrication of amperometric glucose biosensors. Anal Chim Acta 262:13
147. Butler D, Pravada M, Guilbault GG (2006) Development of a disposable amperometric immunosensor for the detection of ecstasy and its analogues using screen-printed electrodes. Anal Chim Acta 556:333.
148. Luangaram K, Boonsua D, Soontornchai S, Promptmas C (2002) Development of an amperometric immunosensor for the determination of methamphetamine in urine. Biocatal Biotransform 20:397.
149. Ferriera AAP, Colli W, Alves MJM, Oliveira DR, Costa PI, Guell AG et al (2006) Investigation of the interaction between Tc85-11 protein and antibody anti-T. cruzi by AFM and amperometric measurements. Electrochim Acta 51:5046.
150. Pemberton RM, Pittson R, Biddle N, Drago GA, Hart JP (2006) Studies towards the development of a screen printed carbon electrochemical immunosensor array for mycotoxins: a sensor for aflatoxin B1. Anal Lett 39:1573.
151. Alarcon SH, Micheli L, Palleschi G, Compagnone D (2004) Development of an immunosensor for ochratoxin A. Anal Lett 37:1545.
152. Alarcon SH, Palleschi G, Compagnone D, Pascale M, Visconti A, Barna-VL (2006) Monoclonal antibody based electrochemical immunosensor for the determination of ochratoxin A in wheat. Talanta 69:1031.
153. Chikkaveeraiah BV, Mani V, Patel V, Gutkind JS, Rusling JF (2011) Microfluidic electrochemical immunoarray for ultrasensitive detection of two cancer biomarker proteins in serum. Biosens Bioelectron 26:4477.

Chapter 5
Electrochemical Glucose Sensors and Their Application in Diabetes Management

Adam Heller and Ben Feldman

5.1 Scope

5.1.1 Coverage

Over 7,000 peer reviewed articles have been published on electrochemical glucose assays and sensors over recent years. Their number makes a full review of the literature, or even of the most recent advances, impossible. Nevertheless, this chapter should acquaint the reader with the fundamentals of the electrochemistry of glucose and provide a perspective of the evolution of the electrochemical glucose assays and monitors helping diabetic people, who constitute about 5% of the world's population. Because of the large number of diabetic people, no assay is performed more frequently than that of glucose. Most of these assays are electrochemical. The reader interested also in nonelectrochemical

---

The present chapter is a slightly revised version of an article by the authors in Chem Rev 2008, 108 2482–2508. Permission was granted by the Authors, the American Chemical Society, and the Publishers which are greatly acknowledged.

A. Heller (✉)
Department of Chemical Engineering, University of Texas at Austin, Austin, TX 78712, USA
e-mail: heller@che.utexas.edu

B. Feldman
Abbott Diabetes Care, 1360 South Loop Road, Alameda, CA 94502, USA

assays used in, or proposed for, the management of diabetes is referred to a 2007 excellent review of Kondepati and Heise [1].

5.1.2 Exclusion of Studies on Glucose Electrooxidizing Anodes of Cardiac Assist Devices, Pacemakers, Waste-Utilizing Electrical Power Generators, and Bioelectronic Devices

Historically, glucose electrooxidizing anodes have been studied not only because of their importance in diabetes management, but also in the context of glucose-O_2 biofuel cells. The objectives of biofuel cell research were overambitious. After almost 50 years of research, there is still not a single biofuel cell in use. Originally, the biofuel cells were intended to power cardiac assist devices ("artificial hearts") [2, 3], then cardiac pacemakers [4–6], then to supply electrical power to homes or electrical grids by electrooxidizing glucose in, or derived of, wastes [7–9]. The earliest studies already identified insurmountable power density and stability associated limitations, but these were not recognized by all investigators. Today, the power density of the glucose-O_2 biofuel cells remains about 10^4-fold below that required for a cardiac assist device and about 10^3-fold below that necessary to competitively supply power to the electrical grid. Furthermore, the operational lives of low-power-density biofuel cells for cardiac pacemakers are about 10^3 times shorter than required. Biofuel cell research undertaken in the context of unspecified bioelectronic devices for undefined applications [10–16] are not considered in this chapter. Hence, this chapter covers only the glucose anodes of those disposable biofuel cells that might provide for a few weeks the low power required by subcutaneously implanted glucose sensors.

5.2 Roots and Fundamentals

5.2.1 Direct, Nonenzymatic, Electrooxidation, and Electroreduction of Glucose

Glucose was directly electrooxidized to gluconic acid in a sulfuric acid solution at a lead anode in 1909 by Loeb [17]. In 1937, the Atlas Powder

Company manufactured sorbitol commercially by electroreducing glucose in an NaOH–Na_2SO_4 solution at an amalgamated lead electrode in a diaphragm cell [18]. Studies of direct electrooxidation [19] and electroreduction [20] of glucose in basic (pH >11) and acidic (pH <1) solutions continue to date. At pH 7.4 glucose has been directly electrooxidized, at a current density of 1 mA/cm, on an electrode coated with a 4,4′,4″,4‴-tetrasulfo-phthalocyanine complex of molybdenum oxide [21]. Nevertheless, partial oxidation products of glucose irreversibly adsorb on and poison most electrocatalysts [22]. Hence, electrochemical assays of biological glucose solutions utilize glucose oxidation-catalyzing enzymes.

5.2.2 The Enzymes of Glucose Electrooxidizing Anodes

The two families of enzymes that are widely used in the electrooxidation of glucose are glucose oxidases (GOx) [23] and PQQ-glucose dehydrogenases (PQQ-GDH). The wild-type enzymes were originally derived, respectively, from *Aspergillus niger* and *Acinetobacter calcoaceticus*. The wild-type enzymes were replaced by engineered enzymes, produced in other organisms. The purpose of their mutation and expression in different organisms was to increase enzyme yield, facilitate enzyme purification, increase specific activity, improve the enzyme stability, and enhance selectivity for glucose [24–35].

The two enzyme families differ in their redox potentials, the strengths of the bonds between their protein-devoid apoenzymes and their cofactors, their cosubstrates, their turnover rates, their Michaelis constants (K_m), and their selectivity for glucose.

The FAD cofactor of GOx is strongly bound to apo-GOx and $FADH_2$-GOx reacts with O_2 to yield FAD-GOx and H_2O_2. The apparent formal redox potential of GOx at 25°C at pH 5.3 is -0.063 ± 0.011 V vs. SHE; at pH 9.3 it is -0.200 ± 0.010 V vs. SHE [36]. Nevertheless, according to a recent re-estimate the apparent formal redox potential of GOx at pH 7.2 is -0.048 V vs. SHE [37]. GOx is relatively specific for glucose. In the electrochemically relevant half-reaction in which glucose is oxidized by FAD-GOx, about 5×10^3 glucose molecules are oxidized per second [15].

PQQ-GDH catalyzes not only the oxidation of glucose, but also of other sugars; the PQQ cofactor is moderately well bound to the apoenzyme in the presence of excess Ca^{2+}, which also stabilizes the binding of the PQQ cofactor by the apoenzyme [38]. Its redox potential at pH 7.0, in the presence of excess Ca^{2+}, is 10.5 ± 4 mV vs. SHE [38]. Unlike the $FADH_2$ of GOx, the $PQQH_2$ of GDH is not oxidized by O_2 [39]. In the half-reaction of PQQ-GDH, in which glucose is oxidized, 11,800 glucose molecules are oxidized per second.

In addition to PQQ-GDH, two other members of the dehydrogenase family have begun to see application in electrochemical glucose detection. These are NAD-dependent GDH [40] and FAD-dependent GDH [41, 42]. These enzymes combine the oxygen independence of PQQ-GDH with the specificity (toward nonglucose sugars) of GOx, and it is likely that they will be more widely used in the future.

5.2.3 Enzyme-Catalyzed O_2-Oxidation of Glucose

In a 1932 study, Otto Warburg and Walter Christian showed that the "yellow enzyme" from yeast was rendered colorless upon reduction by its then still undefined substrate(s) and that its color was restored upon its reoxidation by shaking with gaseous O_2 [52]. In 1936, Hugo Theorell showed that reaction of the reduced "yellow enzyme" with O_2 produced H_2O_2. Franke and Deffner isolated in 1939 yellow glucose oxidase (GOx) from *Aspergillus niger* and established that the GOx reaction center contained flavin. They also showed that the glucose-reduced flavin of GOx was oxidized by O_2, by cytochrome C, and by quinonoid dyes like toluylene blue, thionine, methylene blue, pyocyanine, and safranine T [23]. The dependence of the Pt electrode potential on O_2 partial pressure was well-known at the time, as was the electrooxidation of H_2O_2 on platinum. Thus, the basis for building GOx-based electrodes that could have monitored glucose concentrations potentiometrically through measuring the drop in O_2 partial pressure upon its consumption in reactions 1 and 2, or amperometrically, upon the electrooxidation of the H_2O_2 produced (reaction 3), existed already in 1939.

$$\text{Glucose} + \text{yellow FAD-GOx} \rightarrow \text{gluconolactone} + \text{colorless } FADH_2 - \text{GOx} \quad (5.1)$$

$$\text{Colorless } FADH_2\text{-GOx} + O_2 \rightarrow \text{yellow FAD-GOx} + H_2O_2 \quad (5.2)$$

$$H_2O_2 \rightarrow 2H^+ + O_2 + 2e^- \tag{5.3}$$

Nevertheless, two decades passed before such electrodes were built. The first electrochemical glucose assay, based on reactions 1–3, was described in 1961 by Malmstadt and Pardue. They added to reactions 1–3 a fourth reaction, that of molybdate-catalyzed H_2O_2-oxidation of I^- to I_2 (reaction 4), to enable determination of the H_2O_2 concentration by I^-/I_2 potentiometry [43, 44].

$$H_2O_2 + 2H^+ + 2I^- \rightarrow 2H_2O + I_2 \tag{5.4}$$

Five years later, Kajihara and Hagihara, then Makino and Konno, monitored glucose concentrations through O_2 consumption by reactions 1 and 2, first without, then in combination with, catalase, decomposing the H_2O_2 produced to water and O_2 (reaction 5) [45, 46].

$$2H_2O_2 \rightarrow 2H_2O + O_2 \tag{5.5}$$

Next, Updike and Hicks [47, 48] simplified the electrochemical glucose assay by immobilizing and thereby also stabilizing GOx. Like Makino and Konno, they coupled the GOx membrane with Leland C. Clark's polarographic O_2 electrode, the membrane of which enabled its use in biological fluids [49, 50]. Note that in the presence of catalase ½ of an O_2 molecule is consumed per glucose molecule (reaction 6), while in the absence of catalase one O_2 molecule is consumed per glucose molecule (reaction 7).

$$2\,\text{glucose} + O_2 \rightarrow 2\,\text{gluconolactone} + 2H_2O \tag{5.6}$$

$$\text{Glucose} + O_2 \rightarrow \text{gluconolactone} + H_2O_2 \tag{5.7}$$

5.2.4 Enzyme-Catalyzed Redox Couple-Mediated Electrooxidation of Glucose

Direct electron tunneling from the $FADH_2$ of GOx to an electrode is much too slow, because the $FADH_2$ is buried at a depth of about 13–15 Å below the electrode-contacting periphery of its glycoprotein

[51]. This is the case even when the enzyme is uniquely oriented to minimize the distance between one of its two $FADH_2$-centers and the surface of an electrode. For a 13–15 Å tunneling-distance the tunneling rate is much slower than the rate of glucose oxidation by FAD-GOx, even when the glucose concentration is much less than its physiological 4–8 mM concentration in blood and other tissues of nondiabetic people. In fact, the rate of FAD-GOx-catalyzed glucose electrooxidation is too slow to be measured in the absence of redox mediators.

Warburg and Christian discovered in 1932 not only that the reduced "yellow enzyme" is oxidized by O_2, but also that it is oxidized, even more rapidly, by the quinoid dye methylene blue [52]. Franke and Deffner similarly showed, in 1939, that the glucose-reduced flavin of GOx, which they discovered, was oxidized not only by O_2, but also by cytochrome C, and by quinoid dyes such as thionine, methylene blue (MB), pyocyanine, and safranine T [23]. Again, their observations could have opened the way to redox couple-mediated electrooxidation of glucose (reactions 8 and 9).

$$FADH_2\text{-}GOx + \text{blue MB} \rightarrow FAD\text{-}GOx + \text{colorless } MBH_2 \quad (5.8)$$

$$\text{colorless } MBH_2 \rightarrow \text{blue MB} + 2H^+ + 2e^- \quad (5.9)$$

It was, however, 30 years later, in 1970, that Silverman and Brake [53] described redox couple-mediated electrooxidation of glucose. They showed that MB/MBH_2, 2,6-dichloroindophenol, indigo disulfonate, phenosafranin, and phenazine methosulfate effectively mediated the electrooxidation of glucose by oxidizing glucose-reduced GOx (reaction 8), that is, $FADH_2$-GOx, and by being electrooxidized (reaction 9) [53].

In many of the studies, but not in the actual home blood-glucose monitoring strips in use, GOx or PQQ-GDH was immobilized on electrodes, most often gold or vitreous carbon, or was immobilized within carbon pastes [54–85]. In other studies the enzyme was deposited as an organized electrocatalytic multilayer [86–90], to facilitate the modeling of the transport of glucose and the mediator [86, 87]. Theoretical models now fully account for the glucose concentration dependence, the pH dependence, the redox mediator concentration dependence, and the GOx loading dependence of the glucose electrooxidation current of electrodes on which GOx is immobilized [91, 92].

5.2.4.1 Organic Mediators

Examples of organic mediators include quinoid dyes [70, 93–97], quinones [54–56, 98–102], oxidized viologens [103–107], and quinone and quinone derivatives [98, 100, 101, 108–112], including polymeric quinones [102, 113–115]. All oxidize glucose-reduced GOx. Mediating quinones have also been synthesized in situ by tyrosinase-catalyzed oxidation of phenols [116, 117]. Among the quinoid dye mediators [118], methylene blue has been the longest studied [15, 16, 23, 53, 119]. Measurements of the bimolecular rate constants for the electron transfer from reduced glucose oxidase to oxidized members of this family of mediators provided the exemplary values of 1.6×10^4 M/s for thionine, 4.0×10^2 M/s for brilliant cresyl blue, 9.8×10^2 M/s for azure A, 9.0×10^3 M/s for daunomycin, and 1.2×10^6 M/s for dopamine [110]. A drawback of some, but not all, quinones is their reaction with cysteine and other protein residues, which destabilizes the mediator-enzyme systems [120–122].

The bimolecular rate constants of the steady-state oxidation of GOx by phenothiazines, phenoxazines, Wurster's salts, dithia- and tetrathia-aromatic compounds, measured at pH 7.0, vary between 10^3 and 10^8 M/s. For phenothiazines, phenoxazine, and Wurster's salts, the rate constants depend on the redox potential and vary according to the outer sphere electron transfer theory of Marcus and Sutin [123]. The rate constants of some of the substituted thiaaromatic compounds with the GOx reaction channels or centers differ, however, from those predicted by the classical electron transfer theory [124].

Glucose electrooxidation, catalyzed by GOx and heterocyclic dihydropolyazines such as 5,10-dihydro-5,10-dimethylphenazine and 1,4-dihydro-1,3,4,6-tetraphenyl-s-tetrazine adsorbed on graphite, has been attributed to the radical cation of the dihydropolyazines [94]. N,N'-Di(4-nitrobenzyl)viologen dichloride, poly(o-xylylviologen dibromide), and poly(p-xylylviologen dibromide) mediate electron transfer from $FADH_2$-GOx to carbon paste electrodes, whereas the redox potentials of other viologens are too reducing to oxidize $FADH_2$-GOx [104].

Quinones and quinoid dyes (Q) also catalyze the electrooxidation of NADH generated when NAD^+-GDH is reduced by glucose (reactions 10–12) [40, 125–129]. Of these, 1,10-phenanthroline quinone is in use in glucose monitoring strips [130].

$$\text{Glucose} + \text{NAD}^+\text{-GDH} \rightarrow \text{gluconolactone}$$
$$+ \text{NADH-GDH} + \text{H}^+ \quad (5.10)$$

$$\text{NADH-GDH} + \text{Q} \rightarrow \text{NAD}^+\text{-GDH} + \text{QH}_2 \quad (5.11)$$

$$\text{QH2} \rightarrow \text{Q} + 2\text{H}^+ + 2\text{e}^- \quad (5.12)$$

5.2.4.2 Inorganic Mediators

The O_2/H_2O_2 couple itself can be considered as the natural mediator for GOx. It is, however, far from being the optimal mediator of GOx electrooxidation, because of the low (~0.2 mM) solubility of O_2 in physiological solutions at ambient temperature. Mediation by the hexacyano-complexes of iron [131, 132], cobalt, and ruthenium has been extensively studied [133–135]. Of these, the $Fe(CN)_6^{3-/4-}$ couple is used in home blood-glucose monitoring strips.

5.2.4.3 Metal–Organic Mediators

The redox potentials of complexes of pentacyanoferrate(III) with pyridine, pyrazole, imidazole, histidine, and aza- and thia-heterocycles or benzotriazole, benzimidazole, and aminothiazole as their sixth ligand span the potential range from 300 to 470 mV vs. SHE at pH 7.2. Study of their rates of oxidation of $FADH_2$-GOx established that the parameters controlling the rate constants included the mediator's self-exchange rate constant, its charge, and its steric fit into and binding by the GOx reaction center [37].

Families of metal–organic redox mediators actually used in blood-glucose monitors include ferrocene-derivatives and $Os^{2+/3+}$-complexes [136–138]. Because Exactech, the first electrochemical home blood-glucose monitor introduced by Genetics International/Medisense, utilized PQQ-GDH and a ferrocene-derivative [139], ferrocenes were extensively studied as electron-shuttling mediators between both GOx and PQQ-GDH and electrodes [140–146]. Ferrocene is small enough to penetrate the reaction channel of GOx [147].

Mediation rates were determined for 42 ferrocene-derivatives and explained by their redox potentials and structure. Their bimolecular rate constants for $FADH_2$-GOx oxidation range from 3×10^4 to 8×10^6 M/s [148]. The most intensively studied ferrocene-derivatives are ferrocenemethanol [88, 90, 91, 149–161] and ferrocenecarboxylic acid [155, 162–164]. Several of the ferrocenes meet the requirements for application in blood-glucose analyzers, which include high solubility in water, fast electron-shuttling, stability, and pH-independence of the redox potential. One of the best mediating ferrocenes for $FADH_2$-GOx electrooxidation is 1,1′-dimethyl-3-(2-amino-1-hydroxyethyl) ferrocene, the mediator of the MediSense ExacTech and Precision QID blood-glucose meters [165–167].

Nickelocene, adsorbed on pyrolytic graphite, has a redox potential of 115 mV vs. SHE. It also mediates the electrooxidation of $FADH_2$-GOx at 220 mV [168]. Manganese cyclopentadienyl (Cp) half-sandwich complexes are comparable in their electron-shuttling rates to ferrocene-derivatives, an exemplary bimolecular rate constant, of $[(hMeC_5H_4)Mn(NO)(CN)_2]Na$, being 2.1×10^5 M/s [167]. The chromium half-sandwich complex $[h-C_6Me_4(NH_2)_2Cr(CO)_3]$, the redox potential of which is +270 mV vs. SHE, displays reversible electrochemistry, but is a relatively slow mediator of $FADH_2$-GOx electrooxidation [167].

The redox potentials of tris-(4,4′-substituted-2,2′-bipyridine) complexes of the group VIII metals $Fe^{2+/3+}$, $Ru^{2+/3+}$, and $Os^{2+/3+}$ range from −100 to +840 mV vs. SHE and their bimolecular rate constants for $FADH_2$-GOx oxidation reach 10^7 M/s. $Os^{2+/3+}$ complexes with amino substituted bipyridines and tris-(4,4′-dimethoxy-2,2′-bipyridine) have particularly high rate constants, suggestive of their binding within the GOx reaction channel or reaction center [138]. The bimolecular rate constants for $FADH_2$-GOx oxidation by substituted 1,10-phenanthroline complexes of $Ru^{2+/3+}$ and $Os^{2+/3+}$ are of 10^6–10^7 M/s when the substituents are electron-donors [169]. cis-$[Ru(LL)_2XY]^{n+}$ complexes (LL = 2,2′-bipyridyl (bpy), 1,10-phenanthroline, and 4,4′-dimethyl-2,2′-bipyridyl (X, Y = Cl^-, B^-, CO_3^{2-}, NO_2^-, SCN^-, N_3^-, H_2O, and DMSO) mediate electrooxidation of $FADH_2$-GOx [170]. Cyclometalated $Ru^{2+/3+}$ compounds are also efficient mediators of $FADH_2$-GOx electrooxidation [171].

5.2.5 Electrical Wiring of GOx by Electron-Conducting Redox Hydrogels

The mediating redox couples can be freely diffusing, protein-bound [172–174], or bound to a peripheral oligosaccharide of GOx via a long, typically 8–13 atom, spacer arm [175]. Redox hydrogels constitute, however, the only known electron-conducting phase in which glucose, gluconolactone, and water-soluble ions dissolve and diffuse. Redox hydrogels, in which GOx is immobilized, catalyze the electrooxidation of glucose [176]. Unlike the electrocatalysts formed of GOx or PQQ-GDH and diffusing redox mediators, the glucose oxidation-catalyzing redox hydrogels have no leachable constituents. Their redox centers are tethered to the insoluble, but water-swollen, cross-linked polymer network of the gel [177–179]. Because the redox hydrogels envelope the redox enzymes, they electrically connect the enzymes' reaction centers to electrodes irrespective of the spatial orientation of the enzyme at the electrode surface and also connect multiple enzyme layers. Hence, the attained true current densities are usually about tenfold higher, and in some cases 100-fold higher, than they are when enzyme monolayers are packed onto electrode surfaces and when most of their redox centers are electrically connected to the electrode surfaces. Specifically, the current densities of glucose electrooxidation on smooth, nonporous, electrodes exceed 1 mA/cm already at 0.0–0.1 V vs. Ag/AgCl [177–180].

5.2.5.1 Mechanism of Electron Conduction in Redox Hydrogels

Redox hydrogels conduct electrons by self-exchange of electrons or holes between rapidly reduced and rapidly oxidized redox functions tethered to backbones of cross-linked polymer networks. Although the networks, which are formed by cross-linking of water-soluble redox polymers, swell in water, they do not dissolve. The redox polymers conduct electrons, or holes, through self-exchange in the water-swollen hydrogels [181]. The self-exchange results from Marcus-type collisional electron transfer [123], which physicists know as phonon-assisted tunneling. Here a reduced redox-species collides with an oxidized redox-species, the reduced species transferring its electron, or the oxidized species transferring its hole. Although, in theory, electrons or holes could also propagate by hopping between fixed-position redox centers [182], trap-to-trap hopping of solid state physics is rarely seen in redox hydrogels.

Because electron transfer by self-exchange requires collisions between reduced (electron-loaded) and oxidized (hole-loaded) redox centers [181, 183], electron diffusion slows when an overwhelming majority of the redox centers are either oxidized or reduced. Thus, the electronic conduction is poor when the hydrogel is poised at a potential far positive or far negative of its redox potential. It is highest when the density of reduced and oxidized centers is about equal, that is, when the hydrogel is poised at its redox potential. The rate of self-exchange of electrons or holes decays exponentially with distance. It is fastest when the redox functions are tethered to the cross-linked polymer networks by long and flexible spacers which are, optimally, between 10 and 15 atom long [178, 180]. The long and flexible spacers increase the amplitude of the displacement of the tethered redox centers. They enable thereby effective electron-transferring collisions, even when the time-averaged distance between the oxidized and the reduced redox centers is between 1 and 3 nm, that is, when the concentration of the tethered redox centers in the fully swollen hydrogel is between 1 and 0.1 M.

The apparent electron diffusion coefficients, D_e, of the redox hydrogels depend on, and are predominantly determined by, the segmental mobility, which increases with hydration and decreases upon excessive cross-linking [181]. Increasing the cationic charge of the redox polymer backbone, either by quaternizing part of the pyridines of a poly(4-vinylpyridine) (PVP)-based redox polymer, or by heavy coordination of the PVP with [Os(bpy)$_2$Cl]$^{2+/3+}$, ensures adequate hydration at any pH and provides a D_e of 3.9×10^{-8} cm/s [183].

Charge neutrality is maintained in any volume element of the hydrogel that contains a redox center. Hence, either anions diffuse upon the transfer of an electron from a reduced to an oxidized redox center in the direction opposite to that of the movement of the electron, or cations diffuse in the direction of the movement of the electron [184], or both anions and cations diffuse in their respective directions. For this reason, the value of D_e may approach, but can never exceed, the diffusion coefficient of the most rapidly diffusing anion or cation present [185]. This limiting value is closely approached when the tethers binding the redox centers to the redox polymer backbone are long and flexible. For a 13 atom-long flexible tether, D_e reaches 5.8×10^{-6} cm/s [178].

5.2.5.2 Mechanism of Direct Glucose Electrooxidation

Catalysis of the direct electrooxidation of glucose involves (a) transfer of electrons (and protons) from glucose to FAD reaction centers of glucose oxidase, which are reduced to $FADH_2$, (b) transfer of electrons from the $FADH_2$ centers to the "wiring" electron-conducting hydrogel, and (c) transport of electrons through the hydrogel to the electrode. The second and third steps are enabled by motion of tethered segments of the cross-linked redox polymer network, allowing the redox centers to approach each other sufficiently ("collide") to transfer electrons or holes. Through such collisions, electrons flow from the protein-buried $FADH_2$ centers to redox centers of the polymer network. The collisions also lead to exchange of electrons between reduced and oxidized centers of the network, which is the underlying cause of the electron conduction by the water-swollen networks. Thus, hydration of the cross-linked redox polymer has two effects, both essential for glucose electrooxidation. It enhances the movement of segments, which are, in the absence of hydration, tightly held by electrostatic (ionic and dipolar) interactions, and it makes the cross-linked polymer permeable to the water-soluble reactant, glucose, and to gluconolactone, the electrooxidation product. Hydration, and with it the electron diffusion coefficient, generally increases with the charge density on the polymer in the wired enzyme systems [176, 186].

5.2.5.3 Organic and Metal–Organic Redox Centers in Electron-Conducting Hydrogels

The redox centers of the hydrogels can be organic or metal–organic. An example of a hydrogel with organic redox centers is the hydrogel formed upon reacting templated, linear-chain, polyaniline, cross-linked, and made hydrophilic by a water-soluble, extended chain diepoxide, like poly(ethylene glycol) diglycidyl ether [187].

When GOx is immobilized in such an organic polymer-based hydrogel, the hydrogel catalyzes the direct electrooxidation of glucose [187].

The most extensively studied direct glucose electrooxidation-catalyzing redox hydrogels are, made however, of GOx and water-soluble polycationic polymers, like PVP, poly(N-vinylimidazole), or poly(acrylamide)-copoly(N-vinylimidazole), with tethered complexes of $Os^{2+/3+}$. The $Os^{2+/3+}$ ligands are typically substituted or unsubstituted pyridine (py), 2,2′-bipyridine (bpy) [188–194], di-*N*-alkylated-2,2′-di-imidazole [178].

5.2.5.4 Mechanical Properties: Balancing the Strength against the Electronic Conductivity

Because the redox hydrogels are formed of water-soluble polymers, their mechanical properties, ranging from soft jellies to tough, leather-like materials, are defined by the molecular mass of the starting polymer and by the extent of cross-linking. And because the segmental mobility, on which the electron diffusion depends, increases with hydration, and because cross-linking limits hydration and thus segmental mobility, mechanical strengths and high apparent electron diffusion coefficients are difficult to achieve simultaneously. They are, nevertheless, simultaneously achieved when the tethers are long and flexible because, in this case, even if the backbones are highly cross-linked, the redox functions at the ends of the tethers can still swing and exchange electrons.

The shear strength of the redox hydrogel films on rotating disk electrodes is conveniently measured by determining the angular velocity at which a drop in the voltammetric peak is first observed [195]. The shear stress, τ, resulting from the rotation, is $\tau = 0.616 \rho^{1/2} \Omega^{3/2} r$. Here ρ is the density of the solution; ν is its kinematic viscosity, μ/ρ; μ is the viscosity of the solution; Ω is the rotation rate; and r is the distance from the center of the rotating disk. Above a critical angular velocity, the part of the hydrogel closest to the rim of the rotating electrode is sheared off, and a drop in the current is observed. Adequately cross-linked redox hydrogels withstand shear stresses of 10^{-2} N/m^2, but are sheared off above 0.1 N/m^2 [195]. Water-soluble cross-linkers, with reactive functions separated by long and flexible spacers, such as 400 Da polyethylene glycol diglycidyl ether, are preferred for the mechanical strengthening of the hydrogels [195]. Films of some of the Os^{3+} complex-containing redox hydrogels can, however, also be cross-linked by ligand exchange [196].

5.2.5.5 Electrodeposition of Glucose Electrooxidation-Catalyzing Electron-Conducting Hydrogels by Ligand Exchange

While inner coordination sphere halides, for example, chlorides, are not exchanged in Os^{3+} complexes, where they are electrostatically strongly bound, they are exchanged by pyridine, imidazole, or primary amine functions if the complex is of electroreduced Os^{2+}, where the electrostatic bond is weaker. Thus, when an electrode is densely covered by adsorbed redox polymer and the Os^{3+} is electroreduced to Os^{2+}, inner coordination sphere halides of one strand are exchanged by backbone pyridine, imidazole, or primary amine functions of proximal adsorbed strands, coordinatively binding the two strands. In effect, the polymer is electrodeposited from its aqueous solution by the reductive cross-linking [196]. Because GOx is a polyanion and forms an electrostatic adduct with the electrodeposited redox polymer, and because the protein of GOx has ligand-exchanging amines at its periphery, the redox hydrogel and GOx can be coelectrodeposited. The electrodeposited films catalyze the electrooxidation of glucose [196–198].

The shelf life of concentrated solutions of the redox polymers that can be electrodeposited by ligand exchange is short when the oxidized redox centers are reduced by a codissolved organic constituent, the reduction causing now unwanted cross-linking and precipitation. Solutions and hydrogels of redox polymers having complexes that cannot exchange ligands are, however, stable. For example, solutions and hydrogels comprising Os^{3+} complexes with six heterocyclic nitrogen ligands, such as tethered $Os(bpy)_3^{2+/3+}$, are particularly stable.

5.2.5.6 Redox Potentials of the Electron-Conducting Hydrogels

The redox potentials of the electron-conducting hydrogels are defined primarily by the transition metal ion of their complex and by its ligands [199]. The reported redox potentials of $Os^{2+/3+}$ comprising hydrogels range from about -0.2 V vs. Ag/AgCl (for the tris N, N'-dialkylated-2,2'-di-imidazole complex-based gels) [178, 180] to $+0.55$ V vs. Ag/AgCl (for the $(4,4'$-dimethyl-2,2'-bipyridine$)_2$(4-aminomethyl-4'-methyl-2,2'-bipyridine)]$^{2+/3+}$ complex comprising gels) [194, 200]. Because the half-cell potential for the exemplary electrode reaction

$[Os(bpy)_3^{3+}][Cl^-]_3 + e^- \leftrightarrow [Os(bpy)_3^{2+}][Cl^-]_2 + Cl^-$ is chloride anion concentration-dependent, the concentration of chloride (or of other anions) in the hydrogels affects the redox potential. When the cross-linked redox polymer is a polycation, for example, partly quaternized PVP, the chloride concentration in the hydrogel can be as high as about 1 M when in equilibrium with a physiological solution, in which the chloride concentration is only 0.14 M. Hence, the redox potential is upshifted by about 50 mV. Because the density of cationic sites increases when the redox polymer is cross-linked and water is squeezed out, excessive cross-linking also upshifts the redox potential, though the shift is typically small, only 10–20 mV.

5.2.5.7 Charge of the Polymer Backbones of the Electron-Conducting Hydrogels

The GOx-wiring redox hydrogels are tailored to be polycations, to avoid partial phase separation from glucose oxidase, which is a polyanion at physiological pH [201]. The redox center of the exemplary polymer I of Fig. 5.1 is tailored to have a redox potential of −0.20 V vs. Ag/AgCl, just slightly oxidizing relative to the FAD/FADH$_2$ center of glucose oxidase.

5.2.5.8 Applications of Glucose Oxidation Electrocatalysts Based on Electron-Conducting Redox Hydrogels

The electron-conducting redox hydrogels serve to electrically connect the redox centers of enzymes to electrodes, enabling their use whenever leaching of electron-shuttling diffusional redox mediators must be avoided. This is the case in glucose concentration monitoring electrodes implanted in diabetic people [202, 203], and in membraneless biofuel cells, the anodes and cathodes of which would be shorted if the mediator could diffuse, and in flow cells for the electroanalysis of glucose.

FreeStyle Navigator of Abbott Diabetes Care of Alameda, CA [204] is a glucose monitoring system for diabetes management, measuring and transmitting the glucose concentration to a PDA-like device about every minute. Its core component, a disposable, miniature subcutaneously implanted amperometric glucose sensor, comprises

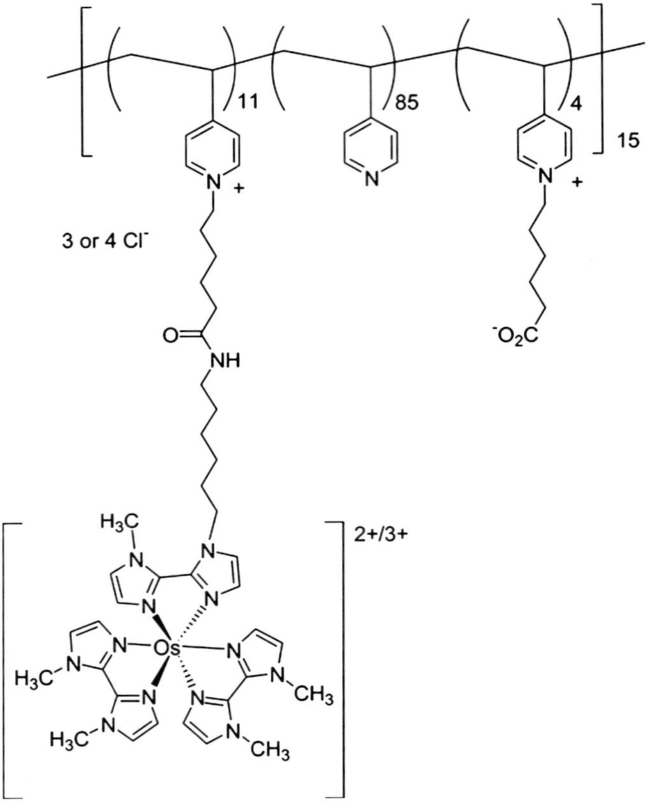

Fig. 5.1 Structure of polymer I, a redox polymer designed to electrically connect the reaction centers of glucose oxidase to electrodes. A 13 atom long tether binds the redox center to the poly(4-vinylpyridine) backbone, which is partially quaternized to make the polymer–water soluble. The apparent electron diffusion coefficient of the redox hydrogel formed upon cross-linking the polymer with polyethylene glycol diglycidyl ether and hydration is 5.8×10^{-6} cm^2/s, and its redox potential is -0.2 V vs. Ag/AgCl. The functions bound to the poly (4-vinylpyridine) backbone are randomly distributed [178, 180]

redox hydrogel-wired glucose oxidase [203]. Comparison of 20,362 measurements of glucose with the Navigator Continuous Glucose Monitoring System in the interstitial fluid, with measurements of venous blood glucose with the Yellow Springs Instrument laboratory reference glucose analyzer, showed a median absolute relative difference (ARD) of 9.3%. The percentage of the FreeStyle Navigator measurements that were in the clinically accurate Clarke error grid A zone was 81.7% and the percentage in the benign error grid B zone was 16.7%. In the first of the recommended 5 days of its use 82.5% of the measurements were in the A zone, and on the fifth day 80.9% were in the A zone [205]. The clinical performance of the FreeStyle Navigator Continuous Glucose Monitoring System in children and adults has been analyzed and reported [206–209].

5.2.6 Metal-Particle GOx-Plug Relay-Based Glucose Electrooxidation Catalysts

In 2003 Xiao et al. reconstituted apo-glucose oxidase using FAD bound to 1.4-nm gold nanocrystals showing that it is possible to electrooxidize glucose when the electrons are relayed to an electrode via a gold-nanoplug in the GOx. The contact to the electrodes was, however, poor and glucose was electrooxidized only at high overpotentials. Current densities were not reported [210]. In 2007 Dagan-Moscovich et al. wired GOx, by modifying the periphery of the enzyme through its reaction with poly(glutaraldehyde), reacting the aldehyde-functions with alanine and reducing by the resulting Schiff-bases Ag^+ to form silver nuclei, on which more silver was precipitated. The Ag^+ ions did not substantially affect the functioning of GOx [211]. Again, the overpotential was high and the current density was not reported.

5.3 Electrochemical Monitoring of the Glucose Concentration by Its GOx-Catalyzed O_2-Oxidation

5.3.1 O_2-Depletion Monitoring upon GOx-Catalyzed O_2-Oxidation of Glucose

The concentration of glucose can be monitored through the GOx-catalyzed oxidation of glucose by O_2 in combination with amperometric monitoring of the rate of decline of the solution O_2-concentration (reaction 6 in the presence of catalase, or reaction 7 in its absence). The O_2 concentration has been monitored either with a Pt electrode or with an indium tin oxide (ITO) electrode, coated with a polymer that was highly O_2 permeable and sufficiently proton-permeable, so as to allow reaction 13 or 14, following, respectively, reaction 6 or 7.

$$O_2 + 4H^+ + 4e^- \rightarrow 2H_2O \tag{5.13}$$

$$O_2 + 2H^+ + 2e^- \rightarrow H_2O_2 \tag{5.14}$$

Rigorous chemical engineering modeling of glucose sensors based on amperometric O_2 monitoring pointed to their performance limits and preferred, but usually difficult to reproducibly manufacture, structures. The parameters affecting their dynamic range, response time, and sensitivity include the membrane thickness, the O_2 permeability of the membrane, the thickness of the GOx-containing film, and the specific activity and loading of the GOx, which have been related through the Damköhler and Biot numbers and the Thiele O_2 moduli [212, 213]. Miniature, low drift and low-cost sensors, made by techniques used in the manufacture of integrated circuits and their interconnects, reaching 90% of the ultimate current in about a minute and responding in the 20–1.4 mM glucose concentration range, have been designed [214–219]. Amperometric probes with transparent and flexible ITO electrodes, monitoring glucose concentrations in the 60–1.2 mM range, were also reported [220]. Furthermore, the high O_2-solubility in, and the resultant high O_2-permeability of, elastomeric silicone was exploited in a poly(dimethylsiloxane)-based carbon-paste glucose sensor, which allowed glucose assays at concentrations as high as 40 mM, with linearity maintained up to 20 mM [221].

5.3.2 Electrooxidation of the H_2O_2 Produced upon Enzyme-Catalyzed O_2-Oxidation of Glucose

The assay of glucose through enzyme-catalyzed reactions in which H_2O_2, the product of reactions 1 and 2, is amperometrically monitored has been the subject of 400 publications. In most, the catalyst of the O_2 oxidation of glucose has been GOx, although the use of pyranose oxidase [222] also has been explored. The amperometric H_2O_2 assay is carried out in one of three formats: (a) catalytic electrooxidation of the H_2O_2 at +0.3–0.8 V vs. SCE, commonly at about 0.6 V vs. SCE; (b) catalytic electroreduction of the H_2O_2, typically near −0.1 V vs. SCE; or (c) H_2O_2-oxidation of a peroxidase, usually horseradish peroxidase, followed by mediated or direct electroreduction of the oxidized peroxidase.

Although Pt on graphite is most often used as the H_2O_2 electrooxidation catalyst, other catalysts and substrates have been studied. The catalyst-substrates [223, 224] studied include carbon pastes, highly oriented pyrolytic graphite, diamond, carbon nanotubes, and conducting polymers, such as polypyrrole and polyaniline. The nonplatinum-based H_2O_2 electrooxidation catalysts include palladium, nickel cyclam, ruthenium, ruthenium–platinum alloys, iridium dioxide, single-walled carbon nanotubes, and polypyrrole functionalized multiwalled-carbon nanotubes [62, 225–235]. Part of the catalyst research has been aimed at reducing the potential at which the H_2O_2 is electrooxidized, that is, at better selectivity in the presence of electrooxidizable interferants [236]. H_2O_2 has been typically electrooxidized on Pt about 600 mV vs. SCE, where ascorbate, urate, and acetaminophen are also electrooxidized. When present, they were excluded through the use of cation exchange or other permselective membranes.

5.3.3 *Electroreduction of H_2O_2 Produced upon Enzyme-Catalyzed O_2-Oxidation of Glucose*

The H_2O_2-electroreduction catalysts studied include Pd, Pd–Pt, Pt-nanowires; Au-nanoparticles distributed in a porous silicate; DNA-Cu^{2+}; DNA-Ag^+; Cu^{2+}, Fe^{2+}; Zn^{2+} and Ce^{3+}-modified silicate xerogels and $Co(CN)_6^{3-/4-}$-chitosan-modified carbon nanotubes [237–244]. Miniaturization of glucose sensors based on H_2O_2-electroreduction provided for their implantation and integration in arrays [245–247]. Membranes improving their selectivity for glucose and extending their dynamic range have been described [248].

5.3.3.1 Peroxidase-Catalyzed H_2O_2 Electroreduction

Amperometric assay of the H_2O_2 produced in reactions 7 or 13 through its oxidation of a peroxidase [163, 236, 249–261] (reaction 15), where HRP^{2+} is Fe(IV)-HRP, that is, peroxidase with four-valent iron, followed by mediated (reactions 16, 17) [163, 262, 263] or direct electroreduction [125, 250] of the peroxidase, has been widely studied. The peroxidase most commonly used is horseradish peroxidase (HRP), though other peroxidases, such as thermostable soybean peroxidase [196, 264–266], have also been used.

$$H_2O_2 + HRP \rightarrow HRP^{2+} + 2OH^- \tag{5.15}$$

$$HRP^{2+} + 2M^{n+} \rightarrow HRP + 2M^{(n+1)+} \tag{5.16}$$

$$2M^{(n+1)+} + 2e^- \rightarrow 2M^{n+} \tag{5.17}$$

5.3.4 Monitoring the Drop in pH upon Enzyme-Catalyzed O_2-Oxidation of Glucose with Field Effect Transistors

Because H$^+$ is released in the electrooxidation of H_2O_2 (reaction 3), glucose concentrations can be monitored by measuring the pH, as long as the influx of glucose and the outflux of protons are well-defined. Hydrolysis of gluconolactone, the oxidation product of GOx or NADH-GDH catalyzed reaction of glucose and oxidants, produces gluconic acid. When the gluconolactone generated (reactions 5 and 7) is rapidly hydrolyzed to gluconic acid, the glucose concentration can be similarly monitored through the associated pH change. Potentiometric glucose monitoring is, however, rarely practiced because of the many parameters affecting the local pH and because the scaling of the signal with the logarithm of the H$^+$ concentration, rather than its scaling linearly, as is the case in amperometric monitoring. Thus, when the rates of generation and neutralization of the acid are known, the concentration of glucose can be related to the local pH at a pH-sensitive device, usually an unencapsulated field effect transistor (FET), referred to as a CHEMFET, proton-sensitive FET, ion-sensitive FET (ISFET), or enzyme-FET (ENFET), all of which are usually made of silicon. Here the current between their source and drain is a function of the surface density and type of charge of the adsorbed ions, hence the pH sensitivity. Typically, the currents of two matched FETs are compared, with GOx immobilized on one and a noncatalytic protein on the other [267–281].

5.4 Central Laboratory and Desktop Glucose Analyzers

5.4.1 The First Central Laboratory Glucose Analyzers

On the basis of the work of Kadish and his colleagues, Beckman Instruments of Fullerton, CA, introduced in 1968 the first commercially available clinical glucose analyzer [282, 283]. In the Beckman glucose analyzer, GOx was added to the analyzed sample, and the decrease in dissolved O_2 concentration was monitored (see Sect. 3.1), initially with the Beckman Polarographic Oxygen sensor electroreducing O_2 on a gold cathode, then with the polarographic O_2-electrode of Clark [284].

The YSI model 23 glucose analyzer was introduced by Yellow Spring Instruments (Yellow Springs, Ohio) in 1974. This model and later YSI glucose analyzers served historically as gold standards with which the accuracy of other glucose analyzers was compared. Even though Hardy Trolander, founder of YSI, and Clark were closely associated, the model 23 and the later YSI glucose analyzers assayed, unlike the Beckman glucose analyzer, not the O_2 consumed, but the H_2O_2 produced in reaction 2 by its electrooxidation on Pt. The core of the YSI Model 23 and of later YSI glucose analyzers was its glucose monitoring probe, which had two polymer layers and an inner Pt electrode, on which the H_2O_2 was electrooxidized to O_2 (reaction 3). The membrane contacting the analyzed solution, which was 30 times diluted when blood or serum was analyzed, was made of polycarbonate. Other than preventing fouling, this membrane reduced the glucose influx, extending thereby the dynamic range and the useful life of the probe. The GOx, which catalyzed the glucose conversion, that is, the generation of H_2O_2, was immobilized on the Pt-side of this outer polycarbonate membrane. The inner membrane was made of cellulose acetate. It excluded most interferents, but was permeable to H_2O_2, allowing its diffusion to the Pt electrode.

The YSI glucose analyzers introduced two significant design principles, which were adopted in later flow-analyzers and in implantable glucose electrodes. First, the influx of glucose was reduced (by dilution of the analyzed solution and by the outer polycarbonate membrane) sufficiently for all of the glucose passing the outer membrane to react with dissolved O_2, even though the concentration of O_2 in water is at saturation only about 0.2 mM at 25°C, while the glucose concentration in the undiluted blood of a diabetic patient can be 30 mM, 150 times higher. Second, reduction of the glucose influx and immobilization (through glutaraldehyde cross-linking) of an initially large excess of GOx on the Pt-side of the outer membrane ensured complete conversion of the glucose that passed the outer membrane, even after most of the initial GOx activity was lost. This extended the operational life of the probe and provided for linear dependence of the current on the glucose concentration as long as all of the glucose influx was converted.

5.4.2 Contemporary Central Laboratory Electrochemical Glucose Analyzers

The current YSI 2300 STAT glucose analyzer utilizes a 25 µL sample, has a throughput one sample per 100 s, measures glucose concentrations up to 50 mM, and its error is the larger of ±2% or 0.2 mM. The working life of its membrane is 21 days.

Eppendorf-Netheler-Hinz GmbH of Hamburg, Germany, and PGW Prüfgeräte-Werk Medingen, Germany, introduced amperometric H_2O_2 electrooxidation-based clinical glucose analyzers in 1986. These analyzers were followed by those of EKF Diagnostic GmbH, Magdeburg, Germany, and of CARE Diagnostica GmbH Voerde, Germany. All use a GOx-comprising membrane, made by BST Bio Sensor Technologie GmbH Berlin, Germany. They combine periodically replaced GOx membranes with permanent, built in, Pt-based O_2 and Ag/AgCl electrodes. The BST GOx membranes are designed to provide selectivity for glucose, low drift, and long operational life. Furthermore, for the recent CARE Diagnostica systems, as well as those of other companies, BST introduced and now manufactures a sensor that is long-lived and comprises the GOx-containing membrane, a Pt–O_2 electrode and an Ag/AgCl electrode, all integrated on a ceramic substrate. The novel BST membranes and sensors are so stable that they are replaced only a few times a year.

Unlike in the Beckmann or YSI systems, which measure the approach of a current plateau, the Eppendorf, CARE, and Voerde systems measure the first derivative of the H_2O_2 electrooxidation current, which allows completion of the assays in less than 5 s. The blood samples are diluted 50-fold, the dilution disrupting the red blood cells, interrupting their glycolysis. Because of the interruption of glycolysis, the diluted blood samples are stable for at least 24 h, contributing to the accuracy of the assays and making these convenient for use by central hospital and clinical laboratories, allowing efficient usage when many blood samples arrive in a narrow time-window and a few samples arrive in others.

A&T Co. of Kanagawa, Japan, produces a GOx/O_2 electrode-based glucose analyzer for hospital and central laboratory usage requiring about 30 µL of blood or serum, with a throughput of about 200 samples/h, providing about ±0.3 mM reproducibility.

5.4.3 Hand-Held Electrochemical Glucose Analyzers for Hospital Wards, Emergency Rooms, and Physician's Offices

In 2004 BST introduced the first hand-held, relatively low-cost electrochemical clinical glucose analyzer (its Glukometer 3000), made with a biosensor required replacement only after 30 days of use or after assay of as many as 1,000 samples. Following assay of a whole blood sample, the fluidics of the BST analyzer allows its cleaning with a few microliters of rinsing solution, reducing the biohazardous fluid-disposal burden. Glucose flux reduction and selectivity for glucose are provided by a film topping of the GOx layer, obviating the need for YSI's inner H_2O_2-selective polymer layer on the H_2O_2 electrooxidizing Pt electrode. The glucose concentrations measured are 0.5–33.3 mM. The system is used in wards of hospitals, emergency rooms, and physician's offices, filling, at low cost, the gap between large central laboratory glucose analyzers and single-use strip-based home blood-glucose monitors, used by self-monitoring diabetic people.

5.5 Home Blood-Glucose Monitors Used by Self-Monitoring Diabetic People

5.5.1 The Need for Glucose Monitoring in Diabetes Management

In diabetes, a disease of which about 150 million people suffer [285], the blood and tissue glucose concentrations are not maintained in their normal range by the controlling feedback loops of the body. The blood-glucose concentration of people not afflicted by diabetes is usually in the 70–120 mg/dL, or 4–8 mM range; it is lower when a person is hungry and higher after a meal. In people with diabetes the range is much wider, 30–500 mg/dL or 2–30 mM. Diabetic people perform annually ~10^{10} glucose assays, far more than all other assays performed by humanity. While a physician treating a sick person needs to know the results of most of the diagnostic assays (s)he prescribes within hours or days, the diabetic self-monitoring diabetic patient needs to know his or her blood-glucose concentration in less than 20 min in order to avoid life-endangering episodes of hypoglycemia. In extreme cases the rate of decline of the blood-glucose concentration of a type 1 diabetic person is as fast as 5 (mg/dL)/min [286].

Type I diabetes affects about 20 million people worldwide and is most frequently diagnosed in children and in young adults. In type 1 diabetes the pancreatic production of insulin, the hormone promoting the uptake of glucose by cells, is impaired. The lives of type I diabetic patients can be sustained only with injections of insulin, which, after meals, lowers the glucose concentration in the blood. In order to maintain their health type I diabetic people need to monitor their blood-glucose concentration 5–6 times a day.

Type II diabetes is much more prevalent than type I diabetes: it afflicts about 5% of the people of the world, most often people who are mature, overweight, and physically inactive. In type II diabetes the insulin produced does not adequately accelerate the uptake of glucose by cells and the insulin-stimulated skeletal-muscle glycogen-synthesis is decreased. The decrease results from reduced insulin-stimulated trans-membrane Glut-4-mediated active glucose transport, caused by intracellular lipid-inhibition of insulin-stimulated insulin receptor substrate (IRS)-1 tyrosine phosphorylation, which reduces the IRS-1-associated phosphatidyl inositol-3 kinase activity [287].

Diabetes has acute and chronic effects. The dramatic, but rare, acute effects, including fainting, coma, and death, result from hypoglycemia. The chronic effects, which are debilitating, result from persistently high glycemia, maintained by many diabetic people in order to avoid the acute effects of hypoglycemia. Persistent maintenance of higher-than-normal blood-glucose concentrations damages the retina, kidneys, nerves, and circulatory system. It is the dominant cause of the reduced longevity of diabetic people and is the leading cause of blindness among the US adults 20–74 years of age, with 12,000–24,000 new cases recorded each year, and also of end-stage renal disease, with about 20,000 cases recorded annually. It accounts for 35% of dialyses and kidney transplants and is the cause of the majority of limb amputations. By controlling his/her glucose concentration within tight limits the diabetic person can avoid the acute hypoglycemia risks and can drastically reduce the likelihood of the devastating complications of diabetes. Tight control is possible only when the diabetic person monitors his or her glucose concentration as often as required by its swings.

5.5.2 Roots of the Electrochemical Glucose Assays Performed by Self-Monitoring Diabetic People

In 1970 Williams, Doig, and Korosi demonstrated the first amperometric assay of blood glucose by a redox couple-mediated, GOx-catalyzed, reaction. They also assayed blood lactate with the FAD-lactate dehydrogenase, FAD-LDH, using $Fe(CN)_6^{3-/4-}$ as redox mediator (reactions 18–20) [98]. $Fe(CN)_6^{3-/4-}$ was subsequently very widely used in electrochemical blood-glucose monitoring strips for diabetes management.

$$FAD\text{-}LDH + lactate \rightarrow FADH_2\text{-}LDH + pyruvate \quad (5.18)$$

$$FADH_2\text{-}LDH + 2Fe(CN)_6^{3-} \rightarrow FAD\text{-}LDH \\ + 2Fe(CN)_6^{4-} + 2H^+ \quad (5.19)$$

$$2Fe(CN)_6^{4-} \rightarrow 2Fe(CN)_6^{3-} + 2e^- \quad (5.20)$$

5.5.3 Gradual Shift from Photonic to Electrochemical Monitoring of Blood Glucose by Self-Monitoring Diabetic People

The 1970 study of Williams, Doig, and Korosi did not lead to rapid application of amperometry in home blood-glucose monitoring. Until 1987 diabetic people assayed their blood glucose by enzyme-based photonic methods, mostly by measuring change in the light reflectance of a dye-containing strip, resulting in an enzyme-catalyzed glucose oxidation reaction [286, 288, 289].

The first electrochemical blood-glucose monitor for self-monitoring diabetic people, which was pen-sized, was disclosed by Higgins, Hill, and Plotkin [290, 291]. It was launched in 1987 as the ExacTech Blood Glucose Meter by Genetics International Inc. of Cambridge, MA. The company subsequently changed its name to MediSense Inc. and was acquired by Abbott Laboratories. In ExacTech, glucose was amperometrically assayed by GOx-catalyzed [292] electroreduction of a ferricinium cation to a ferrocene, which was electrooxidized on a screen-printed carbon-paste electrode of a strip [293, 294]. The ferrocene/ferricinium mediator was based on a study of Cass et al., who showed that ferrocene and its derivatives rapidly shuttled electrons from GOx to electrodes [145].

Today, the majority of the $\sim 10^{10}$ annual glucose assays performed by self-monitoring diabetic people are electrochemical. Unlike the photonic assays, the electrochemical assays do not require conversion of an electron current to a photon flux, and reconversion of a photon flux to an electrical current. In addition, the electrochemical strips require smaller blood sample volumes than the photonic strips: all of the presently available strips utilizing sub-microliter blood sample volumes are electrochemical. Also, electrochemical strips can be more easily integrated with automatic and simple fill-detectors, ensuring that an appropriate volume of blood has been applied to the strip.

The home blood-glucose monitors use plastic or paper strips comprising electrochemical cells and contain PQQ-GDH, NAD-GDH, FAD-GDH, or GOx and a redox mediator. The cells and assays differ in the volume of blood they require, in their structure, in their electrode materials, in their redox mediators, and in their measurement method. Their measurement method can be amperometry or chronoamperometry or coulometry [101]. An example of an amperometric monitor is the Precision Xtra of Abbott/MediSense. The measurement of FreeStyle, the monitor of TheraSense Inc., now also part of Abbott Laboratories, is microcoulometric. FreeStyle measures the blood-glucose concentration in a blood sample as small as 300 nL, which is painlessly obtained [295–299].

Considerations in mediator choice include solubility and rate of dissolution, stability in mixtures with proteins, redox potential, availability and cost, and intellectual property rights. Solubility and fast dissolution are important because the fastest assays are now completed

Table 5.1 Enzyme/mediator combinations of selected electrochemical test-strips

Strip	Enzyme	Mediator
One Touch Ultra	GOx	Ferricyanide
Arkray	PQQ-GDH	Ruthenium hexamine
Ascensia Contour	FAD-GDH	Ferricyanide
BD Test-Strip	GOx	Ferricyanide
FreeStyle	FAD-GDH	Os complex
Precision Xtra	NAD-GDH	Phenanthroline quinone
TrueTrack Smart System	GOx	Ferricyanide
Accuchek Aviva	PQQ-GDH	Proprietary

in 5 s, the mediator dissolving in the blood sample applied in a second or less.

For example, an older family of home blood-glucose monitors of Roche utilizes enzyme-catalyzed oxidation of all of the glucose in the cell by $Fe(CN)_6^{3+}$ and chrono-amperometrically assays the $Fe(CN)_6^{4+}$ produced. An Abbott/MediSense system monitors amperometrically the ferrocene-derivative-mediated electrooxidation of glucose. In the coulometric assay of FreeStyle of Abbott/TheraSense, glucose is electrooxidized through an $Os^{2+/3+}$-mediated reaction. Further examples of mediators used in the most recent home blood-glucose monitors are described in Table 5.1, Above.

5.5.4 Practical Considerations in Home Glucose Test-Strip Design

The modern commercial electrochemical blood-glucose test-strip has a small volume electrochemical cell, utilizes capillary fill, and comprises a stable enzyme and redox mediator. It is accurate (5–10% RMS error vs. a laboratory standard), fast (5–15 s assay time), and requires a small blood volume (0.3–4.0 µL). It is produced in high volume (ca. six billion total electrochemical strips/year in 2007), at high manufacturing yield, and at low cost (5–15 cents/test-strip), with a defect rate of less than 0.1%. It can be stored for at least 18 months at room temperature. It fills reproducibly with blood in less than 3 s, typically in 1 s. The commercially available strips also provide a plethora of additional features, including (1) automatic (nonvisual)

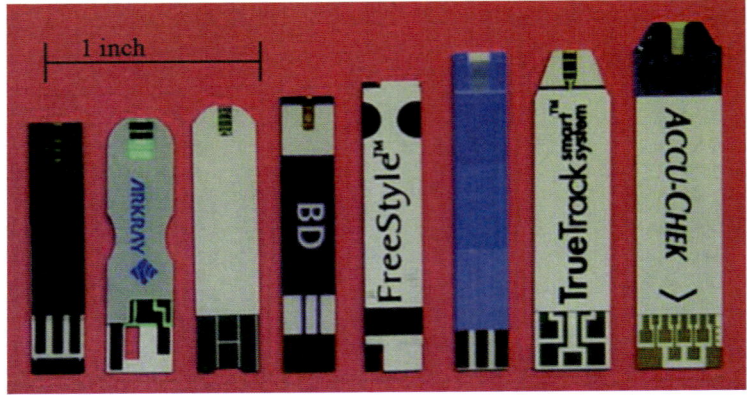

Fig. 5.2 A sampling of electrochemical blood-glucose strips with analyzed blood volumes of ca. 1 µL or less. From left to right, One Touch Ultra, Arkray, Ascensia Contour, BD Test-Strip, FreeStyle, Precision Xtra, TrueTrack Smart System, and Accuchek Aviva

fill detection, (2) code-free operation, (3) the option to fill the strip with multiple blood aliquots over a period of time, (4) automatic control solution detection, and (5) on-strip hematocrit compensation. This section describes strip designs providing such performance. Figure 5.2 shows a sampling of the commercially available electrochemical glucose test-strips.

Typical test-strips are about an inch long and a quarter inch wide. This size is dictated by ergonomic considerations, as the actual sample chamber (visible at the top for many of the strips) is quite small and occupies only a small fraction of the strip area; all of the shown strips require blood sample volumes of 1µL or less. At their bottom, the strips have electrical contact pads, connecting to their respective meters.

The inner workings of a representative strip are pictured in Fig. 5.3, showing six elements common to all electrochemical test-strips. These elements include (1) a plastic substrate material, comprising at least (2) a working electrode and (3) a counter/reference electrode. (The working and counter/reference can also be contained separately on facing plastic substrates, as depicted in Fig. 5.2.) A small volume (ca.1µL) capillary chamber (4) is formed over the plastic substrate(s) and its attached electrodes, often by means of a spacer such as a pressure-sensitive adhesive and a cover layer. The strip

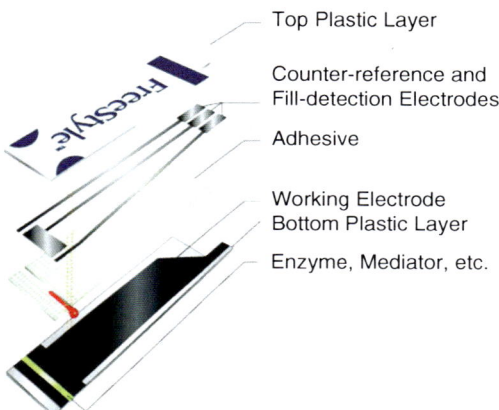

Fig. 5.3 Expanded view of an electrochemical blood-glucose monitoring strip used for diabetes management. The capillary chamber is shown partially filled with a droplet of blood. More than a billion of the strips shown are produced annually

chemistry (5), consisting of an enzyme, a redox mediator, and other components, is distributed (in dry form) within the capillary chamber and generally covers at least the working electrode, and often an entire side of the capillary chamber. Most strips include, in addition, (6) fill detection electrodes, which are vestigial electrodes enabling the meter to detect that the strip is sufficiently filled with blood and to initiate, as soon as the cell is filled, the assay. Some strips employ fill detection strategies that do not require additional electrodes.

Ingenious combinations of these six elements allow construction of electrochemical cells precisely measuring the glucose concentration in near- or sub-microliter blood samples. The six common components of strips are described in the following subsections.

5.5.4.1 Plastic Substrates for Home Glucose Test-Strips

The strip body is generally constructed of a thin (ca. 0.005–0.015 in.) piece of plastic. It serves as a foundation for the electrodes, which are generally deposited by either screen printing or vapor deposition. The substrate material has a high glass transition temperature, so that high temperature process steps (e.g., drying after application of the strip reagents in liquid form) do not cause distortion of the plastic or its electrodes. Its mechanical strength allows physical handling (e.g., insertion into the glucose meter), yet provides for machinability, such that small sections can be rapidly and accurately cut from a large sheet ("web") of material, during strip production. The most widely used materials are polyesters, such as Melinex and Mylar from DuPont.

5.5.4.2 Working Electrodes for Home Glucose Test-Strips

The working electrode is the portal through which glucose-derived electrons exit the sample and enter the meter. It is most commonly constructed of screen-printed carbon ink (a mixture of carbon particles and a polyester binder) or of vapor-deposited Au or Pd. Common electrode configurations are illustrated in Fig. 5.4. The working electrode area must be known and constant for the strips to be reproducibly sensitive to glucose. This area is generally defined by the electrode deposition process (i.e., a reproducible area is deposited or scribed), by an insulating dielectric overlayer which masks a reproducible fraction of the working electrode, or by a combination of the two. Generally the active reagents are deposited over the working electrode, but sometimes they are admixed into the conducting materials. The distance between working and counter/reference electrode is minimized, both to reduce the sample volume, and the interelectrode electrolytic resistance.

5.5.4.3 Counter/Reference Electrodes for Home Glucose Test-Strips

The commercially available electrochemical strips are usually two electrode devices; the counter and reference electrode functions are combined in a single electrode. This counter/reference electrode is the portal through which glucose-derived electrons exit the glucose meter and re-enter the sample. Common configurations are pictured in Fig. 5.4. The counter/reference electrode can be coplanar with the working electrode (in which case it often lies "upstream" of the working electrode) or it can be located on an opposite wall of the capillary

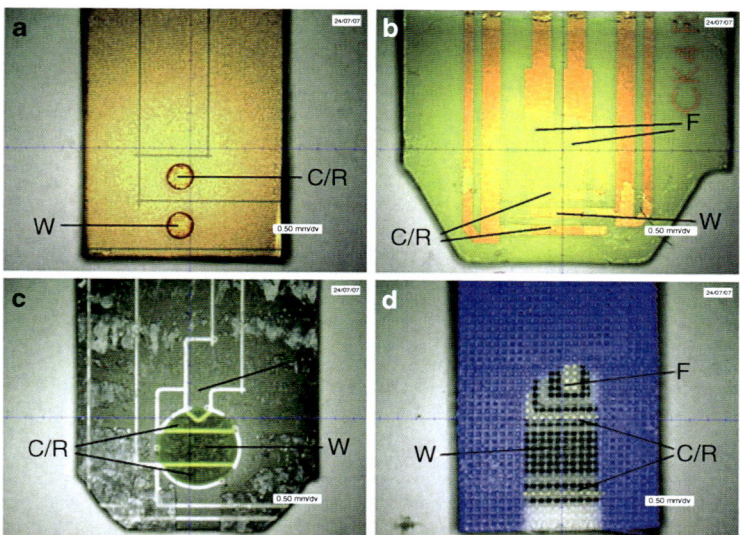

Fig. 5.4 Small volume electrochemically blood-glucose monitoring strips with their top cover layer removed. W, C/R, and F indicate working, counter/reference, and fill detection electrodes, respectively. (**a**) BD test-strip, with electrodes in recessed wells in an insulating layer. (**b**) Accuchek Aviva, its fill electrodes doubling as hematocrit compensation electrodes. (**c**) Ascensia Contour. (**d**) Precision Xtra with electrodes covered by a mesh

cell, such that it "faces" the working electrode. The two often-used types of counter/reference electrodes comprise Ag/AgCl or an inert conductor.

The Ag/AgCl counter/reference electrodes are formed by screen printing an Ag/AgCl ink, which consists of a mixture of Ag particles and AgCl particles in a polyester binding material (some use a single type of particle with an Ag center and an AgCl periphery). Here glucose-derived electrons react with AgCl to produce Ag, thereby ejecting chloride ions into the sample chamber. Such electrodes are generally designed to have an available Coulombic capacity of reducible AgCl which exceeds by ca.1 order of magnitude, the greatest charge the strip will pass in an actual glucose assay; this available charge is typically a few millicoulombs.

The inert conductor counter/reference electrodes are generally made of the material of the working electrode of a particular strip, such as screen-printed carbon or vapor-deposited Au or Pd. This reduces the cost, because the working and counter/reference electrodes are deposited in the same manufacturing step. The counter/reference electrode functions by reducing part of the excess of oxidized mediator in which it is bathed. Thus, initially, in the dry state, both the working and the counter/reference electrodes are coated with a large excess of oxidized mediator. When glucose-containing blood fills the strip, a small fraction of the excess mediator is reduced via the enzyme-catalyzed reaction with glucose. The working electrode oxidizes the reduced mediator, and the counter-reference reduces additional oxidized mediator from the large available pool. Thus, inert conductor counter/reference electrodes are feasible only in the presence of a large stoichiometric excess of the oxidized mediator over the glucose.

5.5.4.4 Capillary Chamber for Home Glucose Test-Strips

The capillary chamber is the "beaker" of the miniature electrochemical cell. It is formed on at least one broad face by the electrode substrate, and on the other by a cover plate (in facing electrode designs, both surfaces are formed by electrode substrates). A spacer forms the edges. Generally its shape is a rectangular solid, with a width of approximately 1 mm, a length of a few millimeters, and a thickness on the order of 0.1 mm, corresponding to a volume of ca. 1 µL. It is open at one end to admit the liquid sample and has an additional opening to allow displaced air to escape as the strip fills. Strip filling can be modeled by the Washburn equation [300], which can be modified to describe capillary flow between two parallel plates:

$$t = 3\mu x^2 / [\sigma \cos(\theta_w)]s$$

where t=filling time, μ=viscosity, x=length (along fill axis), σ=liquid surface tension, θ_w=wetting angle, and s=capillary thickness.

The important result here is that the filling time varies with the square of the chamber length and the inverse of the thickness and the wetting angle. Fill time can, therefore, be decreased greatly by reducing the chamber length. The wetting angle can be decreased by surfactant coating, as described below. In practice, fill times are generally somewhat longer than predicted by the Washburn equation, possibly because of surface discontinuities and formation of the capillary chamber of dissimilar opposite surfaces.

5.5.4.5 Reagents for Home Glucose Test-Strips

The reagents are deposited in dry form over at least the working electrode, but they may cover the entire capillary chamber. In some strips (e.g., Precision Xtra) reagents are mixed directly with the conducting carbon ink, and the mixture is codeposited onto the strip. The reagents include at least an enzyme and a mediator for oxidizing glucose and may further include surfactants to minimize the strip filling time, enzyme stabilizers, and film forming agents, among others. Table 5.1 lists the enzyme/mediator combinations for some of the available electrochemical strips.

Typical surface-active agents enhancing strip-filling include fluorosurfactants, such as DuPont Zonyl, block copolymers of ethylene oxide, and propylene oxide such as BASF Pluronic, or Union Carbide Triton X-100, a nonionic surfactant with a hydrophilic polyethylene oxide group coupled to a hydrocarbon hydrophobic group.

Enzyme stabilizers may include compounds such as monosodium glutamate, trehalose, bovine serum albumin, and buffers (e.g., HEPES, PIPES, etc.) which maintain the enzyme at a favorable pH during storage.

Film forming agents are reputed to improve hematocrit performance, described below. An example of a film-forming agent is silica, used in One Touch Ultra.

5.5.4.6 Fill Detection in Home Glucose Test-Strips

It is of essence that the strip be completely filled with blood when the electrochemical assay commences. This is obvious for coulometric strips which rely upon a precisely known sample volume to calculate a glucose concentration. It is equally true for amperometric strips, where partial working electrode coverage by blood introduces an error. Formerly, filling was visually confirmed by the user; today it is automated to reduce the likelihood of user error and to make the assay faster. Fill detection is realized in electrochemical strips by:

1. Positioning the counter/reference electrode downstream (in the sample flow path) from the reference electrode. This condition guarantees that the working electrode is completely covered with blood before the electrochemical circuit is completed by exposure of the counter/reference to blood.
2. Using identical dual working electrodes, one downstream from the other in the sample flow path. The signal from the two electrodes can then be compared, and the measurement rejected if not sufficiently similar.
3. Adding another "sensing" electrode downstream from the working electrode. When current is detected at the sensing electrode, the strip is deemed full. There are many variations on this theme; typically the sensing electrode is a small additional working or counter/reference electrode.

5.5.5 *Calibration and Characterization of Home Blood-Glucose Test-Strips*

The strips pass in quality-contol laboratories a comprehensive battery of tests before their release. Strip lots are first calibrated and are tested to characterize linearity, coefficient of variation (CV), hematocrit dependence, and response to electrochemical interferents, among others. Tests are generally performed with whole blood, but in some cases a formulated blood substitute ("control solution") is used.

5.5.5.1 Calibration of Home Blood-Glucose Test-Strips

Strips are "factory" calibrated by (1) spiking blood samples to low, medium, and high glucose concentrations, then (2) testing these blood samples by both a reference method and by blood-glucose test-strips. Strip response (current or charge) is plotted vs. the glucose reference value, as shown in Fig. 5.5, above. The resulting slope and intercept are used to select a strip code for use in the meter, or to accept or reject the strip lot for "codeless" systems. Strips (ca. 300 in Fig. 5.5) are tested from throughout the production lot, which may consist of several hundred thousand strips. "Plasma calibration" is generally used, meaning plasma (prepared by centrifuging the whole

5 Electrochemical Glucose Sensors...

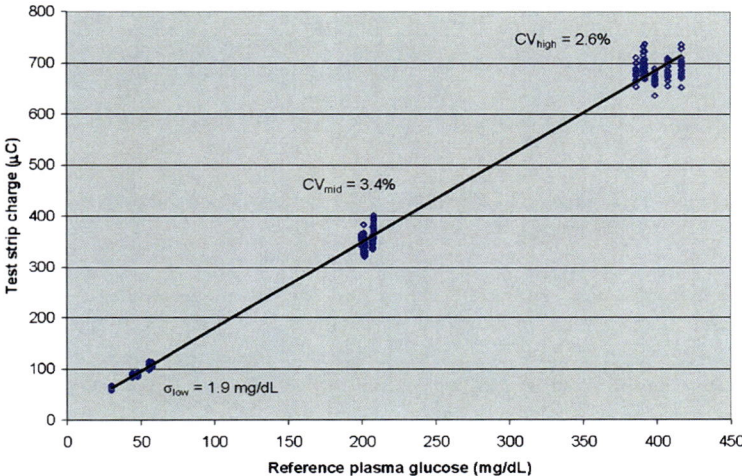

Fig. 5.5 Typical lot calibration plot for a coulometric test-strip, 298 strips were tested

blood sample) is tested on the reference device, while whole blood is tested in the strips. Strips so calibrated report an estimated plasma glucose value, even though they are filled with whole blood. This is done to improve agreement with clinical laboratory glucose measurements, typically performed on plasma.

5.5.5.2 Linearity and Coefficient of Variation of Home Blood-Glucose Test-Strips

Strips should perform accurately over the entire range of clinical interest, generally at least 20–500 mg/dL (ca. 1–30 mM). This performance is gauged by testing multiple (10–20) blood-glucose concentrations spanning the range of interest on both a reference analyzer and the calibrated blood-glucose strips. The coefficient of variation (CV) can be derived from the same data; they are particularly important in gauging a strip's performance, since, unlike non-linearity, data-scatter is difficult to eliminate by adjusting the calibration algorithm. Commercially available strips generally show a CV of 2–5%. The CV is generally higher at very low glucose concentrations, where strip background (signal in the absence of glucose) makes a significant and variable contribution. The CV is lessened by selecting adjacently produced strips from within a given batch: the within-lot CV is generally larger than the within-vial (the unit of commercial packaging) CV.

5.5.5.3 Hematocrit Dependence of Home Blood-Glucose Test-Strips

Hematocrit, the percentage of blood volume occupied by erythrocytes, has a marked effect on the outcome of the strip-based assay, for a number of reasons. Oxygen from erythrocytes can compete with the redox mediator for glucose-derived electrons in strips when the enzyme used is GOx. Also the viscosity of blood increases with hematocrit, the increase slowing the diffusion of all components and reducing the current in amperometric sensors. In general, there is an inverse relationship between hematocrit and strip response. The theoretical minimum hematocrit dependence for a plasma calibrated strip is ca. 0.25%/hematocrit unit, based on the 25% solids composition of an erythrocyte. Commercial electrochemical strips generally have a hematocrit dependence in the range of 0.25–1%/hematocrit unit, although this number can be reduced by hematocrit compensation based on interelectrode impedance measurements. The allowable hematocrit range for electrochemical blood-glucose test-strips varies with strip design; strips with high hematocrit dependence are useful for 25–55% hematocrit, while those with a low innate dependence or hematocrit correction capability can operate over ranges as wide as 0–70% hematocrit.

5.5.5.4 Electrochemical Interferents in Home Blood-Glucose Test-Strips

Electrochemical interferents in blood can cause a false high glucose reading by donating nonglucose-derived electrons. A list of suggested "standard" interferents, developed by the FDA, is shown in Table 5.2, along with interferent concentrations obtained from the National Center for Clinical Laboratory Standards. (NCCLS Guideline EP7-P, Interference Testing in Clinical Chemistry). Strips are tested at low glucose concentrations, with and without the specified interfer-

Table 5.2 List of potential electrochemical interferents[a]

Interferent	Suggested test level (mg/dL)	Ref
Acetaminophen	20	1
Salicylic acid	50	1
Tetracycline	4	1
Dopamine	13	1
Ephedrine	10	2
Ibuprofen	40	1
L-DOPA	5	3
Methyl-DOPA	2.5	1
Tolazamide	100	2
Tolbutamide	100	1
Ascorbic acid	3	1
Bilirubin (unconjugated)	20	1
Cholesterol	500	1
Creatinine	30	1
Triglycerides	3,000	1
Urate	20	1

[a](1) From NCCLS Document EP7-P; (2) Calculated assuming that the drug, at ten times the dosage rate, becomes promptly available in 5 L of blood. (3) ten times the maximum plasma concentration [301]

ents, in order to determine the signal increment due to interferents. From the compounds listed in Table 5.2, those most likely to electrochemically interfere are ascorbate, acetaminophen, and urate. Strips using a carbon working electrode poised at potentials between about −0.1 V and about 0.2 V vs. Ag/AgCl, with fast mediators having redox potentials between about −0.2 V and about 0.1 V vs. Ag/AgCl, do not oxidize urate or acetaminophen, the combined blood concentration of which can be as high as 0.6 mM. However, virtually all commercial strips cross-react with ascorbate. Generally, an interference of less than 10% for an interferent concentration at the upper end of the normal physiological range is considered acceptable.

5.5.5.5 Additional Testing of Home Blood-Glucose Test-Strips

Testing for dependence on temperature, humidity, and the like is also performed. Of particular interest are "error-stacking" experiments, in which a combination of contingencies designed to elicit particularly egregious errors are tested. For example, combinations of high temperature, low hematocrit, and high ascorbate concentration would provide a particularly large positive error. Results obtained in such tests must be judiciously evaluated with regard to their plausible frequency of occurrence in the real world.

5.5.6 *Variables Affecting the Outcome of the Glucose Assays Performed by Self-Monitoring Diabetic People*

Because the well-being of self-monitoring diabetic people depends on the accuracy of the blood-glucose assays they perform, the outcome of the assays must have minimal dependence on parameters that are not independently monitored and compensated for. Some important sources of error have been discussed above, including hematocrit dependence (Sect. 5.6.3) and electrochemical interferents (Sect. 5.6.4). Other sources of error include (1) temperature dependence, (2) skin contamination with glucose or other sugars, (3) improper measurement technique, including miscoding, (4) peritoneal dialysis, and (5) site-to-site variations in glucose concentration.

Because the apparent activation energies can be large, variations with temperature are compensated for, usually by measuring the temperature and applying a correction algorithm. Coulometric systems have an intrinsically lower temperature dependence than amperometric ones, but that does not necessarily translate into a significant difference in operating temperature range (typically about 5–40°C), because the temperature compensation in amperometric systems is quite accurate.

Skin surface contamination is a significant problem, because a blood droplet can dissolve glucose left on finger-tips by food, for example, grapes, in which the concentration of glucose is particularly high. The skin-contamination problem might have been magnified by the trend to smaller blood sample sizes. As the area/volume ratio of the blood droplet increases, so does the concentration of a dissolved skin-contaminant. All product labelings request that users wash with soap and water prior to lancing the skin.

Miscoding occurs when the calibration code (assigned to the strip vial during production) is incorrectly entered into the meter. This is addressed by automatic coding "chips" supplied with the strip vial; however, miscoding is still possible if the coding-chip is not used. The best solution is in code-free strips, which rely on either (1) a rigorous selection process during production to limit sensitivity variations, or (2) identifying readable marks on each strip to allow automatic code assignment by the meter.

Peritoneal dialysis involves injection into the peritoneum, an iso-osmotic fluid. Icodextrin, a polymer of maltose, is frequently used, resulting in significant elevation of the blood maltose level. Strips made with PQQ-GDH cross-react with maltose and must not be used by diabetic patients on peritoneal dialysis [302].

A time-lag may exist between venous glucose levels and those at nonfinger sites, such as the forearm, especially noticeable during periods of rapid change [303]. Lag was observed with an unusual and extreme protocol of Jungheim and Koschinsky for type 1 diabetics: after fasting overnight, the tested people omitted the usual prebreakfast insulin. Instead of breakfast, each patient ingested 75 g of glucose, so that the blood-glucose readings would rise to 300–400 mg/dL; this was followed by the usual mealtime (6–15 units) short-acting dose of insulin. Blood-glucose testing was then performed on the unrubbed forearm and compared to venous levels. Testing under more realistic conditions suggests that the results for the well-rubbed forearm are accurate, although it is recommended that testing specifically for hypoglycemia be performed on the finger [304].

5.6 Diabetes Management Based on Frequent or Continuous Amperometric Monitoring of Glucose

Until the root causes of type I and type II diabetes are eliminated, the life-shortening and quality-of-life damaging consequences of the disease need to be avoided through glucose monitoring systems. Beyond the single-use strips, frequent and semicontinuous systems, some monitoring the glycemia minute-by-minute are now available. The recently introduced wearable systems utilize subcutaneously implanted, innocuous, nearly painlessly inserted and removed amperometric sensors.

5.6.1 Bedside Glucose Monitors Measuring the Blood-Glucose Concentration in a By-Stream of Venous Blood

Although there is no evidence to suggest that diabetes is better managed through monitoring the glycemia in the blood than in the interstitial fluid (the fluid between cells) or in the peritoneal fluid, the standard clinical practice of diabetes care has been, and remains, based on monitoring and controlling the blood glycemia. Because it was deemed that diabetes is best managed by monitoring the glycemia of the blood, the earliest intermittent bedside-monitors measured blood-glucose concentrations [305].

The first bedside system, called the Biostator, was engineered in 1977 by Miles Laboratories of Elkhart, IN. It was a hospital bedside unit, routing a by-stream of blood to an external GOx-based (reaction 7) monitoring unit. Even though only a few hundred units were produced, the Biostator was a core diabetes-research instruments in medical schools and hospitals for more than two decades [306–309].

At this time Via Medical manufactures a bedside venous blood-glucose monitoring system, the via blood-glucose monitor, sampling a venous by-stream at 5- to 10-min intervals. The sensor of the system is based on reaction 7, with the decline in O_2-partial pressure monitored by polarography [310, 311].

5.6.2 Surgeon-Implanted Long-Term Glucose Monitors

Surgeon-implanted, transmitter-containing packages, also based on reaction 7, with footprints larger than 5 cm^2, were subcutaneously implanted in animals. Even though some operated for over 100 days, they are not in clinical use. For longevity, their sensors contain a sufficient amount of GOx, stabilized by cross-linking. Their tissue interface comprises a glucose-flux-limiting membrane, ensuring that it is not the enzyme activity, but the glucose permeation-rate through the membrane, that controls the current [312–315]. With an adequate excess of GOx in the membrane-shielded compartment, the usefulness of the implanted sensor-transmitter package depends on the stabilization of the sensitivity. Maintenance of a fixed sensitivity requires fixed glucose-permeability of the membrane, avoidance of adhesion of glucose-metabolizing cells to the membrane, and prevention of encapsulation by glucose-metabolizing tissue. The vascularization of the tissue encapsulating the sensor changes with time. Because change in vascularization perturbs the balance between glucose supply and consumption, the trend has been toward design of sophisticated membranes, many of them modified polyurethanes [316–318]. Those maintained the desired glucose permeation characteristics and either were not encapsulated by tissue, or were overgrown with unchanging well-vascularized tissue, that is not glucose-depleted.

The long-term implants comprise, in addition to their sensor and its associated electronics, a transmitter, consuming most of the power and necessitating a relatively large high energy capacity battery. The battery-life is extended when the distance of the receiver is shortened, and when the transmissions are less frequent. The trend has, therefore, been to minimize transmission power, accept a short reception distance, and to transmit information about the glucose concentration only at ~10 min intervals.

5.6.3 Systems with Subcutaneous Ultrafiltration and Microdialysis Fibers and Externally Worn Sensors

Continuous electrochemical glucose-monitoring systems based on transporting subcutaneous fluid to an external sensor through an implanted sealed-end ultrafiltration fiber [319–328], or based on forcing the flow of a solution through a microdialysis fiber have been extensively studied [67, 325, 329–353]. The SCGM1 system of Roche Diagnostics, Mannheim, Germany [354] and the GlucoDay system of A. Menarini IFR S.r.l, Florence, Italy [331, 345, 346, 349–352, 355] are microdialysis-based. Their microdialysis fibers are implanted in the abdominal adipose tissue, their fluid flowing to an external amperometric H_2O_2-electrooxidizing GOx-utilizing sensor. The readings of SCGM1 lag by 30 min behind the actual blood glycemia [354].

In microdialysis, an isotonic buffer solution is forced to flow through a hollow and microporous fiber. The flowing solution acquires a glucose concentration, which increases with the concentration of glucose in the surrounding adipose tissue. Equilibration with the subcutaneous adipose tissue fluid is only partial, and the glucose concentration difference between the flowing solution and the subcutaneous fluid depends on the flow rate and on the extent of fouling of the fiber by proteins and adhering cells. The useful life of both the ultrafiltration and the microdialysis-based systems is determined primarily by fouling and bacterial contamination of their fibers and other compartments and only secondarily by inflammation, that is, glucose-consuming macrophages, near the fiber. The growth of any organism in any exterior or interior compartment of the ultrafiltration or microdialysis system, or of any cell on the ultrafiltration or microdialysis fiber, lowers the apparent concentration of glucose and makes the relating of the subcutaneous glucose concentration to the measured current particularly problematic under hypoglycemic conditions.

In ultrafiltration fiber-based systems the implanted end of the fiber is sealed, and the distal end is connected to the externally worn sensor compartment, which is connected to an evacuated cylinder, into which the fluid is sucked. The evacuated container is replaced daily. In microdialysis, one fiber terminus resides in the isotonic buffer solution and the other end is connected to the sensor compartment, which is, in turn, connected to an evacuated container. The equilibration and flow-rates, and the volume of the sensor compartment, determine the lag of the measured glucose concentration behind that in the monitored tissue [340–342, 345, 347, 349, 356]. A novel microdialyzer adds a constant glucose concentration to the perfusate and operates in a pulsatile flow mode, eliminating the need for calibration [343].

5.6.4 *Reverse-Iontophoretic Systems*

The GlucoWatch G2 Biographer (GW2B) sold, then discontinued, by Animas Technologies LLC, now part of Johnson & Johnson, iontophoretically transports fluid across the skin to an externally worn glucose monitor [357]. Its sensor consisted of a pair of Pt-graphite working electrodes, each surrounded by an Ag/AgCl iontophoretic electrode, contacting a GOx-containing hydrogel pad, which, in turn, contacted the skin. A current of 0.3 mA was applied for 3 min through an Ag/AgCl electrode pair, resulting in the iontophoretic transport of glucose into the GOx-containing hydrogel pad, where the glucose was air-oxidized according to reaction 7. The iontophoresys-driving current was then switched off, and a 0.42 V vs. Ag/AgCl potential is applied to the working electrode for 7 min to completely electrooxidize the H_2O_2 produced in reaction 7 [357]. The time required for an assay was ~10 min. The concentration of glucose in the iontophoretically derived fluid was well below that in the subcutaneous interstitial fluid, but the glucose concentrations in the two fluids were related. The GW2B measured glucose concentrations and detected trends, identifying hypoglycemic and hyperglycemic events [358]. A study [359] of 89 pediatric patients, who wore 174 GW2Bs, showed a mean relative absolute difference vs. laboratory serum values of 22%, similar to the then available semicontinuous subcutaneously implanted monitors [360]. Utility was limited by iontophoresis-induced skin irritation and missed readings, particularly in periods of perspiration [361].

5.6.5 Subcutaneously Inserted User-Replaced Miniature Amperometric Sensors

Transcutaneous amperometric sensor-based systems are at this time the dominantly used continuous glucose monitors. A thin, sub-1 mm diameter, flexible sensor, having a working electrode with an immobilized enzyme (usually GOx) and an AgCl/Ag counter or counter-reference electrode, is inserted under the skin. The electrooxidation of glucose is mediated by either O_2 (Sect. 6.5.1), or by an immobilized redox mediator (Sect. 6.5.2). A glucose flux-limiting membrane (Sect. 6.5.3) overlays at least the working electrode of the sensor. Although it limits the glucose electrooxidation currents to the nA range, it provides for linear glucose-response over the range of clinical interest, and also provides a biocompatible tissue interface. Representative transcutaneous sensors are shown in Fig. 5.6.

The sensor is inserted into the subcutaneous fat, to a depth of 5–10 mm, usually by a hollow, retractable sharp. Its leads are connected to a small on-skin potentiostat, typically equipped with a

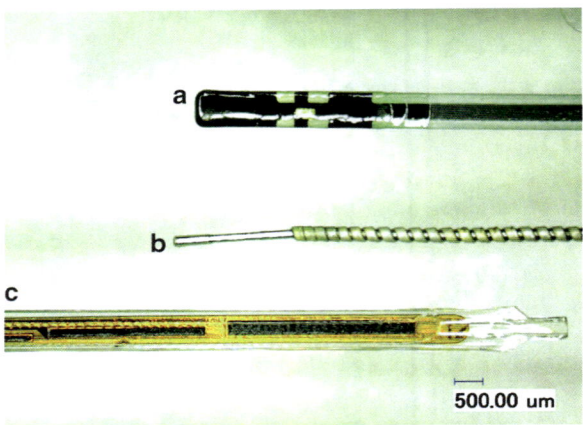

Fig. 5.6 Commercially available transcutaneous sensors. (**a**) FreeStyle Navigator; (**b**) Dexcom STS; (**c**) Guardian RT

wireless transmitter. A suitable break-in period of 1–10 h is then observed, both to equilibrate the sensor with its tissue-environment and to normalize the insertion-wound perturbed site. The sensor is then calibrated *in vivo*, usually by assaying a blood droplet using a strip, and continuous glucose readings then commence, typically every 1–5 min.

5.6.5.1 Subcutaneously Inserted User-Replaced Miniature Sensors Based on GOx-Catalyzed Generation of H_2O_2 and Its Electrooxidation

These systems, reviewed earlier by Wilson and Hu [362], and by Wilson and Gifford [363], contain GOx immobilized on the working electrode and do not contain an immobilized mediator. Their chemical reaction is typically reaction 7 and their electrochemical reactions is typically reaction 3 of section 2.3. The H_2O_2 produced is typically electrooxidized at the working electrode poised near 0.6 V vs. Ag/AgCl, and the H_2O_2-electrooxidation current is monitored. Systems available to diabetic people include Medtronic's Guardian REAL-Time System and DexCom's STS continuous glucose monitor. The sensor of the Guardian REAL-Time System consists of a flexible plastic substrate, less than 1 mm in width, with coplanar working, reference, and counter electrodes formed by a lithographic process [364]. GOx is immobilized on the working electrode, the flux-limiting membrane is coated over the electrode-bearing substrate, and the entire assembly is enclosed in a plastic sleeve equipped with a hole over the working electrode. The system monitors glucose concentrations through the 40–400 mg/dL range and its subcutaneous sensor has a recommended usage time of 3 days between replacements.

The Dexcom STS is a two electrode system, with an Ag/AgCl counter-reference electrode. It comprises an insulated Pt wire, around which a chlorided Ag-wire is coiled. The insulation is stripped off a segment of the Pt, and GOx is immobilized on the wire to form the working electrode. The flux-limiting membrane is deposited around both the working and the counter electrodes. Its monitor reports glucose concentrations in the 40–400 mg/dL range. The sensor is also replaced by the user every 3 days [366]. An improved version of this sensor, with a 7-day wear-time called *Seven*, has been approved by the FDA and is available.

5.6.5.2 Implanted Amperometric Glucose Sensors Built on the Wiring of Glucose Oxidase

Glucose is directly electrooxidized, in the absence of O_2 and without generation of H_2O_2, at electrodes coated with electrocatalysts made by electrically wiring GOx through an electron-conducting redox hydrogel, as described in section 2.5. The first wiring-based amperometric glucose-monitor for diabetes management has been FreeStyle Navigator of TheraSense, now part of Abbott Diabetes Care. Its disposable 5-day wired glucose oxidase glucose sensor is implanted and replaced by the patient with minimal pain and the presence of the miniature sensor under the skin is not felt by the wearer. The surface of the sensor is designed to lessen cell adhesion [365–367]. The sensor consists of a narrow (0.6 mm wide) plastic substrate on which carbon working, Ag/AgCl-reference, and carbon-counter electrodes in a stacked geometry are screen-printed. The active wired enzyme sensing layer covers a small fraction, only about 0.1 mm^2, of the working electrode, and all electrodes are overlaid by a flux-limiting membrane. The sensor resides at about 5 mm depth in the subcutaneous adipose tissue and monitors glucose concentrations over the range 20–500 mg/dL. It is replaced by the user, practically painlessly, after 5 days of use.

5.6.5.3 Flux-Limiting Membranes for Transcutaneous Amperometric Sensors

Membranes on transcutaneous sensors serve two purposes: (1) limiting the glucose flux to the sensing element, and (2) providing a biocompatible interface between the sensor and the body. These functions and their realization are described below.

The current increases linearly with the glucose concentration if all of the arriving glucose molecules are electrooxidized. If the glucose oxidation process is not fast enough to allow the electrooxidation of all of the glucose influx, the electrooxidized fraction of glucose molecules decreases as the glucose concentration increases and a nonlinear current/concentration dependence results. The membranes for the devices described here reduce glucose flux by a factor of approximately 10–100×, enabling the glucose oxidation process to "keep up" with the incoming glucose molecules over the range of roughly 20–500 mg/dL.

Another benefit of encapsulating the sensing layer is the enhancement of its stability. Aging-related decline in sensitivity is avoided by use of the membrane, because even if part of the activity of the glucose electrooxidation catalyst is lost, the residual limiting rate of glucose electrooxidation is still high enough to ensure that all the arriving glucose molecules are electrooxidized. Use of membranes that are excessively resistive to glucose transport is nevertheless avoided, so as to avoid a delay greater than about 3 min between the measured and the true glucose concentration. A longer delay would introduce an excessive measurement error when the glucose concentration rises or declines rapidly.

The glucose transport-limiting membrane defines an apparent Michaelis constant of the sensor, k_M^{app}. Because the current is controlled by the rate of permeation of glucose through the membrane, the temperature dependence of the glucose electrooxidation current is defined by the activation energy for glucose permeation.

The flux of glucose to the sensor membrane surface depends linearly on its solution concentration as long as the viscosity is invariant. This is, however, not the case if the sensor is progressively fouled by a glucose-flux-blocking deposit or if glucose-consuming cells grow on or in the proximity of the sensor. To reduce fouling and adhesion of glucose-consuming cells, the membrane should also be bioinert. Although most membranes were made of polymers, use of cubic-phase lyotropic liquid crystalline membranes having reproducible and uniform thicknesses and easy to reproduce glucose-permeabilities has also been considered [368]. The patent literature discloses proprietary bioinert membrane examples, such as poly(vinylpyridine)-derived hydrogels [369] and swellable polyurethanes [370].

For oxygen-mediated peroxide-measuring sensors, the membranes must be sufficiently permselective for O_2 over glucose for the rate of reaction 7 not to be controlled by the O_2-concentration, which can be in the subcutaneous fluid of a diabetic person 300 times smaller than the glucose concentration (0.1 mM vs. 30 mM) [371–376]. This restriction does not exist for wired enzyme glucose sensors, which do not require oxygen for the glucose-measuring reaction. The latter are therefore compatible with more hydrophilic membrane materials, which do not exhibit as great a preference for oxygen permeation as do some more hydrophobic materials.

5.6.5.4 Calibration of Transcutaneous Amperometric Sensors

When the function relating the signal s and the concentration c is both known and unchanged after implantation, or if it changes predictably after implantation, then a precalibrated sensor does not require recalibration after its implantation. If the function relating s and c changes upon or after implantation, then the sensor must be recalibrated *in vivo*. This last condition generally obtains and "factory calibration" of transcutaneous sensors is not presently offered, for reasons including the following: (1) their more complicated manufacturing process results in greater within-lot coefficient of variation, typically on the order of 10%, as tested *in vitro*, as compared to the 3–5% variation of single-use test-strips; (2) the *in vivo* coefficients of variation being typically larger than the *in vitro* variations, because of the variability of the subcutaneous environment; and (3) sensor output drift and tissue renormalization following insertion-wounding. The *in vivo* calibration is performed by implanting and applying an operating potential to the subcutaneous sensor, allowing a suitable equilibration period of 1–10 h, testing the capillary blood-glucose concentration using a glucose test-strip, and calculating the ratio of the resultant glucose concentration to the transcutaneous sensor current to obtain an operating sensitivity.

Generally, an intercept value (transcutaneous sensor current at zero glucose concentration) must be assumed. The value of the intercept is smaller for wired glucose sensors operating at very low potentials (ca. 40 mV vs. Ag/AgCl), than it is for hydrogen-peroxide electrooxidizing sensors, which operate at ca. 500 mV vs. Ag/AgCl. Note that *in vivo calibration* is inherently less accurate than glucose test-strip factory-calibration, because a consumer glucose meter is usually used for reference in the *in vivo* calibration, and an accurate desktop clinical glucose-analyzer is used for reference in test-strip factory-calibration.

Strategies for improving the accuracy of the *in vivo* calibrations [377, 378] include (1) averaging simultaneous duplicate capillary measurements; (2) rejecting calibrations at extreme glucose concentrations and during periods of extremely rapid glucose concentration change; (3) preassigning a calibration code to the transcutaneous sensor so as to limit the range of sensitivities allowed by the *in vivo* calibration process; (4) incorporating a capillary glucose meter in the transcutaneous sensor hardware to eliminate errors resulting from incorrect user transcription of the capillary glucose value and also to eliminate errors associated with systematic variations between the various commercial glucose meters; and (5) extending the postimplantation waiting period to eliminate erroneous early sensitivities stemming from the insertion-wound-associated trauma.

The *in vivo* calibration process must be repeated during the implantation to correct for possible sensitivity changes. The frequency of these repeat calibrations differs from manufacturer to manufacturer. The FreeStyle Navigator observes a 10 h equilibration period, followed by four *in vivo* calibrations over the following 5 days. The Dexcom STS requires a 1 h equilibration followed by twice daily calibrations over the ensuing 3-day wear, as does the Guardian RT. All systems allow the user to input additional calibration values, if calibration accuracy is deemed to be uncertain.

Calibration-free operation is presently an important research and development objective, both to increase user satisfaction and to eliminate errors in the user-performed calibration process. Its development will require more reproducible transcutaneous sensor manufacturing processes, as well as better understanding of the factors causing sensitivity-variation *in vivo*.

5.6.5.5 The Relationship Between the Glucose Concentrations in Blood and in the Subcutaneous Interstitial Fluid

When the blood-glucose concentration neither rises nor declines, then the glucose concentrations in the blood and in the subcutaneous fluid are similar, or, if not exactly similar, they are related through a proportionality constant. During periods in which the blood-glucose concentration rises or drops, the two concentrations differ.

The relationship of the difference between a sensor-measured glucose concentration and the true glucose concentration is often expressed by a grid proposed by Clarke et al. The zones in this grid are lettered A to E. In zone A the measurement correctly reflects the actual glucose concentration, whereas in zone E the error of the measurement is such that if a clinical decision were based on it, the patient might be harmed. A difference smaller than about 20% between the actual and the measured glucose concentrations is considered as unlikely to lead to an incorrect clinical decision. Greater differences must, however, be avoided, particularly at or near hypoglycemia. The source of the difference between the actual and the measured glucose concentrations can be sensor-related or physiological. A true physiological difference exists when the glucose concentration rises or falls, the difference being a function of the rate at which the glucose concentration changes; the difference is approximately proportional to the slope, dc/dt [379–383].

If, in an animal experiment, the blood-glucose concentration is forced to decline at 6 mg/dL/min, the magnitude of the transient blood-subcutaneous fluid glucose concentration difference can approach the measured concentration. Schmidtke et al. [380] found that when insulin (0.5 U/kg) was intravenously injected in the rat, the subcutaneous concentration was transiently higher, by as much as 84%, than the blood-glucose concentration, the difference reaching its maximum 25 ± 7 min after the injection. Even though the difference was large, they were able to model it mathematically. The modeling provided an algorithm for translating the measured subcutaneous glucose concentration to the blood-glucose concentration at the instant of the subcutaneous measurement. Because the translation required knowledge of dc/dt, its quality depended on the frequency of sampling and thus on the response time, $\tau_{90\%}$, of the sensor. For sensors with $\tau_{90\%}$ of 2 min the average difference between the blood concentrations and the subcutaneous concentrations during the

40-min period after the injection of insulin was reduced from 84 to 29% [381–383].

Under normal conditions, after intramuscular insulin injection in a type I brittle diabetic chimpanzee, the fastest decline in a series of five experiments was 1.8 mg/(dL min) [386]. Although the resulting difference between the blood and the subcutaneous glucose concentrations was readily measurable, the magnitude of the difference was unlikely to lead to an improper clinical decision by the standards of Clarke et al. In these experiments the subcutaneously implanted miniature wired glucose oxidase electrode was operated in conjunction with an on-the-skin ECG Ag/AgCl electrode. The chimpanzee was unconstrained and was trained to wear a small electronic monitor on her wrist and to present her heel for obtaining capillary blood samples. In five sets of measurements with five sensors, averaging 5 h each, 82 capillary blood samples were assayed, their concentrations ranging from 35 to 400 mg/dL. After one-point calibrations were performed at $t=1$ h, the rms error in the correlation between the sensor-measured glucose concentration and that in capillary blood was 17.2%, only 4.9% above the intrinsic 12.3% rms error of the reference strip-monitor. The capillary blood and the subcutaneous glucose concentrations were statistically indistinguishable when the rate of change was <1 mg/(dL min). However, when the rate of decline exceeded 1.8 mg/(dL min) after insulin injection, the subcutaneous glucose concentration was transiently higher.

After considerable research, consensus has emerged on the intrinsic averaged time-lag in humans being about 7 min between the subcutaneous interstitial fluid glucose concentration and the venous blood concentration. Boyne et al. examined 14 patients wearing the Medtronic CGMS and found a mean lag relative to venous blood of about 7 min [384]. A study using the Glucowatch Biographer [385] found a mean delay of 17.2 min, of which 13.5 min were attributed to instrumental delay. Similarly, a study with 30 subjects wearing FreeStyle Navigator found a mean delay of 8 min, of which 3 min could be attributed to sensor response time [378]. A subsequent study [386] with the same sensor, in 58 subjects, found a mean lag of 9.5 ± 4.6 min, again inclusive of a 3-min sensor response time. Standard deviation of this lag was larger between insertions (4.9 min) than within insertions (2.6 min), suggesting real intrasubject and/or intrasite differences.

The continuous glucose monitors Guardian, DexCom, FreeStyle Navigator, and Glucoday had, respectively, during euglycemia, mean absolute relative differences of 15.2, 21.2, 15.3, and 15.6%; during hypoglycemia, 16.1, 21.5, 10.3, and 17.5%. In Clarke error grid analyses, during euglycemia, respectively, 98.9, 98.3, 98.6, and 98.7% of the data points were in zones A+B. Because of frequent loss of sensitivity, the DexCom did not provide sufficient data during hypoglycemia when respectively 84.4, 97.0, and 96.2% of the data points of Guardian, FreeStyle Navigator, and GlucoDay were in Clarke zones A+B [387].

5.6.6 Research Aimed at Integrating a Miniature Power-Source in a 5-Day Patient-Replaced Subcutaneously Implanted Glycemic Status Monitoring and Transmitting Package [388–390]

All presently used batteries contain reactive, corrosive, or toxic components and require cases, usually made of steel. As a battery is miniaturized, the required case dominates its size. Hence, the smallest manufactured batteries are of about 50 mm^3, much larger than the integrated circuits and sensors of functional implantable amperometric glucose sensors for diabetes management.

5.6.6.1 The Potentially Implantable Miniature Zn/AgCl Cell [391]

The presently manufactured small batteries, usually zinc–air, zinc–silver oxide, or lithium–manganese dioxide cells, are difficult to miniaturize because they require a steel case to contain the caustic KOH_{aqu} of Zn batteries or lithium, which reacts rapidly with water. Because the volume fraction of the case increases upon miniaturization, size reduction, though feasible, becomes impractical. The battery could be miniaturized if the anode and the cathode, as well as their reaction products, would be safe enough for implantation in the subcutaneous interstitial fluid so the fluid would serve as the electrolyte, and both the electrolytic solution and case would be obviated.

In a step toward such an implantable battery, the KOH_{aqu} electrolyte and the steel case of a Zn-anode battery were eliminated, opening a route to practical implantable microbatteries. Specifically, the out-diffusing Zn^{2+}, generated in the anode reaction $Zn \rightarrow Zn^{2+} + 2e^-$, was precipitated on the surface of the Nafion-coated Zn anodes by phosphate, to form nonporous lamellae of hopeite-phase $Zn_3(PO_4)_2 \cdot 0.4H_2O$. Surprisingly, the hopeite-phase $Zn_3(PO_4)_2 \cdot 0.4H_2O$ was a Zn^{2+}-cation conducting solid electrolyte. Nonporous inorganic films are usually impermeable to gases, including O_2. Hence, they block or reduce non-Faradaic corrosion. However, because they are rarely ionic conductors, the corrosion-protected Zn anodes can usually not be discharged and are not useful in batteries. The O_2-associated corrosion of the Zn-anode was, nevertheless, reduced in a pH 7.4 physiological (0.15 M NaCl, 20 mM phosphate) buffer solution by growth of a hopeite-phase $Zn_3(PO_4)_2 \cdot 0.4$ H_2O Zn^{2+}-conductive film on the anode. Because the film prevented permeation of O_2 to the electroactive metallic Zn surface, the Zn was efficiently utilized, even when the anode's surface to volume ratio was high and its rate of discharge was slow. For example, the zinc utilization efficiency of 120 μm diameter Zn fiber anodes, discharged over 3-weeks, was 86%. Growth of the nonporous hopeite lamellae and high anode utilization efficiency required precoating with a polyanion, like Nafion, and the addition of a halide like NaCl. The anodes were discharged with little polarization, even when they were overgrown by 100 μm thick hopeite films. At a current density of 0.13 mA/cm^{-2} the excess polarization of the half-discharged, 100 μm hopeite-overgrown anodes was less than 50 mV. At 0.26 mA/cm, the polarization of the half-discharged anodes exceeded their initial polarization only by about 110 mV. The ionic conductance of the hopeite lamellae was $>2 \times 10^{-3}S$.

The zinc anodes, as well as the pH 7.3 physiological buffer electrolyte of the foreseen cells, are harmless enough to be considered for implantation. One cathode explored was the implantable Ag/AgCl cathode, used in FreeStyle Navigator described in section. Because the Ag/AgCl cathode is already used in the body, it is expected that the Ag/AgCl cathode should be safe to implant.

Tests of the Zn(Nafion)-Ag/AgCl cell showed that the nonporous, Zn^{2+} conducting, hopeite-phase $Zn_3(PO_4)_2 \cdot 4H_2O$ grows on the Nafion-coated zinc anode also when the pH 7.4, 0.15 M NaCl, 20 mM phosphate buffer is replaced by serum and that the zinc utilization efficiency remains near 60% when the cell is discharged in serum at a 2-week rate at 1 V. It is projected that the cell will operate in the less fouling subcutaneous interstitial fluid at least as well as it operated in serum.

5.6.6.2 The Potentially Implantable Miniature Zn–O_2 Cell

The energy density of the implantable cell of Sect. 6.6.1 would be increased about tenfold if instead of AgCl/Ag an air-cathode would be used. At physiological pH the cathode could be wired bilirubin oxidase [391–395], overcoated with a bioinert O_2 permeable, proton transporting polymer or hydrogel, or with a cubic-phase lyotropic liquid crystal, which is permeable to O_2, water, and ions but excludes part of the many damaging serum-constituents, including intermediates of urate oxidation [396]. The polarization of the wired bilirubin oxidase O_2 electroreduction catalyst is about 0.3 V less than that of platinum at equal current density, making it superior for use in physiological pH O_2-cathodes [391].

5.6.6.3 The Potentially Implantable Miniature Glucose-O_2 Biofuel Cell

The energy capacity could be further increased by a membrane-less and case-less implanted cell utilizing the glucose and the dissolved O_2 of the subcutaneous interstitial (intercellular) fluid. Because in the glucose-O_2 biofuel neither the cathodic reactant, O_2, nor the anodic reactant, glucose, is cell-contained, but instead diffuse to the electrodes from the subcutaneous fluid, the cell is the smallest. Its laboratory version consists merely of two 7 μm diameter, 2 cm long carbon fibers, one the wired glucose oxidase-coated anode and the other the wired bilirubin oxidase-coated cathode. The cell operates optimally at 0.6 V, where its power density is about 4.8 μW mm^{-2} of carbon fiber area [395, 397, 398].

In its weeklong operation at 37°C such a cell generated ~1 J of electrical energy. The charge passed was 2 C, exceeding about a hundred fold the 0.016 C charge that would have been passed in the discharge of a zinc fiber anode of similar dimension (7-μm diameter, 2 cm long) at 100% utilization efficiency. Unlike other fuel and biofuel cells, the enzyme-wiring-based glucose-O_2 cell is simple and potentially inexpensive, because its anode and cathode compartments need not be separated by a membrane.

The need for the membrane is avoided in biofuel cells when their anodic and cathodic electrocatalysts are wired enzymes. Unlike the platinum alloy electrocatalysts of other fuel cells, the wired enzyme electrocatalysts are so selective for their substrates that neither the crossover of glucose to the cathode compartment, nor the crossover of O_2 to the anode compartment harms the cell. Furthermore, unlike the platinum alloy-utilizing cathodes, which are rapidly poisoned by carbon-containing oxidation intermediates, the wired bilirubin oxidase cathode is not poisoned by glucose nor by its oxidation product, gluconolactone. Because the cell does not have diffusional anodic or cathodic redox mediators, which would short it by the mediator being reduced at the anode and oxidized at the cathode, there is no need for the separation of the compartments. In absence of a membrane, the biofuel cell consists merely of the two wired enzyme-coated electrodes, which would be overcoated in the actually implantable version with a bioinert hydrogels, permeable to glucose, O_2, and ions [389].

The glucose-O_2 biofuel cell operates at this time for about a week in pH 7.3 physiological buffer, containing 0.14 M NaCl and 20 mM phosphate, for about a day in the living grape, the sap of which is particularly rich in glucose [395], but only for hours in serum.

5.7 Concluding Remarks

Electrochemical glucose monitoring has contributed massively to improving the lives of diabetic people. As the prevalence of diabetes is increasing worldwide and as curing of the two types of diabetes remains elusive, humanity is likely to further benefit from advances in the electrochemical monitoring of glycemia.

Following the single-use strips, which now painlessly and accurately monitor glucose concentrations in 300 nL samples of blood, increasingly advanced, continuous glucose monitors are being introduced. Their already small subcutaneous sensors will be further miniaturized, and their 5–7-day operating lives will be extended. Through increasingly accurate and continuous measurement of the first and second derivatives, the diabetic user will be adequately forewarned if and when corrective action is needed to maintain a healthier narrower, closer to normal, glycemic range. The pain of blood-glucose monitoring has been eliminated and the constant worry of diabetic people is now being eliminated through progress in bioelectrochemistry.

References

1. Kondepati VR, Heise HM. Anal Bioanal Chem. 2007;388:545.
2. Appleby AJ, Ng DYC, Weinstein H. J Appl Electrochem. 1971;1:79.
3. Giner, J.; Holleck, G. L.; Turchan, M.; Fragala, R. *Intersociety Energy ConVersion Eng. Conference Proc.* 1971, 256.
4. Rao JR, Richter G. Naturwissenschaften. 1974;61:200.
5. Rao JR, Richter G, Von Sturm F, Weidlich E, Wenzel M. Biomed Eng. 1974;9:98.
6. Weidlich E, Richter G, Von Sturm F, Rao JR, Thoren A, Lagergren H. Biomater Med Devices Artif Organs. 1976;4:277.
7. Turner APF, Aston WJ, Higgins IJ, Davis G, Hill HAO. Biotechnol Bioeng Symp. 1982;12:401.
8. Bennetto, H. P.; Delaney, G. M.; Mason, J. R.; Roller, S. D.; Stirling, J. L.; Thurston, C. F. *Resour. Appl. Biotechnol.* 1989, 363.
9. Chaudhuri SK, Lovley DR. Nat Biotechnol. 2003;21:1229.
10. Skotheim TA, Lee HS, Hale PD, Karan HI, Okamoto Y, Samuelson L, et al. Synth Met. 1991;42:1433.
11. Di Fabrizio E, Gentili M, Morales P, Pilloton R, Mela J, Santucci S, et al. Appl Phys Lett. 1996;69:3280.
12. Willner I, Katz E, Willner B, Blonder R, Heleg-Shabtai V, Buckmann AF. Biosens Bioelectron. 1997;12:337.
13. Willner I. Science. 2002;298:2407.
14. Muguruma H, Kase Y, Uehara H. Anal Chem. 2005;77:6557.
15. Willner I, Baron R, Willner B. Biosens Bioelectron. 1841;2007:22.
16. Willner I, Willner B, Katz E. Bioelectrochemistry. 2007;70:2.
17. Loeb W. Biochem Z. 1909;17:132.
18. Taylor RL. Chem Met Eng. 1937;44:588.
19. Casella IG, Destradis A, Desimoni E. Analyst. 1996;121:249.
20. Fei S, Chen J, Yao S, Deng G, Nie L, Kuang YJ. Solid State Electrochem. 2005;9:498.
21. Arzoumanidis GG, O'Connell JJJ. Phys Chem. 1969;73:3508.
22. Rao MLB, Drake RFJ. Electrochem Soc. 1969;116:334.
23. Franke W, Deffner M. Ann. 1939;541:117.
24. Hodgkins M, Mead D, Ballance DJ, Goodey A, Sudbery P. Yeast. 1993;9:625.
25. Zhu Z, Wang M, Gautam A, Nazor J, Morneu C, Prodanovic R, et al. Biotechnol J. 2007;2:241.
26. Whittington H, Kerry-Williams S, Bidgood K, Dodsworth N, Peberdy J, Dobson M, et al. Curr Gen. 1990;18:531.
27. Park EH, Shin YM, Lim YY, Kwon TH, Kim DH, Yang MSJ. Biotechnology. 2000;81:35.

28. Pulci V, D'Ovidio R, Petruccioli M, Federici F. Lett Appl Microbiol. 2004;38:233.
29. Persson M, Buelow L, Mosbach K. FEBS Lett. 1990;270:41.
30. Sode K, Witarto AB, Watanabe K, Noda K, Ito S, Tsugawa W. Biotechnol Lett. 1994;16:1265.
31. Olsthoorn AJJ, Duine JA. Arch Biochem Biophys. 1996;336:42.
32. Dewanti AR, Duine JA. Biochemistry. 1998;37:6810.
33. Iswantini D, Kano K, Ikeda T. Biochem J. 2000;350:917.
34. Kojima K, Witarto AB, Sode K. Biotechnol Lett. 2000;22:1343.
35. Krishnaraj PU, Goldstein AH. FEMS Microbiol Lett. 2001;205:215.
36. Stankovich MT, Schopfer LM, Massey VJ. Biol Chem. 1978;253:4971.
37. Kulys J, Tetianec L, Ziemys AJ. Inorg Biochem. 2006;100:1614.
38. Sato A, Takagi K, Kano K, Kato N, Duine JA, Ikeda T. Biochem J. 2001; 357:893.
39. Duine JA, Frank J, Jr, Van Zeeland JK, *FEBS Lett.* 1979;108:443–6.
40. Boguslavsky LI, Geng L, Kovalev IP, Sahni SK, Xu Z, Skotheim TA, et al. Biosens Bioelectron. 1995;10:693.
41. Tsujimura S, Kojima S, Kano K, Ikeda T, Sato M, Sanada H, et al. Biosci Biotechnol Biochem. 2006;70:654.
42. Theorell H. Biochem Z. 1936;288:317.
43. Malmstadt HV, Pardue HL. Anal Chem. 1961;33:1040.
44. Pardue HL, Simon RK, Malmstadt HV. Anal Chem. 1964;36:735.
45. Kajihara T, Hagihara B. Rinsho Byori. 1966;14:322.
46. Makino Y, Konno K. Rinsho Byori. 1967;15:391.
47. Updike SJ, Hicks GP. Nature. 1967;214:986.
48. Updike SJ, Hicks GP. Science. 1967;158:270.
49. Clark Jr LC, Sachs G. Ann N Y Acad Sci. 1968;148:133.
50. Clark, L. C., Jr.; Clark, E. W. *Int. Anesthesiol. Clinics* 1987, 25 (1), 29.
51. Hecht HJ, Kalisz HM, Hendle J, Schmid RD, Schomburg DJ. Mol Biol. 1993;229:153.
52. Warburg O, Christian W. Biochem Z. 1932;254:438.
53. Silverman, H. P.; Brake, J. M. U.S. Patent 3,506,544, 1970.
54. Ikeda T, Hamada H, Miki K, Senda M. Agric Biol Chem. 1985;49:541.
55. Ikeda T, Katasho I, Senda M. Anal Sci. 1985;1:455.
56. Ikeda T, Hamada H, Senda M. Agric Biol Chem. 1986;50:883.
57. Miki K, Ikeda T, Todoriki S, Senda M. Anal Sci. 1989;5:269.
58. Gorton L, Karan HI, Hale PD, Inagaki T, Okamoto Y, Skotheim TA. Anal Chim Acta. 1990;228:23.
59. Amine A, Kauffmann JM, Patriarche GJ. Talanta. 1991;38:107.
60. Amine A, Kauffmann JM, Patriarche GJ, Kaifer AE. Anal Lett. 1991;24:1293.
61. Xie Y, Huber CO. Anal Chem. 1991;63:1714.
62. Wang J, Naser N, Angnes L, Wu H, Chen L. Anal Chem. 1992;64:1285.
63. Andrieux CP, Audebert P, Divisia-Blohorn B, Linquette-Maillet SJ. Electroanal Chem. 1993;353:289.
64. Kulys J, Hansen HE, Buch-Rasmussen T, Wang J, Ozsoz M. Anal Chim Acta. 1994;288:193.
65. Taliene VR, Ruzgas T, Razumas V, Kulys JJ. Electroanal Chem. 1994;372:85.

66. Andrieux CP, Audebert P, Bacchi P, Divisia-Blohorn BJ. Electroanal Chem. 1995;394:141.
67. Csoeregi E, Laurell T, Katakis I, Heller A, Gorton L. Mikrochim Acta. 1995;121:31.
68. Kulys J, Hansen HE. Anal Chim Acta. 1995;303:285–94.
69. Kulys J, Krikstopaitis K, Ruzgas T, Razumas V. Mater Sci Eng C. 1995;C3:51.
70. Kulys J, Wang L, Hansen HE, Buch-Rasmussen T, Wang J, Ozsoz M. Electroanalysis. 1995;7:92.
71. Saby C, Mizutani F, Yabuki S. Anal Chim Acta. 1995;304:33.
72. Jezkova J, Iwuoha EI, Smyth MR, Vytras K. Electroanalysis. 1997;9:978.
73. Wang J, Liu J, Cepra G. Anal Chem. 1997;69:3124.
74. Compagnone D, Schweicher P, Kauffman JM, Guilbault GG. Anal Lett. 1998;31:1107.
75. Fernandez JJ, Lopez JR, Correig X, Katakis I. Sens Actuators B. 1998;B47:13.
76. Huang T, Warsinke A, Koroljova-Skorobogat'ko OV, Makower A, Kuwana T, Scheller FW. Electroanalysis. 1999;11:295.
77. Wang J, Chen L, Jiang M, Lu F. Anal Chem. 1999;71:5009.
78. Moscone D, D'Ottavi D, Compagnone D, Palleschi G, Amine A. Anal Chem. 2001;73:2529–35.
79. Rodriguez MC, Rivas GA. Electroanalysis. 2001;13:1179–84.
80. Wang J, Mo JW, Li S, Porter J. Anal Chim Acta. 2001;441:183.
81. Evans RG, Banks CE, Compton RG. Analyst. 2004;129:428.
82. Lawrence NS, Deo RP, Wang J. Anal Chem. 2004;76:3735.
83. Ojani R, Raoof J-B, Salmany-Afagh PJ. Electroanal Chem. 2004;571:1.
84. Laurinavicius V, Razumiene J, Ramanavicius A, Ryabov AD. Biosens Bioelectron. 2004;20:1217.
85. Luque GL, Rodriguez MC, Rivas GA. Talanta. 2005;66:467.
86. Anicet N, Anne A, Moiroux J, Saveant J-MJ. Am Chem Soc. 1998;120:7115.
87. Anicet N, Anne A, Bourdillon C, Demaille C, Moiroux J, Saveant J-M. Faraday Discuss. 2000;116:269.
88. Zhang S, Yang W, Niu Y, Sun C. Anal Chim Acta. 2004;523:209.
89. Ben-Ali S, Cook DA, Bartlett PN, Kuhn AJ. Electroanal Chem. 2005;579:181.
90. Sun Y, Yan F, Yang W, Sun C. Biomaterials. 2006;27:4042.
91. Bourdillon C, Demaille C, Moiroux J, Saveant JMJ. Am Chem Soc. 1993; 115:2.
92. Bartlett PN, Pratt KFEJ. Electroanal Chem. 1995;397:61.
93. Kulys J, Cenas NK. Biochim Biophys Acta. Protein Structure and Molecular Enzymology. 1983;744:57.
94. Kulys J, Palaima A, Urbelis G. Anal Lett. 1998;31:569.
95. Kulys J, Razumas V, Kazlauskaite J, Marcinkeviciene J, Buch- Rasmussen T, Hansen HE, et al. Electroanalysis. 1994;6:740.
96. Kulys JJ. Biosensors. 1986;2:3.
97. Kulys JJ, Samalius AS, Svirmickas GJS. FEBS Lett. 1980;114:7.
98. Williams DL, Doig Jr AR, Korosi A. Anal Chem. 1970;42:118.
99. Ikeda T, Katasho I, Kamei M, Senda M. Agric Biol Chem. 1984;48:1969.

100. Bourdillon C, Laval JM, Thomas DJ. Electrochem Soc. 1986;133:706.
101. Talbott J, Jordan J. Microchem J. 1988;37:5.
102. Arai G, Shoji K, Yasumori IJ. Electroanal Chem. 2006;591:1.
103. Stankovich MT. Anal Biochem. 1980;109:295.
104. Hale PD, Boguslavsky LI, Karan HI, Lan HL, Lee HS, Okamoto Y, et al. Anal Chim Acta. 1991;248:155.
105. Albers WM, Lekkala JO, Jeuken L, Canters GW, Turner APF. Bioelectrochem Bioenerg. 1997;42:25.
106. Liu X, Neoh KG, Cen L, Kang ET. Biosens Bioelectron. 2004;19:823.
107. Ghica ME, Brett CMA. Anal Chim Acta. 2005;532:145.
108. Cenas NK, Pocius AK, Kulys JJ. Bioelectrochem Bioenerg. 1983;11:61.
109. Kajiya Y, Sugai H, Iwakura C, Yoneyama H. Anal Chem. 1991;63:49.
110. Battaglini F, Koutroumanis M, English AM, Mikkelsen SR. Bioconjug Chem. 1994;5:430.
111. Arai G, Yasumori I. Recent Res Dev Pure Appl Chem. 1998;2:179.
112. Cosnier S, Lepellec A, Guidetti B, Rico-Lattes IJ. Electroanal Chem. 1998;449:165.
113. Kaku T, Karan HI, Okamoto Y. Anal Chem. 1994;66:1231.
114. Wang P, Amarasinghe S, Leddy J, Arnold M, Dordick JS. Polymer. 1997;39:123.
115. Piro B, Do V-A, Le LA, Hedayatullah M, Pham MCJ. Electroanal Chem. 2000;486:133.
116. Tatsuma T, Sato TJ. Electroanal Chem. 2004;572:15.
117. Tatsuma T, Sato T. Chem Sensors. 2004;20:782.
118. Gruendig B, Strehlitz B, Krabisch C, Thielemann H, Kotte H, Gomoll M, et al. GBF Monogr. 1992;17:275.
119. Karyakin AA, Strakhova AK, Karyakina EE, Varfolomeyev SD, Yatsimirsky AK. Synth Met. 1993;60:289.
120. Kuhn, R.; Beinert, H. *Ber. - Dtsch. Chem. Ges. [Abteilung] B: Abhandlungen* 1944, *77B*, 606.
121. McDonald TA, Waidyanatha S, Rappaport S. M Carcinogenesis. 1993;14:1927.
122. Crescenzi O, Prota G, Schultz T, Wolfram LJ. Tetrahedron. 1988;44:6447.
123. Marcus RA, Sutin N. Biochim Biophys Acta Rev Bioenerg. 1985;811:265.
124. Kulys J, Buch-Rasmussen T, Bechgaard K, Razumas V, Kazlauskaite J, Marcinkeviciene J, et al. J Mol Catal. 1994;91:407.
125. Gorton L, Bremle G, Csoeregi E, Joensson-Pettersson G, Persson B. Anal Chim Acta. 1991;249:43.
126. Hedenmo M, Narvaez A, Dominguez E, Katakis I. Analyst. 1996;121:1891.
127. Gen L, Boguslavsky LI, Kovalev IP, Sahni SK, Kalash H, Skotheim TA, et al. Biosens Bioelectron. 1996;11:1267.
128. Huan Z, Persson B, Gorton L, Sahni S, Skotheim T, Barlett P. Electroanalysis. 1996;8:575.
129. Kojima S, Tsujimura S, Kano K, Ikeda T, Sato M, Sanada H, et al. Chem Sensors. 2004;20:768.
130. Forrow NJ, Sanghera GS, Watkin JL, Walters S. US Patent. 2004;6:736,957.

131. Mahenc J, Aussaresses H. C R Chim. 1979;289:357.
132. Jaffari SA, Turner APF. Biosens Bioelectron. 1996;12:1.
133. Crumbliss AL, Hill HA, Page DJJ. Electroanal Chem Interfacial Electrochem. 1986;206:327.
134. Lee S-H, Fang H-Y, Chen W-C, Lin H-M, Chang CA. Anal Bioanal Chem. 2005;383:532.
135. Cecchet F, Marcaccio M, Margotti M, Paolucci F, Rapino S, Rudolf P. J Phys Chem B. 2006;110:2241.
136. Degani Y, Heller AJ. Am Chem Soc. 1989;111:2357.
137. Pishko MV, Katakis I, Lindquist SE, Ye L, Gregg BA, Heller A. Angew Chem Int Ed. 1990;102:109.
138. Zakeeruddin SM, Fraser DM, Nazeeruddin MK, Graetzel MJ. Electroanal Chem. 1992;337:253.
139. Matthews DR, Holman RR, Bown E, Steemson J, Watson A, Hughes S, et al. Lancet. 1987;1:778.
140. Frew JE, Hill HA. Philos Trans R Soc Lond B Biol Sci. 1987;316:95.
141. Liaudet E, Battaglini F, Calvo EJJ. Electroanal Chem Interfacial Electrochem. 1990;293:55.
142. Schuhmann W, Wohlschlaeger H, Lammert R, Schmidt HL, Loeffler U, Wiemhoefer HD, et al. Sens Actuators B. 1990;B1:571.
143. Nien P-C, Tung T-S, Ho K-C. Electroanalysis. 2006;18:1408.
144. Ballarin B, Cassani MC, Mazzoni R, Scavetta E, Tonelli D. Biosens Bioelectron. 2007;22:1317.
145. Cass AEG, Davis G, Francis GD, Hill HAO, Aston WJ, Higgins IJ, et al. Anal Chem. 1984;56:667.
146. Green MJ, Hill HAO. J Chem Soc, Faraday Trans 1. 1986;82:1237.
147. Alvarez-Icaza M, Kalisz HM, Hecht HJ, Aumann KD, Schomburg D, Schmid RD. Biosens Bioelectron. 1995;10:735.
148. Forrow, N. J.; Sanghera, G. S.; Walters, S. J. *J. Chem. Soc., Dalton Trans.* 2002, 3187.
149. Jonsson G, Gorton L, Pettersson L. Electroanalysis. 1989;1:49.
150. Bourdillon C, Demaille C, Moiroux J, Saveant J-MJ. Am Chem Soc. 1995;117:11499.
151. Anicet N, Bourdillon C, Demaille C, Moiroux J-M, Saveant JMJ. Electroanal Chem. 1996;410:199.
152. Piro B, Dang LA, Pham MC, Fabiano S, Tran-Minh CJ. Electroanal Chem. 2001;512:101.
153. Yang X, Hall SB, Tan SN. Electroanalysis. 2003;15:885.
154. Ferreyra N, Coche-Guerente L, Labbe P. Electrochim Acta. 2004;49:477.
155. Zhao C, Wittstock G. Anal Chem. 2004;76:3145.
156. Setti L, Fraleoni-Morgera A, Ballarin B, Filippini A, Frascaro D, Piana C. Biosens Bioelectron. 2019;2005:20.
157. Zhang S, Wang N, Niu Y, Sun C. Sens Actuators B. 2005;B109:367.
158. Zhang S, Wang N, Yu H, Niu Y, Sun C. Bioelectrochemistry. 2005;67:15.
159. Zhao C, Wittstock G. Biosens Bioelectron. 2005;20:1277.
160. Zhang S, Yang W, Niu Y, Li Y, Zhang M, Sun C. Anal Bioanal Chem. 2006;384:736.

161. Zhu Z, Wang M, Gautam A, Nazor J, Momeu C, Prodanovic R, et al. Biotechnol J. 2007;2:241.
162. Chen L-Q, Zhang X-E, Xie W-H, Zhou Y-F, Zhang Z-P, Cass Anthony EG. Biosens Bioelectron. 2002;17:851.
163. Tian F, Zhu G. Anal Chim Acta. 2002;451:251–8.
164. Kohma T, Hasegawa H, Oyamatsu D, Kuwabata S. Bull Chem Soc Jpn. 2007;80:158.
165. Green MJ, Hilditch PI. Anal Proc. 1991;28:374.
166. Hilditch PI, Green MJ. Analyst. 1991;116:1217.
167. Forrow NJ, Walters SJ. Biosens Bioelectron. 2004;19:763.
168. Atanasov P, Kaisheva A, Iliev I, Razumas V, Kulys J. Biosens Bioelectron. 1992;7:361.
169. Zhang C, Haruyama T, Kobatake E, Aizawa M. Anal Chim Acta. 2000;408:225.
170. Kurova VS, Ershov AY, Ryabov AD. Russ Chem Bull. 2001;50:1849.
171. Ryabov AD, Sukharev VS, Alexandrova L, Le Lagadec R, Pfeffer M. Inorg Chem. 2001;40:6529.
172. Degani Y, Heller AJ. Phys Chem. 1987;91:1285.
173. Degani Y, Heller AJ. Am Chem Soc. 1988;110:2615.
174. Bartlett PN, Bradford VQ, Whitaker RG. Talanta. 1991;38:57.
175. Schuhmann W, Ohara TJ, Schmidt HL, Heller AJ. Am Chem Soc. 1991;113:1394.
176. Heller A. Curr Opin Chem Biol. 2006;10:664.
177. Ye L, Haemmerle M, Olsthoorn AJJ, Schuhmann W, Schmidt HL, Duine JA, et al. Anal Chem. 1993;65:238.
178. Mao F, Mano N, Heller AJ. Am Chem Soc. 2003;125:4951.
179. Mano, N.; Mao, F.; Heller, A. *Chem. Commun.* 2004, 2116.
180. Mano N, Mao F, Heller AJ. Electroanal Chem. 2005;574:347.
181. Aoki A, Heller AJ. Phys Chem. 1993;97:11014.
182. Blauch DN, Saveant JMJ. Am Chem Soc. 1992;114:3323.
183. Aoki A, Rajagopalan R, Heller AJ. Phys Chem. 1995;99:5102.
184. Andrieux CP, Saveant JMJ. Phys Chem. 1988;92:6761.
185. Surridge NA, Sosnoff CS, Schmehl R, Facci JS, Murray RWJ. Phys Chem. 1994;98:917.
186. Rajagopalan R, Aoki A, Heller AJ. Phys Chem. 1996;100:3719.
187. Mano N, Yoo JE, Tarver J, Loo Y-L, Heller AJ. Am Chem Soc. 2007;129:7006.
188. Forster RJ, Kelly AJ, Vos JG, Lyons MEGJ. Electroanal Chem Interfacial Electrochem. 1989;270:365.
189. Forster RJ, Vos JG. Macromolecules. 1990;23:4372.
190. Forster RJ, Vos JGJ. Electroanal Chem Interfacial Electrochem. 1991;314:135.
191. Forster RJ, Vos JG, Lyons MEGJ. Chem Soc, Faraday Trans. 1991;87:3769.
192. Forster RJ, Vos JG, Lyons MEGJ. Chem Soc, Faraday Trans. 1991;87:3761.
193. Forster RJ, Vos JG. Electrochim Acta. 1992;37:159.
194. Soukharev V, Mano N, Heller AJ. Am Chem Soc. 2004;126:8368.
195. Binyamin G, Heller AJ. Electrochem Soc. 1999;146:2965.
196. Gao Z, Binyamin G, Kim H-H, Barton SC, Zhang Y, Heller A. Angew Chem Int Ed. 2002;41:810.

197. Fei J, Wu Y, Ji X, Wang J, Hu S, Gao Z. Anal Sci. 2003;19:1259.
198. Xie H, Tan XL, Gao Z. Frontiers Biosci. 2005;10:1797.
199. Lever ABP. Inorg Chem. 1990;29:1271.
200. Mano N, Soukharev V, Heller AJ. Phys Chem B. 2006;110:11180.
201. Katakis I, Ye L, Heller AJ. Am Chem Soc. 1994;116:3617.
202. Heller A. Ann ReV Biomed Eng. 1999;1:153.
203. Heller A. AIChE J. 2005;51:1054.
204. Clarke WL, Anderson S, Farhy L, Breton M, Gonder-Frederick L, Cox D, et al. Diabetes Care. 2005;28:2412.
205. Weinstein RL, Schwartz SL, Brazg RL, Bugler JR, Peyser TA, McGarraugh GV. Diabetes Care. 2007;30:1125.
206. Wilson DM, Beck RW, Tamborlane WV, Dontchev MJ, Kollman C, Chase P, et al. Diabetes Care. 2007;30:59.
207. Dassau E, Bequette BW, Buckingham BA, Doyle 3rd FJ. Diabetes Care. 2008;31:295.
208. Weinzimer S, Xing D, Tansey M, Fiallo-Scharer R, Mauras N, Wysocki T, et al. Diabetes Care. 2008;31:525.
209. Wilson DM, Kollman C, Xing D, Fiallo-Scharer R, Weinzimer S, Steffes M, et al. Diabetes Care. 2008;31:381.
210. Xiao Y, Patolsky F, Katz E, Hainfeld James F, Willner I. Science. 2003;299:1877.
211. Dagan-Moscovich H, Cohen-Hadar N, Porat C, Rishpon J, Shacham-Diamand Y, Freeman AJ. Phys Chem C. 2007;111:5766.
212. Lee CW, Chang HN. Chem Eng Sci. 1985;40:873.
213. Gooding JJ, Hall EAH. Electroanalysis. 1996;8:407.
214. Suzuki H, Tamiya E, Karube I. Proc Electrochem Soc. 1987;87–9:393.
215. Suzuki H, Tamiya E, Karube I. Anal Chem. 1988;60:1078.
216. Suzuki H. Fujitsu Sci Techn J. 1989;25:52.
217. Suzuki H, Sugama A, Kojima N, Takei F, Ikegami K. Biosens Bioelectron. 1991;6:395.
218. Suzuki H. Sens Actuators B. 1994;21:17.
219. Suzuki H, Arakawa H, Karube I. Biosens Bioelectron. 2001;16:725.
220. Mitsubayashi K, Wakabayashi Y, Tanimoto S, Murotomi D, Endo T. Biosens Bioelectron. 2003;19:67.
221. Wang J, Li S, Mo J-W, Porter J, Musameh MM, Dasgupta PK. Biosens Bioelectron. 2002;17:999.
222. Silber A, Braeuchle C, Hampp NJ. Electroanal Chem. 1995;390:83.
223. Yacynych AM, Sasso SV, Reynolds ER, Geise RJ. GBF Monogr Ser. 1987;10:69.
224. Armstrong RD, Newton HVJ. Electroanal Chem. 1994;364:87.
225. Gorton L, Joensson GJ. Mol Catal. 1986;38:157.
226. Taniguchi I, Matsushita K, Okamoto M, Collin JP, Sauvage JPJ. Electroanal Chem Interfacial Electrochem. 1990;280:221.
227. Gamburzev S, Atanasov P, Ghindilis AL, Wilkins E, Kaisheva A, Iliev I. Sens Actuators B. 1997;B43:70.
228. Liu J, Lu F, Wang J. Electrochem Commun. 1999;1:341.
229. Mailley P, Cosnier S, Coche-Guerente L. Anal Lett. 2000;33:1733.
230. Elzanowska H, Abu-Irhayem E, Skrzynecka B, Birss VI. Electroanalysis. 2004;16:478.

231. Wang J, Myung NV, Yun M, Monbouquette HGJ. Electroanal Chem. 2005;575:139.
232. Kurusu F, Tsunoda H, Saito A, Tomita A, Kadota A, Kayahara N, et al. Analyst. 2006;131:1292.
233. Tsai Y-C, Li S-C, Liao S-W. Biosens Bioelectron. 2006;22:495.
234. Lee C-H, Wang S-C, Yuan C-J, Wen M-F, Chang K-S. Biosens Bioelectron. 2007;22:877.
235. Yao Y, Shiu K-K. Anal Bioanal Chem. 2007;387:303.
236. Elekes O, Moscone D, Venema K, Korf J. Clin Chim Acta. 1995;239:153.
237. Bharathi S, Nogami M. Analyst. 2001;126:1919.
238. Gu T, Hasebe Y, Uchiyama S. Chem Sensors. 2002;18:52.
239. Prieto-Simon B, Armatas GS, Pomonis PJ, Nanos CG, Prodromidis MI. Chem Mater. 2004;16:1026.
240. Dodevska T, Horozova E, Dimcheva N. Anal Bioanal Chem. 2006;386:1413.
241. De Lara Gonzalez GL, Kahlert H, Scholz F. Electrochim Acta. 2006;52:1968.
242. Wu S, Zhao H, Ju H, Shi C, Zhao J. Electrochem Commun. 2006;8:1197.
243. Yang M, Jiang J, Yang Y, Chen X, Shen G, Yu R. Biosens Bioelectron. 2006;21:1791.
244. Qu F, Yang M, Shen G, Yu R. Biosens Bioelectron. 2007;22:1749.
245. Albers J, Grunwald T, Nebling E, Piechotta G, Hintsche R. Anal Bioanal Chem. 2003;377:521.
246. Piechotta G, Albers J, Hintsche R. Biosens Bioelectron. 2005;21:802.
247. Wang J, Chatrathi MP, Collins GE. Anal Chim Acta. 2007;585:11.
248. Nagy L, Kalman N, Nagy GJ. Biochem Biophys Methods. 2006;69:133.
249. Kulys J. GBF Monogr Ser. 1987;10:51.
250. Gorton L, Bardheim M, Bremle G, Csoeregi E, Persson B, Pettersson G. GBF Monogr. 1991;14:305.
251. Ohara TJ, Vreeke MS, Battaglini F, Heller A. Electroanalysis. 1993;5:825.
252. Tatsuma T, Watanabe T, Watanabe TJ. Electroanal Chem. 1993;356:245.
253. De Benedetto GE, Palmisano F, Zambonin PG. Biosens Bioelectron. 1996;11:1001.
254. Liu H, Ying T, Sun K, Li H, Qi D. Anal Chim Acta. 1997;344:187.
255. Mulchandani A, Pan S. Anal Biochem. 1999;267:141.
256. Patolsky F, Zayats M, Katz E, Willner I. Anal Chem. 1999;71:3171.
257. Kuznetsova LN, Tarasevich MR, Bogdanovskaya VA. Russ J Electrochem. 2000;36:716.
258. Su X, O'Shea SJ. Anal Biochem. 2001;299:241.
259. Castillo J, Gaspar S, Sakharov I, Csoregi E. Biosens Bioelectron. 2003;18:705.
260. Shan D, Cosnier S, Mousty C. Anal Lett. 2003;36:909.
261. Dzantiev BB, Yazynina EV, Zherdev AV, Plekhanova YV, Reshetilov AN, Chang SC, et al. Sens Actuators B. 2004;B98:254.
262. Pizzariello A, Stred'ansky M, Miertus S. Bioelectrochem. 2002;56:99.
263. Asberg P, Inganas O. Biosens Bioelectron. 2003;19:199.
264. Kenausis G, Chen Q, Heller A. Anal Chem. 1997;69:1054.
265. Calvo EJ, Battaglini F, Danilowicz C, Wolosiuk A, Otero M. Faraday Disc. 2000;116:47.

266. Solna R, Dock E, Christenson A, Winther-Nielsen M, Carlsson C, Emneus J, et al. Anal Chim Acta. 2005;528:9.
267. Janata J. Diabetes Care. 1982;5:271.
268. Miyahara Y, Matsu F, Moriizumi T, Matsuoka H, Karube I, Suzuki S. Anal Chem Symp Ser. 1983;17:501.
269. Caras SD, Petelenz D, Janata J. Anal Chem. 1920;1985:57.
270. Murakami T, Nakamoto S, Kimura J, Kuriyama T, Karube I. Anal Lett. 1986;19:1973.
271. Gotoh M, Seki A, Tamiya E, Karube I. Proc Electrochem Soc. 1987;87–9:285.
272. Caras S, Janata J. Methods Enzymol. 1988;137:247.
273. Karube I. GBF Monogr Ser. 1987;10:155.
274. Kimura J, Murakami T, Kuriyama T, Karube I. Sens Actuators. 1988;15:435.
275. Gotoh M, Tamiya E, Seki A, Shimizu I, Karube I. Anal Lett. 1989;22:309.
276. Lee HL, Yang ST, Jung DS, Kim CS, Sohn BK. Anal Sci Technol. 1992;5:177.
277. Shul'ga AA, Sandrovsky AC, Strikha VI, Soldatkin AP, Starodub NF, El'skaya AV. Sens Actuators B. 1992;B10:41.
278. Soldatkin AP, El'skaya AV, Shul'ga AA, Netchiporouk LI, Hendji AMN, Jaffrezic-Renault N, et al. Anal Chim Acta. 1993;283:695.
279. Vering T, Schubmann W, Schmidt H-L, Mikolajick T, Falter T, Ryssel H, et al. Electroanalysis. 1994;6:953.
280. Shul'ga AA, Koudelka-Hep M, de Rooij NF, Netchiporouk LI. Sens Actuators B. 1995;B24:117.
281. Luo X-L, Xu J-J, Zhao W, Chen H-Y. Sens Actuators B. 2004;B97:249.
282. Kadish AH, Hall DA. Clin Chem. 1965;11:869.
283. Kadish AH, Litle RL, Sternberg JC. Clin Chemist. 1968;14:116.
284. Clark, L. C., Jr. U.S. Patent 2,913,386, 1959.
285. Ann. Intern. Med. 2007, 146, ITC1-15.
286. Rosenthal NR, Barrett EJJ. Clin Endocrinol Metab. 1985;60:607.
287. Petersen KF, Shulman GI. Am J Med. 2006;119:S10.
288. Nelson JD, Woelk MA, Sheps S. Diabetes Care. 1983;6:262.
289. Petranyi G, Kyne DA, Alberti KG. Diabetic Med: J Brit Diabetic Assoc. 1986;3:187.
290. Higgins IJ, Hill HAO, Plotkin EV. E V US Patent. 1985;4:545,382.
291. Higgins IJ, Hill HAO. US Patent. 1987;4:711,245.
292. D'Costa EJ, Higgins IJ, Turner APF. Biosensors. 1986;2:71–87.
293. Kyvik KO, Traulsen J, Reinholdt B, Froland A. Diabetes Res Clin Pract. 1990;10:85–90.
294. Ross D, Heinemann L, Chantelau EA. Diabetes Res Clin Pract. 1990;10:281–5.
295. Feldman B, McGarraugh G, Heller A, Bohannon N, Skyler J, DeLeeuw E, et al. Diabetes Technol Therap. 2000;2:221–9.
296. Feldman, B. J.; Heller, A.; E., H.; Mao, F.; Vivolo, J. A.; Funderburk, J. V.; Colman, F. C.; Krishnan, R. U.S. Patent 6,299,757 2001.
297. Feldman, B. J.; Heller, A.; E., H.; Mao, F.; Vivolo, J. A.; Funderburk, J. V.; Colman, F. C.; Krishnan, R. U.S. Patent 6,338,790 2002.

298. Feldman, B. J.; Heller, A.; E., H.; Mao, F.; Vivolo, J. A.; Funderburk, J. V.; Colman, F. C.; Krishnan, R. U.S. Patent 6,592,745 2003.
299. Feldman, B. J.; Heller, A.; E., H.; Mao, F.; Vivolo, J. A.; Funderburk, J. V.; Colman, F. C.; Krishnan, R. U.S. Patent 6,618,934, 2003.
300. Washburn EW. Phys Rev. 1921;17:374.
301. Bierer DW, Quebbemann AJ. Clin Chem. 1981;27:756.
302. Wens R, Taminne M, Devriendt J, Collart F, Broeders N, Mestrez F, et al. Peritonial Dialysis Int. 1998;18:603.
303. Jungheim K, Koschinsky T. Diabetes Care. 2001;24:1303.
304. McGarraugh G, Price D, Schwartz S, Weinstein R. Diabetes Technol Therap. 2001;3:367.
305. Armour JC, Lucisano JY, McKean BD, Gough DA. Diabetes. 1990;39:1519.
306. Clemens AH, Chang PH, Myers RW. Horm Metabol Res Suppl Ser. 1977;7:23.
307. Fogt EJ, Dodd LM, Jenning EM, Clemens AH. Clin Chem. 1978;24:1366.
308. Clemens AH. Med Prog Technol. 1979;6:91.
309. Fogt EJ, Dodd LM, Eddy AR, Jenning E, Clemens AH. Horm Metab Res, Suppl Ser. 1979;8:18.
310. Lucisano JY, Edelman SV, Quinto BD, Wong DK. Polym Mater Sci Eng. 1997;76:564.
311. Wong, D. K.; Lucisano, J. Y. U.S. Patent 5,804,048, 1998.
312. Updike SJ, Shults MC, Rhodes RK, Gilligan BJ, Luebow JO, von Heimburg D. ASAIO J. 1994;40:157.
313. Ward WK, Troupe JE. ASAIO J. 1999;45:555.
314. Updike SJ, Shults MC, Gilligan BJ, Rhodes RK. Diabetes Care. 2000;23:208.
315. Gilligan BC, Shults M, Rhodes RK, Jacobs PG, Brauker JH, Pintar TJ, et al. Diabetes Technol Therap. 2004;6:378.
316. Ahmed S, Dack C, Farace G, Rigby G, Vadgama P. Anal Chim Acta. 2005;537:153.
317. Yu B, Long N, Moussy Y, Moussy F. Biosens Bioelectron. 2006;21:2275.
318. Yu B, Ju Y, West L, Moussy Y, Moussy F. Diabetes Technol Therap. 2007;9:265.
319. Janle-Swain E, Van Vleet JF, Ash SR. ASAIO Trans. 1987;33:336.
320. Ash SR, Baker K, Blake DE, Carr DJ, Echard TG, Sweeney KD, et al. ASAIO Trans. 1987;33:524.
321. Ash SR, Poulos JT, Rainier JB, Zopp WE, Janle E, Kissinger PT. ASAIO J. 1992;38:M416.
322. Janle EM, Ash SR, Zopp WE, Kissinger PT. Curr Sep. 1993;12:14.
323. Ash SR, Rainier JB, Zopp WE, Truitt RB, Janle EM, Kissinger PT, et al. ASAIO J. 1993;39:M699.
324. Ash SR, Zopp WE, Truitt RB, Janle EM, Kissinger PT. Adv Filtration Separation Technol. 1993;7:316.
325. Schmidt FJ, Sluiter WJ, Schoonen AJ. Diabetes Care. 1993;16:695.
326. Moscone D, Venema K, Korf J. Med Biol Eng Comput. 1996;34:290.
327. Tiessen RG, Kaptein WA, Venema K, Korf J. Anal Chim Acta. 1999;379:327.

328. Savenije B, Venema K, Gerritzen MA, Lambooij E, Korf J. Anal Chem. 2003;75:4397.
329. Schmidt FJ, Aalders AL, Schoonen AJ, Doorenbos H. Int J Artif Organs. 1992;15:55.
330. Mascini M, Moscone D, Bernardi L. Sens Actuators B. 1992;B6:143.
331. Moscone, D.; Mascini, M. Ann Biol Chem. 1992;50:323.
332. Moscone D, Pasini M, Mascini M. Talanta. 1992;39:1039.
333. Meyerhoff C, Bischof F, Mennel FJ, Sternberg F, Pfeiffer EF. Int J Artif Organs. 1993;16:268.
334. Moscone D, Mascini M. Analysis. 1993;21:M40.
335. Palmisano F, Centonze D, Guerrieri A, Zambonin PG. Biosens Bioelectron. 1993;8:393.
336. Hashiguchi Y, Sakakida M, Nishida K, Uemura T, Kajiwara K, Shichiri M. Diabetes Care. 1994;17:387.
337. Mascini M, Moscone D, Anichini M. Rev Anal Chem Euroanalysis VIII. 1994;154:298.
338. Osborne PG, Niwa O, Kato T, Yamamoto K. Curr Sep. 1996;15:19.
339. Towe BC, Pizziconi VB. Biosens Bioelectron. 1997;12:893.
340. Vering T, Adam S, Drewer H, Dumschat C, Steinkuhl R, Schulze A, et al. Analyst. 1998;123:1605.
341. Summers LK, Clark ML, Humphreys SM, Bugler J, Frayn KN. Horm Metab Res. 1999;31:424.
342. Lutgers HL, Hullegie LM, Hoogenberg K, Sluiter WJ, Dullaart RP, Wientjes KJ, et al. Neth J Med. 2000;57:7.
343. Hoss U, Kalatz B, Gessler R, Pfleiderer HJ, Andreis E, Rutschmann M, et al. Diabetes Technol Therap. 2001;3:237.
344. Wisniewski N, Klitzman B, Miller B, Reichert WMJ. Biomed Mater Res. 2001;57:513.
345. Maran A, Crepaldi C, Tiengo A, Grassi G, Vitali E, Pagano G, et al. Diabetes Care. 2002;25:347.
346. Poscia A, Mascini M, Moscone D, Luzzana M, Caramenti G, Cremonesi P, et al. Biosens Bioelectron. 2003;18:891.
347. Wientjes KJ, Grob U, Hattemer A, Hoogenberg K, Jungheim K, Kapitza C, et al. Diabetes Technol Ther. 2003;5:615.
348. Ricci F, Moscone D, Tuta CS, Palleschi G, Amine A, Poscia A, et al. Biosens Bioelectron. 1993;2005:20.
349. Wentholt IM, Vollebregt MA, Hart AA, Hoekstra JB, DeVries JH. Diabetes Care. 2005;28:2871.
350. Fayolle C, Brun JF, Bringer J, Mercier J, Renard E. Diabetes Metab. 2006;32:313.
351. Kubiak T, Worle B, Kuhr B, Nied I, Glasner G, Hermanns N, et al. Diabetes Technol Therap. 2006;8:570.
352. Rossetti P, Porcellati F, Fanelli CG, Bolli GB. Diabetes Technol Therap. 2006;8:326.
353. Ricci F, Caprio F, Poscia A, Valgimigli F, Messeri D, Lepori E, et al. Biosens Bioelectron. 2007;22:2032.
354. Schoemaker M, Andreis E, Roeper J, Kotulla R, Lodwig V, Obermaier K, et al. Diabetes Technol Therap. 2003;5:599.

355. Sparacino G, Zanderigo F, Corazza S, Maran A, Facchinetti A, Cobelli C. IEEE Trans Biomed Eng. 2007;54:931.
356. Wang X. Diabetes Technol Therap. 2004;6:883.
357. Kurnik RT, Berner B, Tamada J, Potts ROJ. Electrochem Soc. 1998; 145:4119.
358. Tierney MJ, Tamada JA, Potts RO, Jovanovic L, Garg S. Biosens Bioelectron. 2001;16:621.
359. The Diabetes Research in Children Network (DirecNet) Study Group. Diabetes Technol. Therap. 2003, 5, 791.
360. The Diabetes Research in Children Network (DirecNet) Study Group. Diabetes Technol. Therap. 2003, 5, 781.
361. Chase HP, Beck R, Tamborlane W, Buckingham B, Mauras N, Tsalikian E, et al. Diabetes Care. 2005;28:1101.
362. Wilson GS, Hu Y. Chem Rev. 2000;100:2693.
363. Wilson GS, Gifford R. Biosens Bioelectron. 2005;20:2388.
364. McIvor KC, Cabernoch JL, Branch KD, Van Antwerp NM, Halili EC, Mastrototaro JJ. U.S. Patent 6,360,888; 2002.
365. Csoeregi E, Quinn CP, Schmidtke DW, Lindquist S-E, Pishko MV, Ye L, et al. A; Anal Chem. 1994;66:3131.
366. Csoeregi E, Schmidtke DW, Heller A. Anal Chem. 1995;67:1240.
367. Quinn CP, Pathak CP, Heller A, Hubbell JA. Biomaterials. 1995;16:389.
368. Rowinski P, Rowinska M, Heller A. Anal Chem. 2008;80:1746.
369. Mao, F.; Cho, H. U.S. Patent 6,932,894, 2005.
370. Van Antwerp, W. P.; C D. C.; J., M. J. U.S. Patent 6,784,274, 2004..
371. Clark Jr LC, Duggan C. A Diabetes Care. 1982;5:174.
372. Abel P, Fischer U, Brunstein E, Ertle R. Horm Metab Res, Suppl Ser. 1988;20:26.
373. Clark Jr LC, Noyes LK, Spokane RB, Sudan R, Miller ML. Methods Enzymol. 1988;137:68.
374. Gough DA. Horm Metab Res Suppl Ser. 1988;20:30.
375. Shaw GW, Claremont DJ, Pickup JC. Biosens Bioelectron. 1991;6:401.
376. Wilson GS, Zhang Y, Reach G, Moatti-Sirat D, Poitout V, Thevenot DR, et al. Clin Chem. 1992;38:1613.
377. Shin JJ, Holtzclaw KR, Dangui ND, Kanderian JS, Mastrototaro JJ, Hong PI. U.S. Patent 7,029,444; 2006.
378. Feldman B, Brazg R, Schwartz S, Weinstein R. Diabetes Technol Therap. 2003;5:769.
379. Ishikawa M, Schmidtke DW, Raskin P, Quinn CAJ. Diabetes Complications. 1998;12:295.
380. Schmidtke DW, Freeland AC, Heller A, Bonnecaze RT. Proc Natl Acad Sci USA. 1998;95:294.
381. Freeland AC, Bonnecaze RT. Ann Biomed Eng. 1999;27:525.
382. Bonnecaze, R. T.; Freeland, A. C. U.S. Patent Application 20030100040, 2003.
383. Wagner JG, Schmidtke DW, Quinn CP, Fleming TF, Bernacky B, Heller A. Proc Natl Acad Sci USA. 1998;95:6379.
384. Boyne MS, Silver DM, Kaplan J, Saudek CD. Diabetes. 2003;52:2790.
385. Kulcu E, Tamada JA, Reach G, Potts RO, Lesho MJ. Diabetes Care. 2003;26:2405.

386. Doniger, K.; Budiman, E.; Hayter, G.; M., T.; Rebrin, K. *6th Diabetes Technology Meeting,* Atlanta, GA, 2006.
387. Kovatchev B, Anderson S, Heinemann L, Clarke W. Diabetes Care. 2008. doi:10.2337/dc07-2401; epub March 13.
388. Heller A. Phys Chem Chem Phys. 2004;6:209.
389. Heller A. Anal Bioanal Chem. 2006;385:469.
390. Shin W, Lee J, Kim Y, Steinfink H, Heller AJ. Am Chem Soc. 2005;127:14590.
391. Mano N, Fernandez JL, Kim Y, Shin W, Bard AJ, Heller AJ. Am Chem Soc. 2003;125:15290.
392. Mano N, Kim H-H, Heller AJ. Phys Chem B. 2002;106:8842.
393. Mano N, Kim H-H, Zhang Y, Heller AJ. Am Chem Soc. 2002;124:6480.
394. Mano N, Heller AJ. Electrochem Soc. 2003;150:A1136.
395. Mano N, Mao F, Heller AJ. Am Chem Soc. 2003;125:6588.
396. Rowinski P, Kang C, Shin H, Heller A. Anal Chem. 2007;79:1173.
397. Mano N, Mao F, Heller AJ. Am Chem Soc. 2002;124:12962.
398. Mano N, Mao F, Heller A. Chembiochem. 2004;5:1703.

Chapter 6
Electrochemistry of Adhesion and Spreading of Lipid Vesicles on Electrodes

Victor Agmo Hernández, Uwe Lendeckel, and Fritz Scholz

6.1 Introduction

Biological membranes have developed to separate different compartments of organisms and cells. There is a large number of rather different functions which membranes have to fulfil: (1) they control the material and energy fluxes of metabolic processes, (2) they provide a wrapping protecting the compartments from chemical and physical attacks of the environment, (3) they provide interfaces at which specific biochemical machineries can operate (e.g., membrane bound enzymes), (4) they are equipped for signal transduction, (5) they possess the necessary stability and flexibility to allow cell division, and endo- and exocytosis as well as migration, (6) they present anchoring structures that enable cell-to-cell and cell-to-matrix physical interactions and intercellular communication. These are certainly not all functions of membranes as new functionalities are continuously reported. Since the biological membranes separate essentially aqueous solutions, such separating borders—if they should possess a reasonable stability and also flexibility combined with selective permeability—have to be built up of hydrophobic molecules exposing to both sides a similar interface. It was one of the most crucial and most lucky circumstances for the development and existence of life that certain amphiphilic molecules are able to assemble in bilayer structures (membranes), which—on one side—possess a rather high physical and chemical stability, and—on the other side—are able to incorporate foreign molecules for modifying both the physical properties as well as the permeability of the membranes for defined chemical species. The importance of the chemical function of membranes and all its constituents, e.g., ion channels, pore peptides, transport peptides, etc., is generally accepted. The fluid-mosaic model proposed by Singer and Nicolson [1] is still the basis to understand the biological, chemical, and physical properties of biological membranes. The importance of the purely mechanical properties of membranes came much later into the focus of research. The reasons are probably the dominance of bio*chemical* thinking and bio*chemical* models among biologists and medical researchers, as well as a certain lack of appropriate methods to probe mechanical properties of membranes. The last decades have changed that situation due to the development of techniques like the Atomic Force Microscopy, Fluorescence Microscopy, Micropipette Aspiration, Raman Microspectroscopy, advanced Calorimetry, etc. This chapter is aimed at elucidating how the properties of membranes can be investigated by studying the interaction of vesicles with a very hydrophobic surface, i.e., with the surface of a mercury electrode. This interaction is unique as it results in a complete disintegration of the bilayer membrane of the vesicles and the formation of an island of adsorbed lipid molecules, i.e., a monolayer island. This process can be followed by current-time measurements (chronoamperometry), which allow studying the complete disintegration process in all its details: the different steps of that disintegration can be resolved on the time scale and the activation parameters can be determined. Most interestingly, the kinetics of vesicle disintegration on mercury share important features with the process of vesicle fusion and, thus, sheds light also on mechanisms of endocytosis and exocytosis. Most importantly, not only artificial vesicles (liposomes) can be studied with this approach, but also reconstituted plasma membrane vesicles and even intact mitochondria. Hence, one can expect that the method may provide in future studies also information on the membrane properties of various other vesicles, including exosomes, and may allow investigating various aspects of drug action in relation to membrane properties (transmembrane transport, tissue targeting, bioavailability, etc.), and also the impact of pathophysiological conditions (e.g., oxidative modification) on membrane properties, on a hitherto not or only hardly accessible level.

6.2 Physico-Chemical Properties of Self-Assembled Lipid Membranes

Lipid bilayers are formed by self-assembling of amphiphilic molecules as a result of the so-called hydrophobic effect [2–4]. The hydrophobic effect is mainly entropy-based because the water molecules around a hydrophobic molecule or surface assume a special structure (iceberg structure) in which they lose orientation entropy. Thus, it is a favorable process to liberate these water molecules by self-assembly of the hydrophobic (or amphiphilic) molecules, e.g., in micelles, liposomes, membranes, or the formation of a new volume phase (e.g., oil droplets).

To characterize membranes, the following parameters have to be considered [5, 6]: (a) thickness (L (μm)), (b) density (ρ (g cm^{-3})), (c) shear viscosity (η (N s m^{-2})) (frictional resistance of the membrane to an applied shear force), (d) shear elastic modulus (G (Pa)) (ratio of shear stress to shear strain), (e) interfacial tension (γ (N m^{-1})) (work needed to change the surface area), (f) interfacial potential difference ($\Delta\phi(V)$), (g) transmembrane potential ($\Delta\phi_{TM}(V)$), and (h) Hamaker constant (H_{12} (J)) (these are constants characterizing the *van der Waals* interactions between macroscopic bodies; they have the dimension of an energy). In addition to these parameters, the molecular dynamics of the membrane are very important: thus lateral diffusion coefficients and the flip-flop rates of lipid molecules are crucial for a deeper understanding of membrane functions. Frequently the term "flexibility of a membrane" is discussed, although this is an ill-defined term including flexibility of acyl chains, the lateral and transverse diffusion, etc. [7]. The term "membrane fluidity" is more specific, as it refers to the orientational and dynamic properties of the lipid hydrocarbon chains in the bilayer matrix, and it is experimentally accessible (see, e.g., [8–10]). Membrane fluidity is mainly determined by the quantitative and qualitative lipid composition of the membranes. Generally, a low cholesterol content and a high content of phospholipids with unsaturated acyl chains favor a high fluidity. Recently a novel index has been proposed to calculate membrane fluidity even for the body temperature of mammalians [11].

Cholesterol-dependent changes of membrane fluidity have been recognized as important regulators of the biological functions of membranes and, consequently, of cellular functions. An excellent recent review [12] demonstrates that the activity of members of nearly all major ion channel families is regulated by changing the membrane cholesterol content or the level of partition into cholesterol-rich domains. These effects on ion channel function are, however, heterogeneous. Whereas suppression of channel activity due to increased membrane contents of cholesterol is observed more frequently (e.g., for voltage-dependent Ca-channels), the opposite effect is true for, e.g., Transient Receptor Potential channels (e.g., TRPC). Besides changing the activation thresholds of ion channels, altered cholesterol levels may also affect their coupling to intracellular signaling pathways. According to Levitan et al. [12], three mechanisms could underlie these regulatory effects of cholesterol: (1) direct and specific interaction of cholesterol with the given membrane protein, (2) the maintenance of a scaffold for protein interactions, and, most interestingly here, (3) changes in the physical properties of the membrane bilayer itself. The latter mechanism clearly emphasizes the urgent need to develop and apply reliable methods to measure the physical properties of biological and model membranes/vesicles and, at the same time, to assess their biological functionality. Changes in membrane composition and structural modifications (e.g., peroxidation) have been associated with various physiological and pathophysiological conditions and contribute to the development and progression of diseases. This holds true also for mitochondria, which play a central role not only for the cells' ATP supply, but also regulate various pathways of apoptosis (programmed cell death). As it is outlined below, the interaction of mitochondria with electrode surfaces has been successfully addressed in a recent study [13], which also revealed that they behave like liposomes. Interestingly, changes of mitochondrial membrane fluidity were associated with structural modifications such as the oxidation of cardiolipins. Cardiolipins are a major constituent of the inner mitochondrial membrane. Of note, recent data suggest that a balance of mitochondrial cholesterol to oxidized cardiolipins regulates mitochondrial membrane properties and, thus, mitochondrial function [14].

This example again illustrates that the concomitant analysis of physical and biological parameters of membranes/vesicles is a challenging but inevitable task of future studies.

The current state of the available tools for analyzing the physical characteristics of liposomes/vesicles with electrochemical methods, especially chronoamperometry, is reviewed below.

6.3 Interactions of Vesicles with Surfaces (Hydrophilic and Hydrophobic)

6.3.1 Liposome Adhesion and Spreading on Solid Supports

Liposome adhesion and spreading onto solid supports was pioneered by Brian and McConnell [15], who discovered that phospholipid vesicles spontaneously fuse on treated glass surfaces, forming supported planar membranes. It was soon discovered that the supported bilayer lipid membranes (s-BLMs) thus formed could be designed in order to mimic the composition of real cell membranes. Furthermore, it was proposed that these structures could be used, for example, as biosensors based on ligand–receptor recognition at the lipid surface [16]. Later work showed also that, on hydrophobic supports such as mercury, the adhesion and spreading of liposomes could lead instead to the formation of adsorbed lipid monolayers [17]. These have a higher stability than s-BLMs and allow a wide range of studies on almost defect-free lipid layers.

In order to improve the potential and versatility of supported lipid mono- and bilayers, it has been of the utmost importance to acquire a detailed understanding of the process of adhesion and spreading of liposomes on solid substrates. Several attempts have been made to explain the process in terms of the adhesion-spreading mechanism and the driving forces involved. Most of previous research has been focused on hydrophilic solids such as silica, mica, quartz, and glass (as reviewed by Richter et al. [18]). A broad variety of techniques and instruments have been employed, as, for example, the quartz crystal microbalance with simultaneous frequency and dissipation monitoring

(QCM-D) [19–23], atomic force microscopy (AFM) [24–30], scanning force microscopy (SFM) [31], the surface force apparatus (SFA) [32], fluorescence microscopy [33, 34], microcantilevers [35], surface plasmon resonance (SPR) [23, 36–38], and a range of electrochemical methods (reviewed by Agmo Hernández and Scholz [39]). A relatively general agreement has been reached concerning the mechanism of adhesion and spreading of liposomes on hydrophilic substrates which consists mainly of three stages [40, 41]:

1. Adhesion of the intact vesicle on the surface, causing a deformation of the spherical vesicle shape
2. Rupture of the lipid membrane at the point of maximum curvature
3. Spreading of a lipid bilayer film on the substrate surface

Liposome adhesion (step 1) to a surface involves an interplay between different contributions to the free energy of the system, which must be minimized [42]. The major two contributions are the free energy of liposome/surface interaction, which favors the adhesion, and the free energy of liposome deformation, which opposes the adhesion. For larger liposomes, the surface area of liposome–substrate interaction is greater than for smaller ones. On the other hand, the free energy of liposome deformation is in principle independent of the size of liposomes and depends only on its rigidity. It follows that larger liposomes are more prone to adhere to surfaces. In fact, liposomes below a critical ratio (at which the free energy of liposome/surface interaction is smaller in magnitude than the free energy of liposome deformation) do not adhere.

Step 2, however, will be favored by smaller liposomes, in which the surface tension due to bending is higher than in larger liposomes. Therefore, even though larger liposomes are more prone to adhere to a surface, they will rupture and spread more slowly. Regarding the formation of sBLMs on hydrophilic substrates, it has been shown that sonicated small unilamellar vesicles (SUVs) are optimal [43]. However, one needs to take into consideration that different lipid compositions will assign different properties to the liposomes, and the required conditions for sBLM formation may change accordingly.

In some cases, however, it has been observed that the adhesion and spreading of liposomes on hydrophilic substrates does not follow the mechanism described above. Under certain conditions, it has been observed that liposomes tend to adhere as intact vesicles on the surface. Once a critical liposome surface concentration is reached, the vesicles start to fuse, eventually rupturing and forming a sBLM [19, 23]. On hydrophilic quartz surfaces, it has been observed that a mixed mechanism takes place: some liposomes adhere, rupture, and spread individually while some others accumulate on the surface before fusing and spreading [44]. The reasons behind these differences in mechanism are not yet clear and, thus, remain a matter of investigation.

Adhesion and spreading of liposomes on hydrophobic surfaces is much more complex and less understood. On rough hydrophobic substrates (such as untreated gold), usually only adhesion, and no spreading, occurs, in some cases even with liposomes smaller than the critical radius predicted for adhesion to take place [19]. On smooth hydrophobic surfaces, adhesion and spreading have been reported. The final product of the spreading can be either a supported bilayer (as is the case on smooth gold[1]) [45, 46] or a monolayer (as on mercury or alkyl-modified surfaces) [19, 39, 47, 48]. The mechanisms leading from an intact vesicle to a bilayer or monolayer on hydrophobic substrates is still a matter of discussion. Concerning hydrophobic metals, recent publications have proposed very convincing and widely accepted models for the mechanism leading from an adsorbed intact liposome to the formation of a supported lipid bilayer on gold [45, 46, 49] or an adsorbed lipid monolayer on mercury [47, 48, 50–52]. On gold, the lipid–lipid interaction is stronger than the lipid–gold interaction, and therefore a bilayer is formed. On mercury, on the other hand, the lipid–substrate interaction is the most favored one, leading to the formation of a monolayer. In spite of these differences, both models agree that an initial rearrangement of the lipid molecules in contact with the hydrophobic surface is necessary in order for the rupture and spreading to take place. In the next sections of this chapter, the adhesion and spreading of liposomes on mercury will be described in more detail.

A problem that has been the focus of recent investigations concerns the initial adhesion of the liposomes to the hydrophobic substrate.

---

[1] Whether smooth, clean gold is hydrophilic or hydrophobic is a matter of discussion since the 1930s. Arguments have been presented supporting both hypotheses. However, for the purposes of this discussion, it has been clearly demonstrated in the cited references (among others), that the interaction of liposomes with atomically smooth gold {111} can only be understood if the gold substrate is considered hydrophobic.

It has been shown that not all the liposomes in an apparently homogeneous suspension react in the same way when they encounter a hydrophobic surface. Some of them will adhere and spread immediately, while some others will not attach at all. This observation led to a recently presented model, [52] which proposes that hydrophobic "active" sites must be formed on the liposome before it can adhere to the substrate. This model was further corroborated in a systematic study aiming to describe the kinetics of formation of these active sites and the properties of the liposome that would be affected by their appearance [53]. It was found that, besides the affinity for hydrophobic substrates, active sites are also responsible for the permeability of the membrane. A deeper discussion of this activation–deactivation mechanism and how it can be studied electrochemically will be discussed in the next sections of this chapter.

6.3.2 Electrochemistry on Immobilized Lipid Structures: Liposomes and Supported Lipid Mono- and Bilayers

Several studies involving immobilized liposomes or supported lipid mono- and bilayers have been performed using electrochemistry as a tool to determine the dynamic and elastic properties of these structures, as well as the mechanism underlying the formation of self-assembled structures on electrodes. A review of that topic was recently published by Agmo Hernández and Scholz [39]. However, as the field is developing at an amazing rate, an update on the subject is necessary. In this section, some of the work reviewed in the aforementioned paper is introduced, together with the latest developments in the field.

The relationship between electrochemistry and lipid structures is wide and covers a very broad range of aspects, among them:

(a) Electrochemical techniques as tools to characterize immobilized lipid structures (supported lipid bilayers and monolayers, immobilized liposome layers, etc.) and their interaction with the surroundings. Within this field the study of the effects in the immobilized lipid structure of molecules or particles in solution or embedded in the membrane is included. Examples include the characterization of the properties of the studied lipid structure (capacitance, surface charge density, degree of order of the lipid arrangement as a function of potential) [54–59]; the study of the effects of drugs or other water-soluble molecules on the properties of the membrane (e.g., drug-induced membrane permeabilization) [60]; study the interaction of immobilized lipid layers with nanoparticles in suspension [61]; studies on the effect of the inclusion of peptides or lipid-soluble drugs in the membrane, etc. [54, 55, 62–65].

(b) Electrochemistry for the monitoring of membrane-related *in vivo* processes, such as exocytosis, which can be followed by the so-called patch-clamp technique [66–70], or by chronoamperometry at electrodes located near the site where exocytosis events take place [71–75].

(c) Electrodes modified with lipid bilayers containing electroactive molecules. This kind of experiments has provided great insight into the electrochemical behavior of, e.g., ubiquinone-10 (UQ), in a biomimetic environment [76–80]. Furthermore, the inclusion of electroactive substances in lipid bilayers immobilized on electrodes allows studying changes and alterations of the membrane structure upon its interaction with, for example, surfactants, as has been shown recently by Largueze et al. [81]. This approach also allows interrogating the interactions of redox reaction products with the surrounding aqueous environment [82–84].

(d) Development of electrochemical sensors. An immobilized lipid structure can be modified in order to respond to stimuli received from the surroundings. This response can be monitored as an electrochemical signal. A recent example is the development of a sensor for quaternary ammonium ions based on an immobilized liposome layer or supported lipid bilayer modified with UQ [80].

(e) Electrochemistry applied to the study of the properties of lipid assemblies in suspension and their transition to immobilized structures. Electrochemical techniques such as chronoamperometry and electrochemical impedance spectroscopy allow following the transition from a lipid structure in the bulk solution to a supported lipid bilayer/monolayer or immobilized liposome layer [45, 47, 48, 51, 52, 85–87]. This may allow determining the properties of the original structure found in suspension. Studies on gold and mercury have provided a very clear picture of the liposome-supported lipid bilayer/monolayer transition mechanism. The latter have also provided a model to relate the transition kinetics to the mechanical and dynamic properties of liposomes in suspension [47, 48]. This approach will be the subject of the following sections.

6.4 Chronoamperometry as a Tool to Study the Interactions of Vesicles with Charged Electrode Surfaces

Even though alternate current methods are usually the choice to study non-faradaic processes at electrodes, the development of faster and more sensitive and precise electrochemical instruments allows employing direct current methods to monitor and characterize capacitive processes occurring down to the nanosecond time scale. Chronoamperometry has been the method of choice to study, for example, the interactions of montmorillonite particles [88], oil droplets [89–92], certain biological membranes [13, 93–96], and a wide range of other particles [97–105] with the surface of metal electrodes. The first studies of the adhesion and bursting of liposomes on a static mercury drop electrode (SMDE) and the individual analysis of single adhesion events were performed by Hellberg et al. [50, 106]. The principle of such measurements is the change in the double layer capacitance when a liposome adheres and spreads on the surface of the electrode (Fig. 6.1). On the area covered by the lipid, there is a change in the charge density $\Delta q = (q_{\theta=1} - q_{\theta=0})$, and a capacitive current arising as the excess charge must be removed. In chronoamperometry, this process is recorded as capacitive spikes, each of them arising from the adhesion of a single liposome (Fig. 6.2). If the electrode surface has a positive charge (potentials positive with respect to the potential

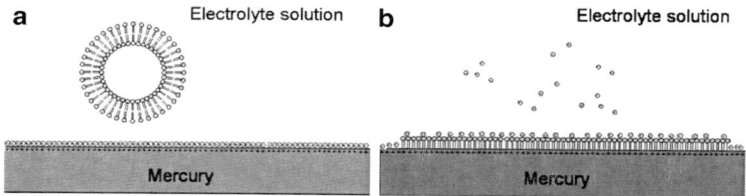

Fig. 6.1 (a) Initial state, with the double layer formed at the interface mercury–electrolyte with a certain charge density $q_{\theta=0}$. (b) State after the adhesion and spreading of a liposome, with an adsorbed lipid monolayer. A mercury–lipid–electrolyte interface is formed, with a lower charge density $q_{\theta=1}$

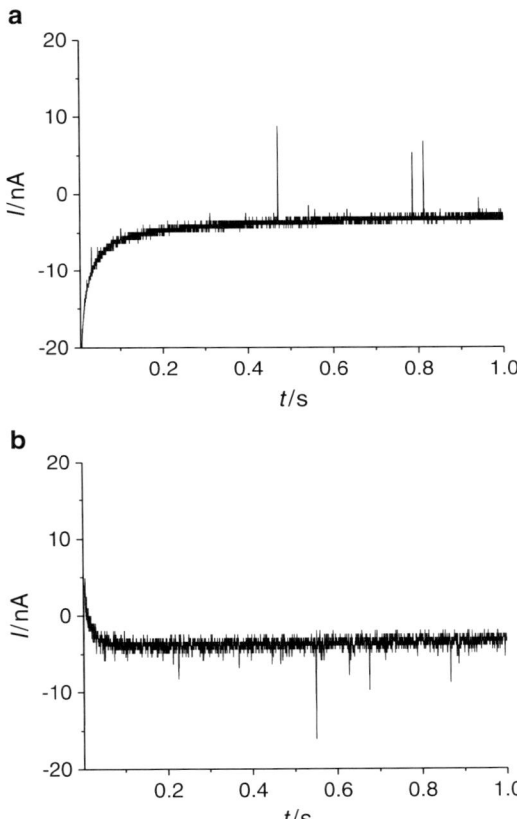

Fig. 6.2 Chronoamperometric signals recorded at a static mercury drop electrode (SMDE) with a 0.1 g/L suspension of multilamellar vesicles (MLVs) in 0.1M KCl solution at 25°C at (**a**) −0.9 V and (**b**) −0.2 V vs. Ag|AgCl 3M KCl (pzc at −0.45 V) (Adapted from Agmo Hernández and [39])

of zero charge (pzc)), the peaks are oriented in the same direction as reduction currents, since electrons must flow to the electrode surface in order to charge the new capacitor. On the contrary, if the surface is negatively charged, the peak currents have an opposite sign, as electrons must be sucked away from the electrode surface.

From the measurements depicted in Fig. 6.2, it is possible to isolate the adhesion-spreading signal of a single liposome. For most lipid compositions and phase states, the general shape of the signal is similar to

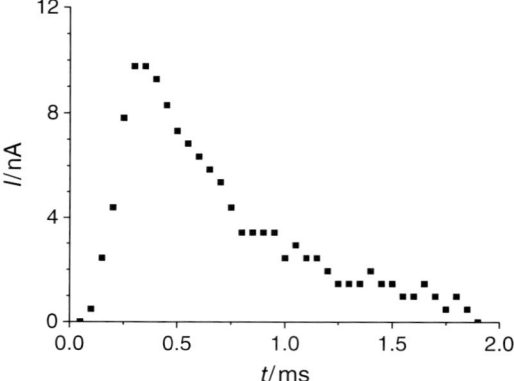

Fig. 6.3 A single adhesion peak produced by the adhesion of a giant unilamellar DMPC vesicle (GUV) at 30°C and a potential of −0.9 V vs. Ag|AgCl 3M KCl (Adapted from Agmo Hernández and Scholz [39])

what is shown in Fig. 6.3: a fast increase in the current is followed by a slow exponential decrease until the baseline is reached. As will be shown in following sections of this chapter, a closer analysis of these signals can provide important information about liposome properties, such as its phase state, its tendency to rupture, and the rigidity of its membrane.

Given that the area under the peak equals the absolute charge difference on the electrode surface before and after the adhesion and spreading (Q_{lip}) of a liposome, the total area covered by the adsorbed lipid island (A_{island}) can be calculated as:

$$A_{island} = \frac{Q_{lip}}{\Delta q} \qquad (6.1)$$

and the total number of lipid molecules forming such island is given by:

$$n_{lecithin} = \frac{A_{island}}{A_{lipid}} = \frac{Q_{lip}}{\Delta q A_{lipid}} \qquad (6.2)$$

where A_{lipid} is the area of one lipid molecule. At the pzc, the adhesion events are not detectable, as no charge is displaced. With an increasing absolute value of the potential with respect to the pzc, Δq increases and so does the total charge displaced by the adhesion-spreading of a liposome. At potentials near the pzc, only liposomes containing a great amount of lipid will displace enough charge to produce a detectable current. At potentials far from the pzc, the detection efficiency increases as the signals arising from the adhesion of small liposomes become larger. However, at potentials too positive or too negative, adsorption of organic material cannot take place anymore, as the interaction of mercury and water becomes stronger. Hellberg et al. [50, 106] determined that the maximum efficiency of the detection of the adhesion of DMPC liposomes suspended in 0.1M KCl (pzc at −0.45 V) is found at −0.2 V vs. Ag|AgCl on the positive side of the pzc and at −1.0 V vs. Ag|AgCl on its negative side.

The study of the adhesion-spreading signals allows determining the size distribution and lamellarity of a given suspension. Equations (6.1) and (6.2) relate the total area under the capacitive peak (Q_{lip}) with the area of the lipid island (A_{island}) and the amount of lipid (n_{lipid}) forming the vesicle. The volume occupied by a lipid molecule can be calculated if we assume a cylindrical form:

$$V_L = A_{lipid} L_L \qquad (6.3)$$

where L_L is the length of a lipid molecule. For a spherical multilamellar vesicle (MLV) with an "onion"-like structure (many layers closely packed), the volume of the vesicle is close to the total volume occupied by the lipid molecules forming it, that is:

$$V_{MLV} = \frac{4}{3}\pi r_{MLV}^3 = n_{lipid} V_L = \frac{Q_{lip} L_L}{\Delta q} \qquad (6.4)$$

and, therefore, the radius of the vesicle is:

$$r_{MLV} = \sqrt[3]{\frac{3 Q_{lip} L_L}{4\pi \Delta q}} \qquad (6.5)$$

For chronoamperometric measurements performed with DMPC ($A_{DMPC} = 0.64 \text{nm}^2$, $L_{DMPC} = 2.5 \text{nm}$) at −0.9V ($\Delta q_{-0.9V} = -7 \times 10^{-6}$ C cm^{-2} according to Hellberg [107]), a peak with an area of 1 pC corresponds to a vesicle ≈410 nm in diameter, which agrees with the data of Hellberg [107].

For the case of unilamellar vesicles, it can be assumed that the area of the adsorbed lipid island is twice the area of the interlayer surface of the vesicle (the surface between the two lipid monolayers). Assuming a spherical shape:

$$A_{island} = 2 A_{int} = 8\pi r_{int}^2 \qquad (6.6)$$

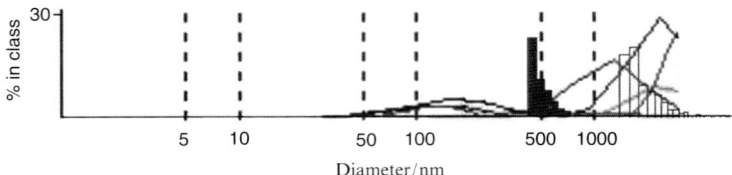

Fig. 6.4 Size distribution of the liposomes in a 0.1 g L^{-1} suspension prepared according to Moscho et al. [108]. *Lines*: different distributions obtained from four consecutive light scattering measurements at 25°C. *Empty bars*: Size distribution obtained from 30 chronoamperometric signals obtained at −0.9 V vs. Ag|AgCl at temperatures above PTT treated as GUVs. *Filled bars*: Size distribution obtained from the same signals, treated as MLVs

where A_{int} is the surface of the interlayer space in the membrane. It follows then that:

$$r_{int} = \sqrt{\frac{Q_{lip}}{8\pi\Delta q}} \qquad (6.7)$$

and the radius of the giant unilamellar vesicle (GUV) is:

$$r_{GUV} = \sqrt{\frac{Q_{lip}}{8\pi\Delta q}} + L_L \qquad (6.8)$$

For giant vesicles, L_L is usually very small in comparison with the first term of (6.8) and can therefore be neglected. For the adhesion of a DMPC GUV displacing 1 pC at a potential of −0.9 V vs. Ag|AgCl, the corresponding diameter is ≈1.5 μm.

Using chronoamperometric measurements, it is possible with the above described relationships to determine the size distribution of the vesicles assuming MLVs or GUVs. The obtained distributions can be compared with size distributions obtained from light scattering measurements and conclusions can be drawn about the kind of vesicles present in the suspension. In the case of the DMPC liposomes produced according to the procedure adapted from Moscho et al. [108], the size distribution treating them as GUVs and as MLVs is compared in Fig. 6.4 with the size distribution obtained with light scattering. It is clearly seen that if the signals are treated as caused by GUVs, the size distribution obtained with the help of chronoamperometry corresponds very well with that obtained by light scattering measurements. If we assume MLVs, the size distribution does not correspond to the light scattering results and, therefore, it is safe to assume that the liposomes produced are mainly GUVs and that the assumptions made in (6.6)–(6.8) are correct. Unfortunately, the chronoamperometric method does not detect liposomes displacing less than a certain charge (given by the instrument sensitivity), and therefore, the left side of the distribution (taking the maximum as reference) cannot be determined. Thus chronoamperometry is very useful to determine the lamellarity of the liposomes, but inaccurate to calculate their size distribution if liposomes smaller than a critical diameter are present.

Equation (6.8) is also applicable to calculate the total (J_c) and gaussian (K_G) curvatures of the liposome as a function of the displaced charge:

$$J_c = 2r_{GUV}^{-1} = \frac{2}{\left(\sqrt{\frac{Q_{lip}}{8\pi\Delta q}} + L_L\right)} \approx \sqrt{\frac{32\pi\Delta q}{Q_{lip}}} \tag{6.9}$$

$$K_G = r_{GUV}^{-2} = \left(\sqrt{\frac{Q_{lip}}{8\pi\Delta q}} + L_L\right)^{-2} \approx \frac{8\pi\Delta q}{Q_{lip}} \tag{6.10}$$

The lipid molecules present on the outer layer of a membrane in the liquid crystalline phase state (defined as $L_{\alpha(d)}$) can be considered as points uniformly distributed on the surface of a sphere or radius r_{GUV}. It is known that a uniform distribution of points on a sphere follows a hexagonal arrangement, with the exception of exactly 12 unit cells (independent of the total number of units), which are pentagons [109]. If the number of points is large, the arrangement can be considered completely hexagonal. The number of hexagons forming the sphere is $N_{hex} = n_{lecithin}/6$, with sides of length ϕr_{GUV}, where ϕ is the angle between two adjacent lipid molecules. The area of a hexagon is:

$$A_{hex} = \frac{3\sqrt{3}}{2}\phi^2 r_{GUV}^2 \tag{6.11}$$

The summed area of all hexagons must be equal to the area of the sphere, and therefore:

$$\frac{3\sqrt{3}}{2}\phi^2 r_{GUV}^2 \left(\frac{n_{lecithin}}{6}\right) = 4\pi r_{GUV}^2 \tag{6.12}$$

and substituting (6.2):

$$\phi = 4\sqrt{\frac{\pi \Delta q\, A_{lecithin}}{\sqrt{3} Q_{lip}}} \tag{6.13}$$

which allows estimating the angle between two adjacent lipid molecules. Furthermore,

$$\phi r_{GUV} = \sqrt{\frac{2}{\sqrt{3}} A_{lecithin}} + 4L_L \sqrt{\frac{\pi \Delta q\, A_{lecithin}}{\sqrt{3} Q_{lip}}} \tag{6.14}$$

represents the distance between the geometrical centers of the polar head groups in the outer layer. The second term has a slight dependence on the size of the liposome and, therefore, by larger liposomes, the head groups are slightly closer together than for smaller ones. The first term in (6.14) is the mean distance between the centers of mass of the ends of the alkyl tails of adjacent molecules and it is independent of the size of the liposome. If the sign of the second term is inverted, one gets the distance between the polar head groups of the inner monolayer, which should then be more closely packed than in the outer layer. It is necessary, however, to be very careful with the obtained results, as they are just a very rough approximation of what happens in actual membranes, where the attractive forces between the alkyl chains and the repulsion of the hydration layers of the polar head groups play an important role. It is also assumed that the lipid molecules in the liposomes occupy the same area as on the mercury electrode. The approach, however, gives already an estimate of the distribution of the lipid molecules on a liquid membrane. Chronoamperometry has the very great advantage of allowing to study single liposomes and to estimate the above-mentioned characteristics of the membrane in a rather simple way.

6.5 The Macro- and Micro-Kinetics of Adhesion-Spreading of Vesicles on Electrodes

A liposome suspension contains liposomes of different sizes, and possibly different structures. It can be characterized by giving the concentration of the lipid in the aqueous suspension in the unit mg L^{-1}. However, this does not say anything about the *number* of liposomes per volume. That can be given in mol L^{-1}, provided that it has been determined experimentally. Even the number of liposomes per volume lacks another important information, viz., that about the size distribution. If the latter is known (e.g., from light scattering measurements), the structure of the liposomes is still undefined. From these considerations follows that a liposome suspension should ideally be characterized by the following set of data:

(a) Concentration of lipid in mg L^{-1}
(b) Concentration of liposomes in mol L^{-1}
(c) The size distribution function
(d) The description of liposome structure, e.g., as unilamellar or multilamellar

As outlined in Sect. 6.4, chronoamperometric measurements allow following the adhesion-spreading of liposomes on a mercury electrode. That process can be studied on two levels: (1) one can count the number of liposomes which produce adhesion-spreading signals per time unit and per surface area unit of the electrode. Of course this is a reaction rate, and it will depend on the concentration of liposomes in the suspension (in mol L^{-1}). Since that reaction rate relates to the concentration of the liposomes, the kinetics behind these measurements is termed the macro-kinetics and measured as the peak frequency J per unit area (see Sect. 6.5.1). (2) The second level to study the adhesion-spreading is to analyze the time dependence of a single adhesion-spreading process. The question behind is, how fast a single liposome undergoes the complete destruction, i.e., the adhesion-spreading starting with an intact liposome shortly before the first contact with the electrode, until the final state of an adsorbed island of lipid molecules. The kinetics of that process is termed micro-kinetics, as it relates to a single liposome. In order to measure the rate of a single adhesion-spreading event, the only experimentally accessible base is given by measuring the capacitive current produced by the attachment of the liposome to the electrode and the follow-up formation of the island of adsorbed lipid molecules. Since the initial mercury–aqueous electrolyte is converted to a mercury–lipid interface, that process is accompanied by a pronounced capacitive current. Most importantly, the micro-kinetics is not a simple one-step reaction, but it can only be described by a sequence of first-order reaction steps (see Sect. 6.5.2).

Of course, the macro-kinetics depends on the micro-kinetics. However, the connection between these two descriptions is not a simple one (see the following sections).

6.5.1 The Macro-Kinetics

By counting the number of adhesion-spreading events recorded over time in a chronoamperometric experiment, it is possible to gain insight into the kinetics of the adhesion mechanism. If the peak appearance frequency at different temperatures is determined, the activation energy of the whole process can be calculated [50] from:

$$\ln J = \ln C_{\text{lip}} + \ln A - \frac{E_a}{RT} \qquad (6.15)$$

where J is the peak frequency per unit area, C_{lip} is the bulk concentration of liposomes, which is assumed to be constant during the measurements, A is the preexponential factor of the Arrhenius equation, and E_a is the activation energy of the process. The slope of the curve $\ln J$ vs. T^{-1}, multiplied by $-R$, is then equal to E_a. As the activation parameters depend on the phase state, i.e., they are different for a liquid ordered liposome and a gel phase liposome, constructing curves over a wide temperature range and with an appropriate temperature spacing allows detecting phase transitions with an accuracy of 0.1 K.

It is necessary to take into consideration that the rate equation corresponding to the Arrhenius equation (6.15) comprises in one single first-order process all the reaction steps leading from a liposome in suspension to an adsorbed liposome island. This means that the determined activation energy does not correspond to the formation of any particular activated state, but it is an apparent value describing the adhesion-spreading process as a whole. A careful analysis of the obtained $\ln J$ vs. T^{-1} plot provides important hints concerning the adhesion-spreading mechanisms. As will be explained in detail in Sect. 6.5.2, the proposed adhesion-spreading mechanism involves a series of first-order processes ("interaction-docking," "bilayer-opening," and "rupture-spreading"). A perfectly linear relationship, e.g., would suggest that only one of these processes is the rate limiting step of the whole reaction. Deviations from linearity, on the other hand, are a clear indication that at least two of the processes occur (either simultaneously or sequentially) with similar rate constants. Finally, a curve with a positive slope (apparent negative activation energy) would imply that an exothermic reversible reaction takes place at some point during the adhesion-spreading process. This first analysis will facilitate the interpretation of the micro-kinetics, as will be discussed in the following section.

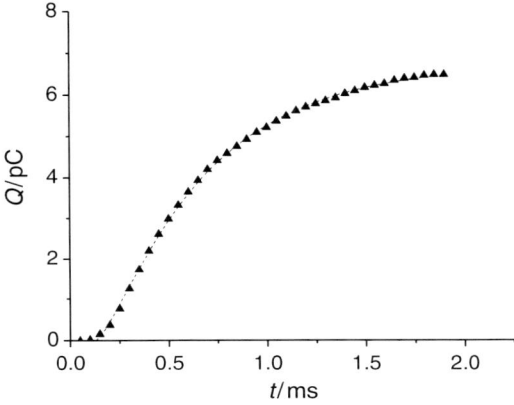

Fig. 6.5 *Solid triangles*: data obtained from the integration of the peak signal depicted in Sect. 6.4, Fig. 6.3. *Line*: fitting with (6.16). (Figure adapted from Agmo Hernández [110])

6.5.2 The Micro-Kinetics

Taking just one peak for analysis, this application of chronoamperometry allows studying the adhesion-spreading event of a single liposome. The peak can be integrated to get a charge vs. time curve (Fig. 6.5) that can be very well-fitted by the empirical equation [50, 52]:

$$Q_{lip} = Q_0(1 - e^{-t/\tau_0}) + Q_1(1 - e^{-t/\tau_1}) + Q_2(1 - e^{-t/\tau_2}) \quad (6.16)$$

As has been shown by Agmo Hernández et al. [52], this empirical equation is equivalent to the analytical solution of the mechanism:

$$L \underset{k_p}{\overset{\text{Trans.}}{\rightleftarrows}} L' \underset{k_b}{\overset{k_a}{\rightleftarrows}} L_D \overset{k_1}{\underset{k_{-1}}{\rightleftarrows}} L_{D^*} \overset{k_1}{\longrightarrow} L_O \overset{k_2}{\rightleftarrows} L_{O^*} \overset{k_2}{\longrightarrow} L_E \overset{k_3}{\rightleftarrows} L_A$$

where L is the free liposome, L' is the liposome in contact with the mercury surface, L_D is the docked liposome, L_{D^*} is the adsorbed docked liposome in a deformed state, L_O is the opened liposome, L_{O^*} the adsorbed opened liposome, L_E, the "deconvoluted" liposome, i.e., a lipid island that is not yet adsorbed, and L_A is the island of adsorbed lipid molecules. According to the model, K_1, K_2, and K_3 are the equilibrium constants of very fast adsorption equilibria. The first reaction represents the transport (defined by "Trans." and "k_p") of the liposomes from the bulk to the electrode surface which is also reversible. The second, reversible reaction (leading from L' to L_{D*}), is termed the "interaction-docking" process. As will be shown in Sect. 6.5.3, this process can be understood as an "activation" of the liposome followed by its "nucleation" on the electrode surface. This means, the liposome (L') if transformed to an active state in which it can then attach to the mercury surface and form a "nucleus" (L_{D*}) for the growing of an adsorbed lipid island. The third process (passing from L_{D*} to L_{O*}) has been coined the "bilayer-opening" step and consists in the rearrangement of the lipid molecules in close contact with the electrode in order to expose their lipophilic tails to the hydrophobic mercury surface. The fourth and last process (passing from L_{O*} to L_A) represents the "rupture-spreading" of the liposome, i.e., the irreversible formation of a hole in the lipid membrane and the spreading of a lipid monolayer on the mercury surface. The whole reaction is represented schematically in Fig. 6.6.

This model assumes that the chronoamperometric signal recorded is caused by the processes leading from L' to L_E. Solving the appropriate kinetic equations, the following solution is found for the response variable (Q_{lip}):

$$Q_{lip} = Q_0\left(1 - e^{-\frac{t}{\frac{2(1+K_1)}{(Z+Y)}}}\right) + Q_1\left(1 - e^{-\frac{t}{\frac{2(1+K_1)}{(Z-Y)}}}\right) + Q_2\left(1 - e^{-\frac{t}{\frac{(1+K_2)}{(k_2 K_2)}}}\right) \quad (6.17)$$

where

$$Y = \sqrt{(k_b + k_a - k_1 K_1 + k_a K_1)^2 + 4K_1 k_1 k_b} \text{ and}$$
$$Z = k_b + k_a + k_a K_1 + k_1 K_1.$$

Comparing (6.16) and (6.17), it follows that:

$$\frac{Z+Y}{2(1+K_1)} = \frac{1}{\tau_0} \quad (6.18)$$

$$\frac{Z-Y}{2(1+K_1)} = \frac{1}{\tau_1}$$

$$\frac{k_2 K_2}{(1+K_2)} = \frac{1}{\tau_2}$$

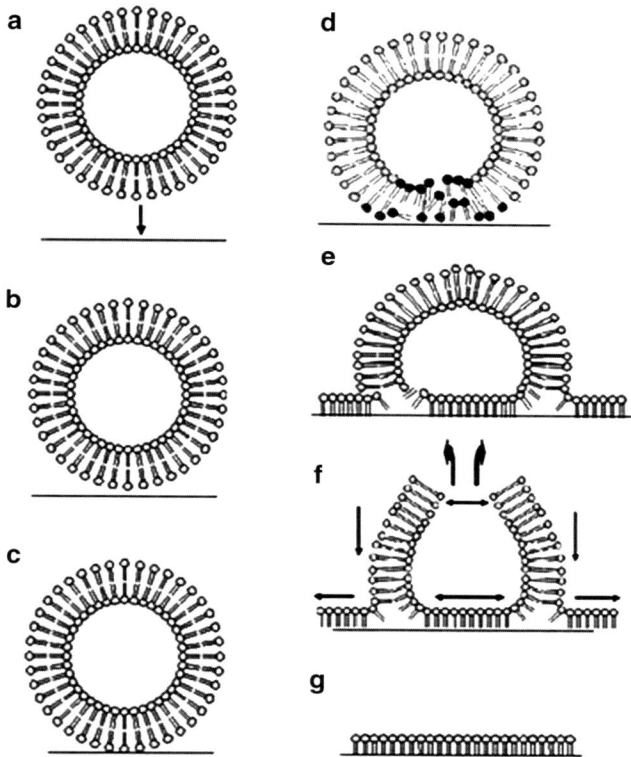

Fig. 6.6 Mechanism of adhesion-spreading of liposomes on a mercury electrode. (**a**) mass transport from the bulk to the electrode surface. (**b**) interaction of the electrical double layers of the liposome and the electrode, (**c**) direct contact of the liposome with the electrode (docking), (**d**) bilayer opening resulting in (**e**) a partially adsorbed liposome with increased lateral tension, (**f**) rupture of the liposome and spreading resulting on (**g**) an adsorbed lipid island. The column to the right represents the processes that can be easily characterized with the chronoamperometric signals obtained on a mercury electrode

Usually, the interaction-step is too fast to be resolved. Therefore, initial reports by Hellberg et al. [50] and Agmo Hernández et al. [47, 48, 87] assumed a zero-order process in their analysis of the charge transients (equivalent to the assumption $k_a = \infty$ and $k_b = 0$). An advantage of this assumption is that the analytical equivalent of time constant τ_1 is simplified as follows:

$$\frac{k_1 K_1}{(1+K_1)} = \frac{1}{\tau_1} \qquad (6.19)$$

Determining the values of τ_1 and τ_2 at different temperatures allowed determining the apparent activation energies of the bilayer-opening and rupture-spreading processes, respectively. For low temperatures, when K_1 and K_2 values can be expected to be much larger than 1, the activation energies determined reflect the real activation barriers of the processes studied. However, at higher temperatures, as the K_1 and K_2 values decrease, the obtained activation energies are only apparent quantities and were, in some cases, even negative. In spite of this not allowing an accurate determination of the real activation energies, it supported the notion of adsorption equilibria being coupled with the first-order steps described.

The use of high resolution potentiostats with high sample frequency allowed estimating values for the time constant of the interaction-docking process (τ_0), which was determined to be in the range of only a few microseconds [51, 52]. Apparent activation energies for the process could also be estimated.

The accessibility of activation parameters allowed Agmo Hernández et al. to propose a hypothesis about the characteristics of activated states of each step [47, 48, 52]. As has been mentioned before, the "interaction-docking" step activation barrier is given by the formation of an "active site" on the liposome that can lead to the "nucleation" of the vesicle on the mercury surface. For the "bilayer-opening" step, the activation parameters were very similar to those previously determined for flip-flop translocation [47], suggesting that the process involved the turning around of lipid molecules in order to show their lipophilic tails to the hydrophobic surface. Finally, the activation barrier for the "rupture-spreading" process was strongly affected by the inclusion of lytic molecules in the liposome solution [47, 48], indication that the formation of a lytic pore in the membrane was the limiting condition for rupture and spreading to occur. Table 6.1 shows a summary of the established steps in the adhesion-spreading process, as well as the proposed activated state on each step.

Table 6.1 Summary of the steps involved in adhesion-spreading, the activated and final states and important remarks for each process

Process	Description	Activated state	Final state	Remarks
Transport	Free liposomes in the bulk diffuse to the electrode surface	–	Liposomes in contact with the mercury surface	No charge displaced. Not measurable. Transport by diffusion. Following a "nucleation time," diffusion becomes the controlling step of the overall reaction
Interaction-docking	The liposome at the surface is "activated," possibly by the electric field and adheres to the electrode surface	Lipid molecules being rearranged, creating an hydrophobic patch on the liposome surface	Adhered (docked) liposome	Controls the overall process. Displaces a small amount of charge. Time dependence detectable only with high resolution measurements. The "activation" of the liposome is spontaneous or induced by the electric field
Bilayer-opening	The molecules in contact with the electrode turn around to expose their hydrophobic tails to the mercury surface. The inner monolayer pushes its way to the electrode	Molecules turning around	Vesicle partially flattened. Produces tension in the membrane. Small area already adsorbed	Size independent process, with the exception of very small vesicles, which have a high activation energy. Related to the flip-flop mechanism
Rupture-spreading	The flattened vesicle breaks up to release the inner solution and spreads to form an adsorbed lipid island	Formation of a critical pore	An adsorbed lipid island	Microscopic rate determining step. The activation energy is related to the bending, Gaussian and compressibility moduli as well as with the line tension of the membrane and the stretching tension arising during bilayer opening. In the gel phase is strongly dependent on the vesicle size

6.5.3 The Nucleation Model: A Connection Between Macro- and Micro-Kinetics

The study of individual adhesion-spreading signal demonstrated that, once a liposome is docked on the mercury surface, it undergoes a very fast transformation to an adsorbed lipid island. The number of adhesion-spreading events (N) during a certain time (t) is, therefore, not controlled by the bilayer-opening or rupture-spreading processes, but by the preceding steps, i.e., the mass transport and the interaction-docking steps. Agmo Hernández et al. demonstrated that the overall adhesion-spreading kinetics exhibit mixed control, i.e., both the mass transport and the docking kinetics play a role [52]. Interestingly, the kinetic part, converting a liposome at the electrode surface (L') to a docked liposome (L_{D^*}) and then, irreversibly and rapidly, to an adsorbed lipid island, formally resembled the kinetic model of metal nucleation on electrodes as proposed by Milchev [111, 112]. In that case, the first reversible step is the formation of active nucleation sites and the second one is the actual nucleation of the metal. In the case of liposome adhesion-spreading of liposomes on mercury, the system is formally analogous, the first step is the "activation" of a liposome (forming a potential "nucleation site") and the second one, is the actual "nucleation": an irreversibly attached liposome, which may be either in the L_O, L_E, or L_A states. Assuming the transport ("Trans." and k_p in the depicted model) to be caused by diffusion, the number of expected adhesion-spreading events on a given surface as a function of time is given by:

$$N(t) = A_{\text{SMDE}} C^*_{\text{lip}} \sqrt{D} \left(\begin{array}{l} A_1 \text{erf}\left(\sqrt{-k_{o1}t}\right) e^{-k_{o1}t} \\ +A_2 \text{erf}\left(\sqrt{-k_{o2}t}\right) e^{-k_{o2}t} + A_3 \sqrt{t} \end{array} \right) \tag{6.20}$$

where

$$A_1 = \frac{Z_o}{X_o(Y_o - X_o)(Y_o + X_o)^2 \sqrt{-2(Y_o + X_o)}}$$

$$A_2 = -\frac{Z_o}{X_o(Y_o - X_o)^2 (Y_o + X_o)\sqrt{-2(Y_o - X_o)}}$$

$$A_3 = \frac{2Z_o}{\sqrt{\pi}(Y_o - X_o)^2(Y_o + X_o)^2}$$

$$k_{o1} = \frac{1}{2}(Y_o + X_o)$$

$$k_{o2} = \frac{1}{2}(Y_o - X_o),$$

$$X_o = \sqrt{k_a^2 + 2k_ak_p + 2k_ak_b - 2k_ak_n + k_p^2 - 2k_pk_b - 2k_pk_n + k_b^2 + 2k_nk_b + k_n^2}$$

$$Y_o = k_a + k_p + k_b + k_n$$

$$Z_o = 16k_ak_n(k_ak_n + k_pk_b + k_pk_n)$$

and k_n is the "nucleation rate" and corresponds to the microscopic rate limiting step (the slowest microscopic process, usually the rupture-spreading).

Studies on the temporal distribution of the adhesion-spreading events showed that all events are independent of each other and are preceded by an stochastic process provoking the docking of the liposome, in complete agreement with the metal nucleation theory cited above [51]. This was proven by comparing the probability distribution of the number of adhesion-spreading signals obtained at $t = 100$ ms with the Poisson distribution given by:

$$P_m = \frac{N^m \exp(-N)}{m!} \qquad (6.21)$$

where N is the average number of signals obtained in a certain time period (from 0 to 100 ms in this case), and P_m is the probability to form exactly m nuclei (attached liposomes) within the same time interval.

The good correlation between the experimental data and (6.21), shown in Fig. 6.7, implies that the attachment of liposomes to the electrode surface can be considered as a time-dependent flux of independent random events, just as the formation of metal nuclei at electrodes. The number of obtained signals follows the same kind of distribution at all time intervals studied.

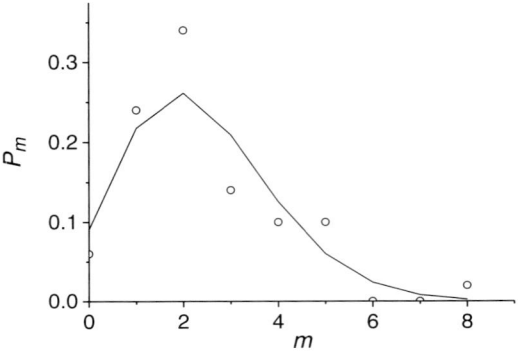

Fig. 6.7 P_m vs. m relationship. *Circles*: Experimental distribution of the number of liposome adhesion-spreading signals obtained after 100 ms at −0.9 V vs. Ag|AgCl (average = 2.4) in a suspension containing 8.03×10^7 GUVs (mL^{-1}). *Line*: Expected Poisson distribution for the given average according to (6.21). (Adapted from Agmo Hernández et al. [51])

A frequently studied stochastic quantity is the probability $P_{\geq 1}$, that is, the probability of forming at least one nucleus. It can be expressed by:

$$P_{\geq 1} = 1 - P_0 = 1 - \exp(-N) \tag{6.22}$$

which can be differentiated to yield:

$$dP_{\geq 1} = \exp(-N) \frac{dN}{dt} dt \tag{6.23}$$

and the average time of expectation to get one nucleus will be given by:

$$\overline{t_1} = \int_0^\infty t \, dP_{\geq 1} \tag{6.24}$$

Equations (6.22) and (6.23) provide information about the most important parameters to study the kinetics of the nucleation process: the rate of nuclei formation dN/dt and the average number of nuclei N.

Substituting (6.20) in the relationship given by (6.22) results in a $P_{\geq 1}(t)$ function that allows testing the proposed micro- and macro-kinetic models. Experimentally, the determination of $P_{\geq 1}$ is easier and more accurate than the determination of the actual average number of attached liposomes (nuclei) N, as, in the latter case, it is necessary to be sure that every signal recorded corresponds to one, and only one,

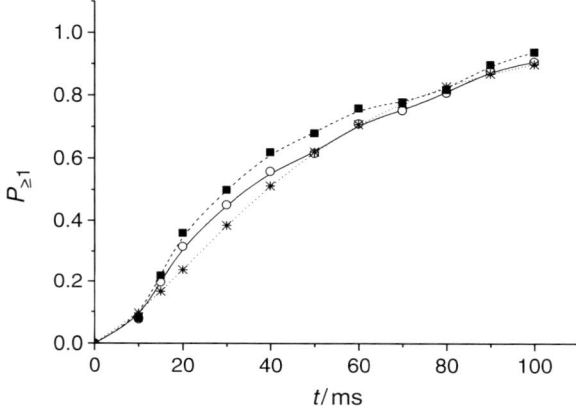

Fig. 6.8 Time dependence of the probability $P_{\geq 1}$. *Solid squares*: Experimentally determined values of $P_{\geq 1}$. *Stars*: Data for $P_{\geq 1}$ evaluated combining the theoretical model predicting the average number N of events and the Poisson distribution ((6.20) and (6.22)). Circles: Data for $P_{\geq 1}$ calculated from the experimental values of N using (6.22). The lines shown are guides to the eye. (Adapted from Agmo Hernández et al. [51])

adhesion-spreading event. It is also necessary to be sure that no adhesion-spreading events occurred in the time span between electrode preparation and the start of the measurement. These requirements can never be guaranteed. In the case of the determination of $P_{\geq 1}$, it does not matter if the recorded signal arises from exactly one or from more than one adhesion events. Previous, unrecorded events are also unimportant. Important is, however, that at least one adhesion-spreading event has taken place.

Figure 6.8 compares experimental values of $P_{\geq 1}$ obtained at different time intervals (solid squares) with a 0.05 g/L suspension of DMPC GUVs, with those evaluated by combining (6.20) and (6.22). The predicted and experimental results match satisfactorily well. Figure 6.8 shows also the data for $P_{\geq 1}$ calculated from the experimental average values of N by means of (6.22). As seen, these data correspond very well with the other curves. At times later than those shown in Fig. 6.8, all curves converge to $P_{\geq 1} = 1$, as expected.

These observations support the hypothesis that the mixed transport-kinetic model given by (6.20) describes the overall adhesion-spreading process of liposomes. Furthermore, it shows that the profound knowledge of the kinetics of metal nucleation at electrodes can be used to study the adhesion-spreading of vesicles at mercury electrodes, and probably also at other hydrophobic surfaces.

The nucleation model, assuming the formation and consumption of "active" liposomes (or, more precisely, of hydrophobic active sites on the liposome surface), can be used to explain several properties of liposomes in suspension, as has been shown in a recent publication [53]. Although on mercury electrodes it is likely that the activation is triggered by the presence of electric fields, the cited report shows that a similar process occurs in undisturbed membranes simply due to the dynamic motion of lipids in the membrane. Although the rate constants in the latter case are much smaller, the formed "active" sites have similar properties. As has been shown in the cited publication [53], they are responsible for the affinity of liposomes for hydrophobic substrates, as well as for their permeability. Current studies suggest that these sites may affect the partition of lipid-soluble molecules to the membrane as well as the resistance of the membrane to deformation and rupture, and the lipid-driven liposome–liposome fusion process. The dynamic nature of the activation and deactivation process gives rise to an intrinsic heterogeneity in liposome suspensions and is the starting point for a new area in liposome research. These series of discoveries were first possible thanks to the chronoamperometric measurements performed on mercury, as they permitted isolating single events and allowed kinetic analysis that would have been impossible otherwise.

6.5.4 Experimental Studies

6.5.4.1 Liposomes

The adhesion and spreading on mercury electrodes of GUVs and MLVs of different compositions, as well as the effect on the adhesion-spreading kinetics of membrane-active molecules (membrane-active peptides and surfactants), have been exhaustively studied by the authors [47, 48, 50–52, 87, 106, 107, 110]. Important conclusions have been drawn from these reports. Among other remarkable findings, it was shown that the kinetic and activation parameters clearly depend on the lamellarity of the liposomes [47]. MLVs presented higher energy barriers for both microscopical processes as well as for the macroscopical process and the steps were proceeding more slowly than in case of GUVs. This was a confirmation of the higher

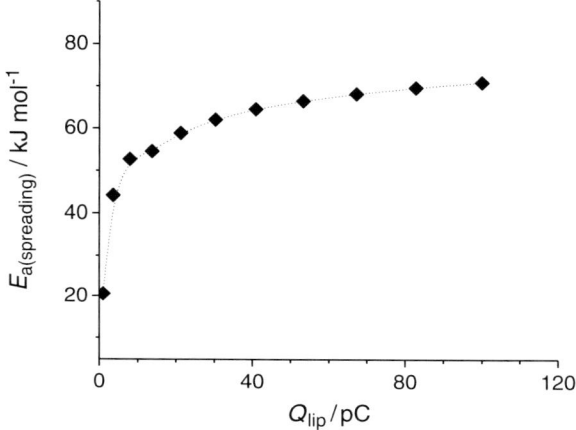

Fig. 6.9 *Curve* showing the dependence of the activation energy of rupture-spreading of DMPC GUVs in the gel phase on the size of the liposome expressed in terms of Q_{lip} (the *dotted line* is a guide to the eye) (Adapted from Agmo Hernández and Scholz [47])

stability of MLVs compared to GUVs. Interesting was also the discovered fact that, in the case of GUVs in the pretransition (rippled) gel phase ($P_{\beta'}$), the time constant and the activation energy of the rupture-spreading process depend on the size of the vesicles [47] as shown in Fig. 6.9. The authors explained this considering that, in order for rupture-spreading to take place, a pore larger than a certain critical size (critical radio $r_c = \Gamma/\sigma$, where Γ is the line tension along the pore rim and σ is the liposome surface tension) must be formed in order to irreversibly rupture the membrane. According to the literature, the energy barrier needed to build such a pore is given by [113]:

$$\delta E = \pi \Gamma^2 / \sigma \qquad (6.25)$$

The surface tension of the membrane arises mainly from the contributions of bending and stretching, which results in the energy barrier being given by:

$$\delta E = \pi \Gamma^2 ((2\kappa + \kappa_G) r_{GUV}^{-2} + 0.5 K (\delta A_s / A_s)^2)^{-1} \qquad (6.26)$$

where the first term inside brackets represents the bending tension and the second term is the stretching tension, κ and κ_G are the bending and Gauss moduli, K is the area compressibility (stretching) modulus, and $\delta A_s / A_s$ is the relative surface area change when the liposome flattens on the electrode surface. According to (6.26), a larger liposome will present a larger energy barrier for pore opening. This dependence will only be clear in cases where the first and the second term of the denominator of (6.26) are of the same order of magnitude. In the case of DMPC liposomes, this happens only when the vesicles are in the gel phase, in agreement with the observations reported by the authors. Furthermore, in order to support the assumption that the formation of a critical pore limits the rupture-spreading process, the authors determined the time constants and activation energies of this step for liposomes treated with lytic peptides [47, 48], observing that the rupture-spreading became much faster and the activation barrier was strongly diminished, in agreement with the hypothesis.

In order to derive even more information about the liposome from the kinetic and activation parameters of the rupture-spreading process, (6.26) can be linearized:

$$\delta E^{-1} = (2\kappa + \kappa_G)\pi^{-1} \Gamma^{-2} r_{GUV}^{-2} + 0.5 K \pi^{-1} \Gamma^{-2} (\delta A_s / A_s)^2 \quad (6.27)$$

and a plot of $1/\delta E$ vs. r_{GUV}^{-2} results in a straight line, the slope and y-intercept of which are related to the bending modulus and the stretching tension, respectively.

The activation and kinetic parameters of the bilayer-opening process can also provide information needed to characterize the liposomes. Agmo Hernández and Scholz [47] found that the activation energy of bilayer opening is independent of the size of the vesicles, with the exception of very small liposomes. The latter exhibit a clear increase of the activation energy of the process. The authors considered that this step involves only the lipid molecules in contact with the mercury surface. These molecules must turn around in order to show their lipophilic ends to the hydrophobic mercury surface, in a process requiring a high energy. In the case of large liposomes, that area of contact must, in any case, have a certain concentration of defects, which can prompt (catalyze) the bilayer-opening process acting as nuclei for lipid molecules to turn around. For small liposomes, the area of contact decreases and it is possible that no defects at all are present, and the process is no longer catalyzed. The values of the activation energy of bilayer opening can be extrapolated to get the activation energy barrier that an infinitely small liposome, completely defect-free, would possess. In that hypothetical case, the activated state would involve only a single lecithin

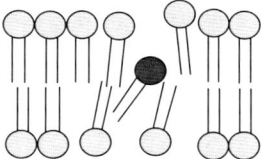

Fig. 6.10 Activated state of bilayer opening, which is very similar to that of the flip-flop translocation

molecule turning around, exposing its hydrophilic headgroup to the hydrophobic part of the membrane (Fig. 6.10), a very similar activated state as for flip-flop translocation (i.e., the exchange of lipid molecules between the two monolayers forming the membrane). In fact, the activation energy and preexponential factor determined by the authors for that limiting case is close to the parameters determined for flip-flop translocation in similar systems by different approaches [114, 115]. Chronoamperometric measurements are then shown to be a useful tool for the study of flip-flop translocation and can be used for the characterization of liposomes in a suspension.

The knowledge about the mechanism of adhesion-spreading of liposomes on a mercury electrode allowed the study of the effect of composition, embedded molecules, and molecules in solution on the properties of the membrane. The well-known stabilizing and crystal breaker effects of cholesterol were characterized and the temperature of formation of superlattices could be determined. The changes in the Arrhenius curves at the different phase transitions allow the construction of phase diagrams for liposomes of different compositions. The mechanism of interaction of lytic peptides and liposomes can be inferred, and the effect of surfactants was studied. A summary of all the published results is presented below. This wide compendium of data shows that chronoamperometric measurements and the detailed analysis of the adhesion-spreading signals are powerful tools for the study of liposomal suspensions.

Kinetics of DMPC Liposome Adhesion on a Static Mercury Electrode

The activation parameters defining the kinetics of the process were determined with the method described above. Table 6.2 shows a compilation of the data published so far. As can be observed, the activation

Table 6.2 Kinetic parameters of the adhesion-spreading on a mercury electrode at −0.9 V vs. Ag|AgCl of DMPC GUVs suspended in 0.1M KCl (A = Arrhenius preexponential factor, E_a = apparent activation energy in kJ mol^{-1})

Phase	$A_{overall}$ /L cm^{-2} s^{-1}	$A_{docking}$ /s^{-1}	$A_{opening}$ /s^{-1}	$A_{spreading}$ /s^{-1}	$E_{a(overall)}$	$E_{a(docking)}$	$E_{a(opening)}$	$E_{a(spreading)}$
$L_{\beta'}$	2.07	–	–	–	49.3 ± 11	–	–	–
$P_{\beta'}$	3.3 × 10^{-13}	1.39 × 10^{15}	5 × 10^{13a}	5.37 × 10^{12a}	−21.7 ± 5.7	58.1 ± 11.8	54.6 ± 1.8^a	52.7 ± 8.5^a
L_{α}	3.65 × 10^{-11}	1.39 × 10^{15}	1.1 × 10^{10}	50	−9.3 ± 14	58.1 ± 11.8	34.7 ± 5	−9 ± 4.9

Data compiled from [47, 52]
^{a}For liposomes displacing 7.9 pC (2.1 μm radio)

energies of the overall reaction and the microscopic steps can be determined with small error margins.

Effect of Lytic Peptides on the Kinetics of Adhesion-Spreading of DMPC Liposomes

Amphiphilic alpha-helical peptides are well-known for disrupting the lipid bilayer integrity and forming pores in the membrane and are therefore termed "lytic" or "pore forming" peptides. Antimicrobial peptides (AMPs) usually belong to this family and have been proposed as a new class of antibiotics against which the target organisms cannot develop immunity or resistance [116–119]. Chronoamperometric measurements of the adhesion-spreading of liposomes have allowed determining the effect of two of these peptides (mastoparan X, a component of wasp venom; and melittin, a component of bee venom) on the membrane properties and allows comparing the effect of concentration and nature of the peptide. They have also served to prove that the rupture-spreading step is controlled by the formation of a critical pore, as stated in previous sections. The process becomes much faster when the peptides are added, as shown in Table 6.3. An important reduction of the activation energy of the rupture-spreading process has also been reported [48]. Remarkably, the results have shown that both mastoparan X and melittin have almost precisely the same quantitative effect on DMPC membranes. This is unexpected, as both peptides are very different structurally. This observation wouldn't have been available without the help of the chronoamperometric method here described.

Effect of Surfactants on Membrane Stability

The study of the kinetics of liposome adhesion-spreading when small amounts of the nonionic surfactant Triton X-100 were added to the suspension allowed determining the effect of the detergent as well as the degree to which it is incorporated in the membrane. It was found that it had no significant effect on the adhesion-spreading kinetics of liposomes in the gel phase, suggesting that the surfactant is not incorporated in the membrane (Table 6.4). On the other hand, when the liposomes were in the liquid crystalline phase, the rupture-spreading process became much faster, implying that the detergent facilitates

Table 6.3 Time constants τ_1 and τ_2 for the adhesion-spreading of unilamellar liposomes on a mercury electrode when lytic peptides (mastoparan X or melittin) are added to a 0.1 g/L GUV suspension

Parameter	Pure DMPC	Mastoparan X			Melittin		
		0.01 μM	0.1 μM	0.5 μM	0.01 μM	0.1 μM	0.5 μM
τ_1/ms at 290 K[a]	0.15±0.05	0.14±0.04	0.10±0.02	0.084±0.013	0.19±0.07	0.11±0.04	0.095±0.023
τ_2/ms at 290 K[a]	0.68±0.11	0.48±0.11	0.43±0.10	0.36±0.06	0.73±0.18	0.41±0.14	0.36±0.07
τ_1/ms at 303 K	0.075±0.017	0.13±0.03	0.068±0.026	0.057±0.022	0.13±0.04	0.066±0.022	0.065±0.027
τ_2/ms at 303 K	0.56±0.11	0.41±0.16	0.28±0.09	0.27±0.09	0.49±0.16	0.25±0.06	0.26±0.04

Data compiled from [47, 48]
[a]Gel phase. Average values

Table 6.4 Comparison of the time constants of the bilayer-opening and the rupture-spreading steps of pure unilamellar DMPC liposomes with unilamellar liposomes treated with 0.15 μM of triton X-100 (GUV concentration = 0.1 g/L)

	T/K	t_1/ms Only DMPC	t_1/ms With 0.15 mM Triton X-100	t_2/ms Only DMPC	t_2/ms With 0.15 mM Triton X-100
Gel phase	284	0.23 ± 0.082	0.18 ± 0.1	0.91 ± 0.45	0.75 ± 0.14
	287	0.19 ± 0.06	0.14 ± 0.08	0.78 ± 0.23	0.59 ± 0.25
	290	0.15 ± 0.054	0.12 ± 0.023	0.68 ± 0.11	0.46 ± 0.13
Liquid crystalline	297	0.088 ± 0.043	0.075 ± 0.022	0.56 ± 0.17	0.27 ± 0.06
	300	0.081 ± 0.028	0.067 ± 0.02	0.56 ± 0.14	0.26 ± 0.027
	303	0.075 ± 0.017	0.059 ± 0.016	0.56 ± 0.11	0.25 ± 0.085

Data compiled from [48]

the formation of pores in the membrane. This also suggests that the mechanism by which larger amounts of Triton X-100 cause the dissolution of liposomes may involve the formation of large holes in the membrane, in agreement with reports in the literature [120].

Detection of the Crystal Breaker and Condensing Effects of Cholesterol in DMPC MLVs

As mentioned in Sect. 6.2 of this chapter, cholesterol is a membrane component found in mammalian and other cells, in which it plays a very important role. Cholesterol disturbs the translational order of the phospholipid molecules in the membrane, and, at physiological concentrations (above 30% mol/mol of the total lipid content), it avoids the formation of the gel phase even at very low temperatures, maintaining the membrane fluidity even in this conditions. This is referred to as the crystal breaker effect of cholesterol [121]. At the same time, cholesterol causes a straightening of the phospholipid acyl chains, reducing the surface area per lipid and condensing the membrane, providing it with increased rigidity, resistance, and impermeability. This is known as the stabilizing or condensing effect of cholesterol [121, 122]. Both the crystal breaker and the condensing effects are related to the formation of a liquid ordered phase ($L_{\alpha(o)}$), in which the lipids maintain lateral mobility but the movement of the acyl chains is limited. This molecule guarantees that the cell membrane will remain fluid and, at the same time, will be stable. The reasons behind this remarkable effect of

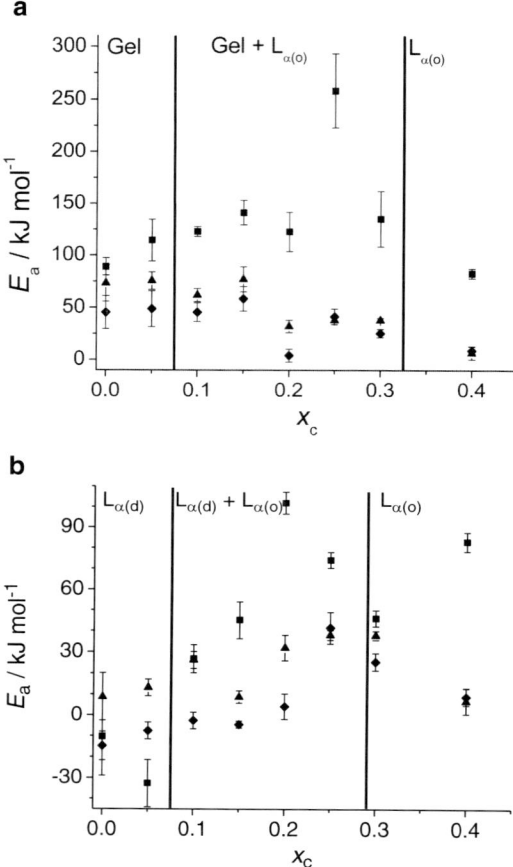

Fig. 6.11 Activation energies of the adhesion-spreading step as function of the molar fraction x_c of cholesterol in multilamellar DMPC liposomes (**a**) at low (2–23°C), and (**b**) at high temperatures (24–45°C): Squares: overall process, Triangles: bilayer-opening process, Rhombus: rupture-spreading process. The vertical bars indicate the standard deviations of measurements (Adapted from Agmo Hernández and Scholz [48])

cholesterol are still a matter of debate, as exemplified in a recent publication [123]. The chronoamperometric measurements described in this chapter are sensitive enough to characterize the crystal breaker and condensing effects of cholesterol, which are reflected in the kinetic parameters of the adhesion-spreading, as shown in Fig. 6.11.

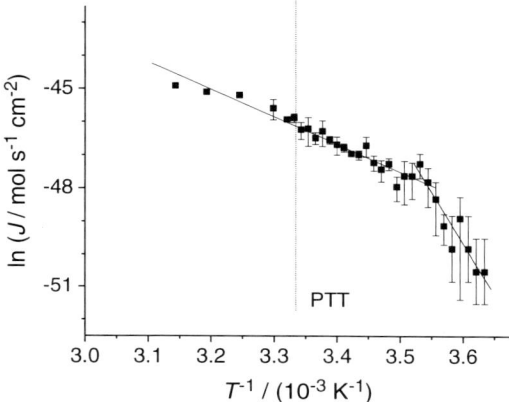

Fig. 6.12 Arrhenius plot for multilamellar DMPC liposomes containing 25 mol-% of cholesterol (a superlattice composition). Error bars were calculated from the standard deviation of the peak frequency. (Adapted from Agmo Hernández and Scholz [48])

How cholesterol distributes in the membrane is also not well established. One hypothesis, referred as the "superlattice view," states that highly ordered regions (superlattices) in which discrete amounts of lipid associate with cholesterol are formed. At certain cholesterol concentrations, the whole liposome membrane should be composed of a single superlattice. This hypothesis is still discussed and very difficult to prove experimentally, as cholesterol in a sample will not distribute evenly in all liposomes and, therefore, just a few will present the right concentration for superlattice formation. The chronoamperometric measurements, however, are able to isolate the responses from single liposomes. Studying the adhesion-spreading events with liposomes with a cholesterol content predicted to create superlattices, it was observed that some adhesion-spreading signals followed different kinetics. An analysis of these signals revealed the vanishing of the main phase transition which could be interpreted as evidence for the formation of superlattices (Fig. 6.12). The possibility of isolating signals arising from single liposomes proves to be a great advantage of the chronoamperometric method.

Table 6.5 Preexponential factors A and activation energies E_a of the bilayer-opening step of liquid crystalline GUVs of different lecithin compositions

	$A_{(opening)}/s^{-1}$	$E_{(opening)}/kJ\ mol^{-1}$
DMPC	1.1×10^{10}	34.7 ± 5
DMPC:DPPC 1:1	1.2×10^{8}	25.5 ± 7.3
DPPC	4.7×10^{6}	17.2 ± 2.7

The Bilayer-Opening Step and its Relationship with the Flip-Flop Translocation Mechanism

Chronoamperometric experiments demonstrated that the kinetics of the bilayer-opening process are dependent on the length of the acyl chains of the lipids. This dependence is due to significant changes in the preexponential factor of the Arrhenius equation when the acyl chain length is modified. The activation energy, on the other hand, does not show significant variations. This is the same behavior observed for flip-flop kinetics [124]. Table 6.5 shows a comparison between the adhesion-spreading kinetic parameters for DMPC (14 carbon chain length) and DPPC (16 carbon chain length) GUVs in which these effects are clearly observed. Furthermore, as stated in Sect. 6.5.4, the activation energies determined for phosphatidylcholine liposomes are very similar for both the bilayer-opening process and the flip-flop translocation [114, 115]. As has been discussed in previous sections, the activated state of both processes is the same: a lecithin molecule with its hydrophilic head embedded in the hydrophobic matrix ("turning around"). This shows that the electrochemical experiments described in this chapter can provide information concerning also the flip-flop process.

6.5.4.2 Studies of Biologically Relevant Systems

Although liposomes are certainly excellent model systems for biological membranes, it is a temptation to study systems with "real" biological membranes, possibly even real cell organelles. Here we shall survey these attempts. The following biologically relevant systems have been studied: (1) The interaction of nitrogen oxide (NO˙), an important messenger molecule, with DMPC and DOPC liposomes [125]. (2) The adhesion-spreading of thrombocyte vesicles [96]. (3) The adhesion-spreading of intact mitochondria [13]. (4) The adhesion-spreading of giant unilamellar liposomes of cardiolipins [126].

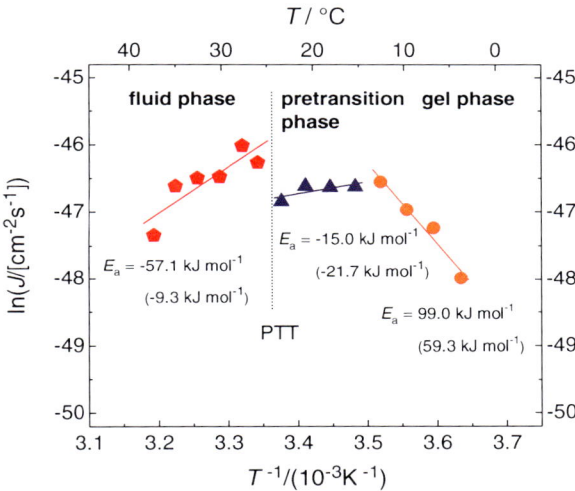

Fig. 6.13 Arrhenius plots of DMPC GUV liposomes, resulting from experiments with in situ application of nitric oxide (see text). The values in parentheses give the respective activation energies, but without nitric oxide action (Adapted from Hermes et al. [125])

The Interaction of Nitrogen Oxide (NO•) with DMPC and DOPC Liposomes

As mentioned earlier, the fluidity of membranes is important for the functions of cellular membranes. Thus, the membrane fluidity of red blood cells has been found to be related to cardiovascular and metabolic disorders, as hypertension, diabetes mellitus, and others [127–129]. Diffusion measurements in cell membranes support the idea that membrane fluidity is also linked to membrane compartmentalization [130]. Membrane fluidity is mainly regulated by the phospholipid and cholesterol content and it is modulated by temperature changes. There is a causal link between lipid rafts inside the plasma membrane and membrane fluidity [131–134]. Microdomain dynamics and membrane fluidity can be studied by ESR [135], fluorescence correlation spectroscopy [136], fluorescence recovery after photobleaching [137], single particle tracking [137], NMR [135], and dual color FRAP [137]. Recently it has been shown that measuring the adhesion-spreading behavior of DMPC and DOPC liposomes allows assessing the effect of nitrogen oxide on membrane fluidity [125]. Figure 6.13 gives the activation data for the macro-kinetics of adhesion-spreading of giant unilamellar DMPC liposomes in the absence and presence of NO•. Like in the absence of NO•, the Arrhenius plots exhibit three different ranges for the three phases of the liposomes: Above the phase transition temperature (PTT) of 23.6°C, the DMPC is in the fluid or liquid crystalline phase. Below the PTT, it is in a gel phase, and it is in a pretransition phase in-between the two other phase states. The fluid phase shows a steeper curve than the pretransition curve, both with *negative apparent activation energies*, while the gel phase gives a positive E_a. However, the resulting E_a values are distinctly different in the presence and absence of NO•. NO• induces in DMPC liposomes (fluid and pretransition phases) increasingly negative apparent activation energies of the macro-kinetics. This may point to an increasing temperature coefficient of adsorption (less strong adsorption with increasing temperature). In the micro-kinetics of DMPC liposomes, one observes for the fluid phase an unchanged activation energy of adhesion and a strongly increased activation energy of spreading. This indicates a stabilizing effect of NO•. The same is true for the gel phase: the activation energy of the macro-kinetics is strongly increased, and the apparent activation energies of the two steps of the micro-kinetics are very negative, showing that an increase in temperature strongly decreases the effect of the involved adsorption steps. The overall stabilizing effect of NO• on lipid bilayer membranes observed here supports earlier findings for the fluidity of inner mitochondrial membranes [138] and human erythrocyte membranes [139]. In this work [125] the authors also determined the partition coefficient of NO• between DMPC liposomes and an aqueous electrolyte solution: $K_{DMPC|H_2O}^{NO} = 10.0 \pm 2.1$. This value supports the conclusions made from the chronoamperometric measurements.

The Adhesion-Spreading of Thrombocyte Vesicles

Thrombocytes (blood platelets) are small cells which quickly form a haemostatic plug at sites of vascular injury in order to prevent hemorrhage. Thrombocytes are also involved in the occlusion of blood flow by formation of a primary thrombus and are involved in tissue injury, inflammation, and wound healing, partly via attracting and binding of leucocytes. The adhesion of thrombocytes on different substrates has been recently the subject of intensive studies, as this is highly impor-

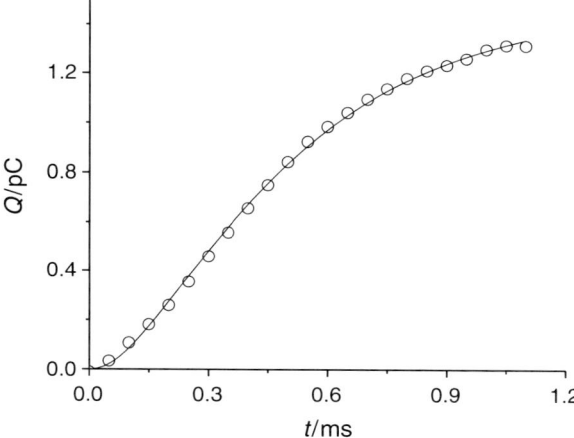

Fig. 6.14 Charge vs. time dependence of the adhesion-spreading of a thrombocyte vesicle at −0.6 V vs. Ag|AgCl. Empty circles represent the experimental data, and the solid line is the fitted curve (Adapted from Agmo Hernández et al. [96])

tant to assess the bio-compatibility of blood with materials exhibiting different physical surface properties [140–144]. An interesting and promising approach to study the properties of thrombocytes is to follow the adhesion-spreading of thrombocyte vesicles: These vesicles are closed unilamellar shells formed from intracellular granule membranes and outer cell plasma membranes when membranes are mechanically disrupted [145, 146]. They form spontaneously when the membrane is broken. The vesicles retain little or no cytosol and lack intracellular organelles. Very similar to liposomes, thrombocyte vesicles dispersed in isotonic aqueous potassium chloride solutions also produce adhesion-spreading signals on mercury electrodes [96]. Figure 6.14 depicts a charge vs. time curve for the adhesion-spreading of a thrombocyte vesicle. The experimental dependence was fitted with the equation

$$Q = Q_0 + Q_1(1 - \exp(t/\tau_1)) + Q_2(1 - \exp(t/\tau_2)) \qquad (6.28)$$

which is a simplified (assuming instant interaction-docking) form of (6.16) (see Sect. 6.5.2).

The fitting of the experimental data with (6.28) was only successful for potential negative with respect to the potential of zero charge. For positive potentials, the errors of the fitting parameters were unacceptable, which has to be taken as an indication that the model on which the equation is based is not valid for the process occurring at these positive potentials. Since the thrombocyte vesicles possess a negative surface charge [147], their attachment on a positively charged electrode surface will certainly involve an attractive coulombic interaction, which may slow down the *disintegration* process and completely change the kinetics of adhesion and spreading. For a negatively charged electrode, an initial repulsion of the negatively charged thrombocyte vesicles must be effective; however, the disintegration of the cell membrane and the adsorption of the hydrophobic tails of the lecithin molecules will quickly and irreversibly initiate the interaction-docking, bilayer-opening, and rupture-spreading of the membrane on the mercury surface. The time constant of bilayer opening τ_1 is in the range of 8.5×10^{-5} to 2.2×10^{-4} s, and the time constant of rupture-spreading τ_2 is in the range of 3.8×10^{-4} to 8.5×10^{-4} s. The data are rather similar to the time constants found for unilamellar and multilamellar DMPC liposomes.

The Adhesion-Spreading of Intact Mitochondria

Mitochondria, the "power plants" of biological cells, are for several reasons interesting objects for the electrochemical studies described in this chapter: (a) they play a key role in the metabolism of cells, (b) they are very rich in membrane structures, i.e., they possess a high content of lipids spreadable on a mercury electrode, and finally (c) their size (500–1,000 nm) is just perfect for signal generation. Not at least, the involvement of mitochondria in human diseases fuels the interest to characterize mitochondria under physiological or near to physiological conditions. In a recent study it could be demonstrated that isolated functionally intact mitochondria interact with the surface of a static mercury electrode in a way which is similar to the adhesion-spreading of liposomes and thrombocyte vesicles, i.e., they attach to the hydrophobic mercury surface and disintegrate by forming islands of adsorbed molecules [13]. Figure 6.15 shows chronoamperograms of a mitochondria suspension at two different temperatures. Figure 6.16a depicts the current transient of a single adhesion-spreading event, and Fig. 6.16b the corresponding charge

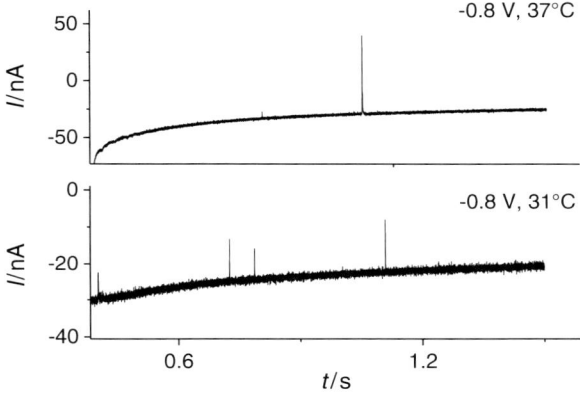

Fig. 6.15 Current-time traces measured in mitochondria suspension at two different temperatures

transient and the curve fitted with (6.28). Figure 6.17 shows typical and reproducible Arrhenius plots of the *macro*-kinetics for mitochondria isolated from BRIN-BD11 cells grown under two different conditions, i.e., (a) normoglycemic (5.5 mmol L^{-1} glucose), and (b) hyperglycemic (25 mmol L^{-1} glucose) conditions. Figure 6.17 exhibits a pronounced break of the straight lines at a temperature of ca. 27.5°C, indicating a phase transition. The PTT is in accordance with literature data of submitochondrial particles [148]. These electrochemically obtained results relate to *intact* mitochondria and reflect the PTT of the intact mitochondrial membrane. The plots in Fig. 6.17 also yield the activation energies of the *macro*-kinetics: the mitochondria grown under the two conditions are behaving differently below the PTT. Essentially the same change of the disintegration kinetics has been observed for mitochondria stressed by a palmitic acid diet (0.1 mM). In this work, an attempt was made to calculate the size of the mitochondria from the overall charge displaced at the electrode upon adhesion-spreading of the mitochondria. The results are in very good agreement with light scattering data. This shows that the interpretation of signals as being due to a complete disintegration of the mitochondrial membranes, i.e., the outer and the inner membranes, is obviously correct. Since mitochondria have an exceptional high content of cardiolipins, an additional study was performed with cardiolipin liposomes (see next part).

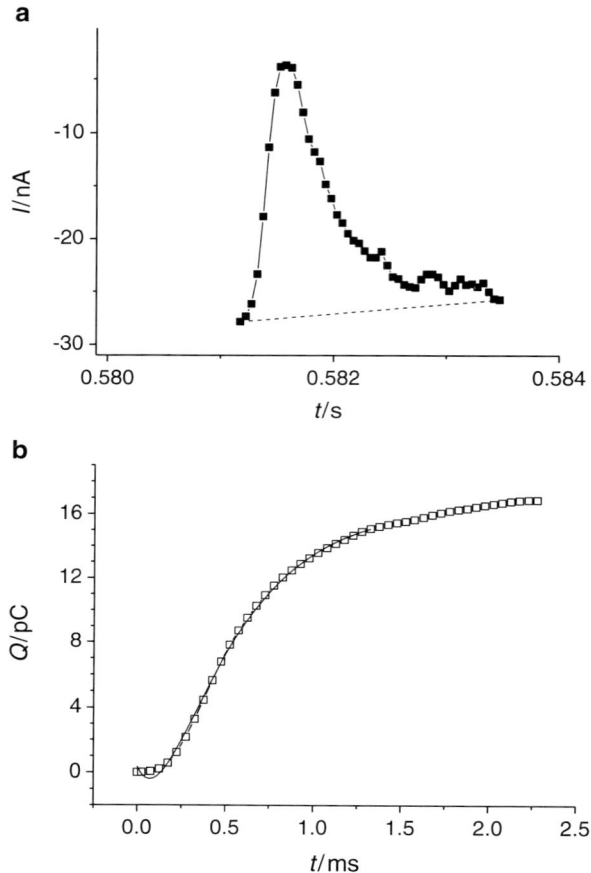

Fig. 6.16 (a) Adhesion-spreading peak as current-time trace. (b) The same peak after integration as charge-time trace. The solid line is the fitting curve using (6.28)

The Adhesion-Spreading of Giant Unilamellar Liposomes of Cardiolipins

Following the study of mitochondria, an attempt was made to study cardiolipin liposomes, as cardiolipins are a major constituent of mitochondrial membranes: Giant unilamellar liposomes of the synthetic cardiolipin 1′,3′-bis [1,2-dimyristoyl-sn-glycero-3-phospho]-*sn*-

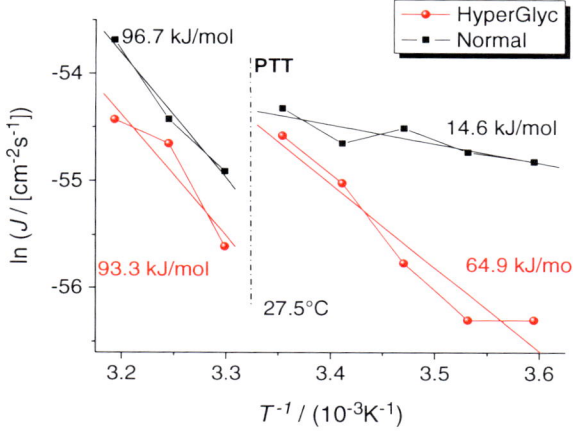

Fig. 6.17 Arrhenius plots of the macro-kinetics of the adhesion-spreading of mitochondria on mercury

glycerol (tetramyristoyl cardiolipin = TMCL) give chronoamperometric current peaks at a stationary mercury electrode [126] (Fig. 6.18). The signals are due to the adhesion and spreading of the liposomes on the hydrophobic mercury surface. The potential dependence shows a minimum of the peak frequency at the point of zero charge, a large maximum of peak frequency at about −0.2 V, and a second, however, smaller maximum at −0.8 V (cf. Fig. 6.19). The electrochemical behavior of the liposomes indicates phase transitions of the cardiolipin (Fig. 6.20) which could be also observed in differential scanning calorimetry. Future studies have to be focused on the relations between functional parameters of mitochondria under various (patho)-physiological conditions and the electrochemical behavior of mitochondria and liposomes reconstituted from mitochondrial cardiolipin fractions that show disease-specific compositions.

6.5.5 *Common Features of Adhesion-Spreading of Vesicles on Electrodes and Vesicle Fusion*

The three-step model of liposome adhesion-spreading on mercury shares several characteristics with the most accepted models for

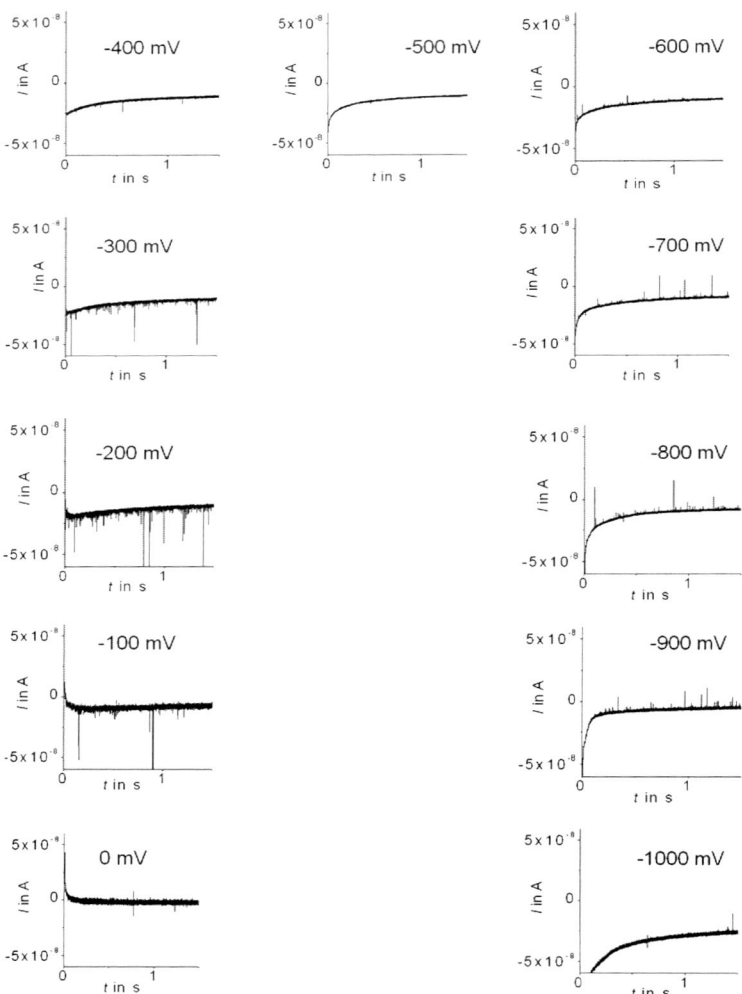

Fig. 6.18 Chronoamperometric traces recorded at different electrode potentials in a suspension of unilamellar TMCL liposomes in 0.1M KCl solution at 40°C (Adapted from Zander et al. [126])

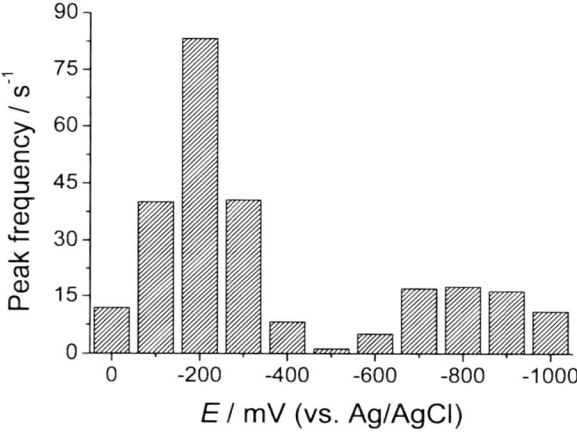

Fig. 6.19 Number of adhesion-spreading events (chronoamperometric traces) per time at different electrode potentials for TMCL GUV liposomes (Adapted from Zander et al. [126])

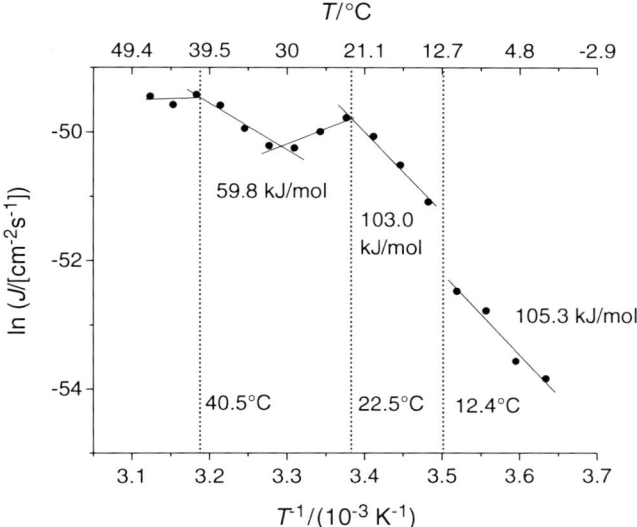

Fig. 6.20 Arrhenius plot for a suspension of TMCL GUV liposomes. Electrode potential: −200 mV vs. Ag/AgCl (Adapted from Zander et al. [126])

liposome–liposome fusion, and analogies between the processes can be built as follows:

(a) *Interaction-docking* vs. *initial liposome–liposome contact*
    In order for two lipid membranes to fuse, they first need to come in close contact with each other. That means overcoming a very high energy barrier given by the strong hydration of the lipid headgroups. Recent theories on membrane fusion have proposed that this barrier can be avoided if the contact between two liposomes occurs at a point on which hydrophobic patches are found [149–151] or formed [152]. As stated above, a recent publication has shown that hydrophobic active sites are indeed spontaneously formed and consumed in liposome membranes, causing heterogeneity in liposome suspensions, which appear then to be formed by two populations with different properties: (1) an "active" population with a critical amount of hydrophobic defects, characterized for its affinity for hydrophobic surfaces and its high permeability towards hydrophilic solutes, and (2) an "inactive" population which behaves like a hydrophilic sphere and which is almost impermeable to hydrophilic solutes [53]. Both populations are in dynamic equilibrium with each other, making it impossible to isolate them for long periods of time. Current, unpublished studies by one of the authors of this chapter (VAH) have shown that the active population is also more fusogenic than the inactive one, with most of the liposome–liposome fusion occurring between active liposomes. This activation–deactivation process is equivalent to the reversible formation of "nucleation sites" in liposomes that precedes their attachment (the interaction-docking step) and spreading on a mercury electrode, as has been discussed in previous sections. The main difference is that, in the case of mercury, the formation and consumption of these sites is driven by the electric fields present [52], while in free liposomes the process is likely to occur due to the dynamic motion of the lipids. In both systems, however, the process controls the overall reaction, as nothing can start without the liposome first being activated and attached.

(b) *Bilayer opening* vs. *hemifusion stalk formation*
    In most fusion models it has been agreed that, once two liposomes are in contact with each other, a "bridge" will form by means of which lipid molecules will be exchanged between them. This results in the formation of a "stalk" binding two lipo-

somes. In order to start the lipid-exchange lipid molecules at the area of liposome–liposome contact rearrange to allow the flow of lipids. This means that they turn partially around. This is a similar rearrangement like the one proposed for the bilayer-opening process, during which the lipid molecules turn around to show their lipophilic tails towards the mercury electrode, causing a lipid flow that will eventually result in the rupture and spreading of the vesicle on the surface [153].

(c) *Rupture-spreading* vs. *fusion pore formation*
Once the stalk is formed, a fusion pore appears that allows the lipids to reacquire a bilayer configuration. This implies the formation of a liposome containing the lipids from the two original vesicles. The inner contents of the individual liposomes are also mixed [153]. The energy barrier for this process is given by the formation of the fusion pore. In a similar way, the rupture-spreading step is controlled by the formation of a critical pore on the membrane. This causes the rupture of the membrane and its spreading on the mercury surface.

As can be seen from this brief comparison, there are mechanistic similarities between the membrane–membrane and the membrane-mercury fusion processes. The detailed study of the latter may shed some light on the many remaining questions concerning the former.

6.6 Conclusions and Perspectives

It was the aim of this chapter to present a new direction of research which is directed on the investigation of the interaction of vesicles with a hydrophobic electrified surface (mercury electrode), in order to get information about the physical properties of the membranes of the vesicles. This is an emerging research field; however, from the first studies of biological and medical applications [13, 96, 125, 126] it can be expected that the developed methodology can considerably contribute to a deeper understanding of membrane alterations and their implications for the metabolism of healthy organisms and pathological processes. The following directions of research are envisaged:

1. It is possible to construct vesicles (liposomes) in a well-defined way, and by studying the effects of deliberate changes of membrane

compositions on the adhesion-spreading behavior, conclusions can be made concerning the membrane properties
2. It is possible to subject vesicles (liposomes) to certain environmental stress factors, and to follow the changes of membrane properties by studying the adhesion-spreading on mercury
3. It is possible to study reconstituted vesicles prepared from cell cultures, cell culture supernatants, or cells isolated from organisms which have been subjected to certain stress factors; again to find out what were the effects on membrane properties
4. It is a special challenge to study the adhesion-spreading events of various cell organelles, also in order to investigate relations between stress factors and membrane properties
5. The methodology outlined in this chapter gives a unique access to the PTTs of intact organelles and reconstituted membrane vesicles. These data are otherwise very hardly accessible, and so it can be expected that the new electrochemical approach may produce a lot of new data needed to characterize membranes

Finally we like to briefly discuss the necessity to study the adhesion-spreading of vesicles with the help of mercury electrodes. It is true that using mercury is not anymore very popular because of well-reasoned general health concerns. However, this metal is the only one which provides a perfect hydrophobic surface while equally allowing to keep that electrode in an ideally isotropic surface state (liquid–liquid interface), and also keeping it extremely pure, i.e., without the usual contaminations which all solid electrodes have, unless they are prepared in an ultra-high vacuum. Further, the charge density of a mercury aqueous electrolyte interface can be determined with highest precision, so that the charges measured upon adhesion-spreading of vesicles can be translated to meaningful size data. Hence, there is simply no alternative to the use of mercury [154], and it should be understood that modern mercury electrodes can be handled without any health risk in the laboratory.

References

1. Singer SJ, Nicolson GL. Fluid mosaic model of structure of cell-membranes. Science. 1972;175:720–31.
2. Tanford C. The hydrophobic effect: formation of micelles and biological membranes. New York: Wiley; 1980.

3. Hunter CA. Quantifying intermolecular interactions: guidelines for the molecular recognition toolbox. Angew Chem Int Ed. 2004;43:5310–24.
4. Gill SJ, Wadso I. Equation of state describing hydrophobic interactions. Proc Natl Acad Sci USA. 1976;73:2955–8.
5. Dimitrov DS, Jain RK. Membrane stability. Biochim Biophys Acta. 1984;779:437–68.
6. Espinosa G, Lopez-Montero I, Monroy F, Langevin D. Shear rheology of lipid monolayers and insights on membrane fluidity. Proc Natl Acad Sci USA. 2011;108:6008–13.
7. Goldstein DB. The effects of drugs on membrane fluidity. Annu Rev Pharmacol. 1984;24:43–64.
8. Kuhry JG, Duportail G, Bronner C, Laustriat G. Plasma-membrane fluidity measurements on whole living cells by fluorescence anisotropy of trimethylammoniumdiphenylhexatriene. Biochim Biophys Acta. 1985;845:60–7.
9. Kubina M, Lanza F, Cazenave JP, Laustriat G, Kuhry JG. Parallel investigation of exocytosis kinetics and membrane fluidity changes in human-platelets with the fluorescent-probe, trimethylammonio-diphenylhexatriene. Biochim Biophys Acta. 1987;901:138–46.
10. Feijge MAH, Heemskerk JWM, Hornstra G. Membrane fluidity of nonactivated and activated human blood-platelets. Biochim Biophys Acta. 1990;1025:173–9.
11. Fajardo VA, McMeekin L, LeBlanc PJ. Influence of phospholipid species on membrane fluidity: a meta-analysis for a novel phospholipid fluidity index. J Membr Biol. 2011;244:97–103.
12. Levitan I, Fang Y, Rosenhouse-Dantsker A, Romanenko V. Cholesterol and ion channels. In: Harris JR, editor. Cholesterol binding and cholesterol transport proteins: structure and function in health and disease. New York: Springer; 2010.
13. Hermes M, Scholz F, Hardtner C, Walther R, Schild L, Wolke C, et al. Electrochemical signals of mitochondria: a new probe of their membrane properties. Angew Chem Int Ed. 2011;50:6872–5.
14. Montero J, Mari M, Colell A, Morales A, Basanez G, Garcia-Ruiz C, et al. Cholesterol and peroxidized cardiolipin in mitochondrial membrane properties, permeabilization and cell death. Biochim Biophys Acta. 2010;1797:1217–24.
15. Brian AA, McConnell HM. Allogeneic stimulation of cyto-toxic t-cells by supported planar membranes. Proc Natl Acad Sci USA. 1984;81:6159–63.
16. Sackmann E. Supported membranes: scientific and practical applications. Science. 1996;271:43–8.
17. Stauffer V, Stoodley R, Agak JO, Bizzotto D. Adsorption of DOPC onto Hg from the G vertical bar S interface and from a liposomal suspension. J Electroanal Chem. 2001;516:73–82.
18. Richter RP, Bérat R, Brisson AR. Formation of solid-supported lipid bilayers: an integrated view. Langmuir. 2006;22:3497–505.
19. Keller CA, Kasemo B. Surface specific kinetics of lipid vesicle adsorption measured with a quartz crystal microbalance. Biophys J. 1998;75:1397–402.
20. Lüthgens E, Herrig A, Kastl K, Steinem C, Reiss B, Wegener J, et al. Adhesion of liposomes: a quartz crystal microbalance study. Meas Sci Technol. 2003;14:1865–75.

21. Reiss B, Janshoff A, Steinem C, Seebach J, Wegener J. Adhesion kinetics of functionalized vesicles and mammalian cells: a comparative study. Langmuir. 2003;19:1816–23.
22. Rodahl M, Hook F, Fredriksson C, Keller CA, Krozer A, Brzezinski P, et al. Simultaneous frequency and dissipation factor qcm measurements of biomolecular adsorption and cell adhesion. Faraday Discuss. 1997;107:229–46.
23. Keller CA, Glasmastar K, Zhdanov VP, Kasemo B. Formation of supported membranes from vesicles. Phys Rev Lett. 2000;84:5443–6.
24. Jass J, Tjärnhage T, Puu G. From liposomes to supported, planar bilayer structures onn hydrophilic and hydrophobic surfaces: an atomic force microscopy study. Biophys J. 2000;79:3153–63.
25. Jenkins ATA, Bushby RJ, Evans SD, Knoll W, Offenhäusser A, Ogier SO. Lipid vesicle fusion on µCP patterned self-assembled monolayers: effect of pattern geometry on bilayer formation. Langmuir. 2002;18:3176–80.
26. Liang XM, Mao GZ, Ng KYS. Mechanical properties and stability measurement of cholesterol-containing liposome on mica by atomic force microscopy. J Colloid Interface Sci. 2004;278:53–62.
27. Liang XM, Mao GZ, Ng KYS. Probing small unilamellar eggPC vesicles on mica surface by atomic force microscopy. Colloids Surf B. 2004;34:41–51.
28. Tero R, Watanabe H, Urisu T. Supported phospholipid bilayer formation on hydrophilicity-controlled silicon dioxide surfaces. Phys Chem Chem Phys. 2006;8:3885–94.
29. Teschke O, de Souza EF. Liposome structure imaging by atomic force microscopy: verification of improved liposome stability during adsorption of multiple aggregated vesicles. Langmuir. 2002;18:6513–20.
30. Winger TM, Chaikof EL. Synthesis and characterization of supported phospholipid monolayers: a correlative investigation by radiochemical titration and atomic force microscopy. Langmuir. 1998;14:4148–55.
31. Kunneke S, Kruger D, Janshoff A. Scrutiny of the failure of lipid membranes as a function of headgroups, chain length, and lamellarity measured by scanning force microscopy. Biophys J. 2004;86:1545–53.
32. Wong JY, Park CK, Seitz M, Israelachvili J. Polymer-cushioned bilayers. II. An investigation of interaction forces and fusion using the surface forces apparatus. Biophys J. 1999;77:1458–68.
33. Nissen J, Gritsch S, Wiegand G, Radler JO. Wetting of phospholipid membranes on hydrophilic surfaces—concepts towards self-healing membranes. Eur Phys J B. 1999;10:335–44.
34. Yuan J, Parker ER, Hirst LS. Cationic lipid absorption on titanium: a counterion-mediated bilayer-to-lipid-tubule-network transition. Langmuir. 2007;23:7462–5.
35. Liu KW, Biswal SL. Using microcantilevers to study the interactions of lipid bilayers with solid surfaces. Anal Chem. 2010;82:7527–32.
36. Hubbard JB, Silin V, Plant AL. Self assembly driven by hydrophobic interactions at alkanethiol monolayers: mechanism of formation of hybrid bilayer membranes. Biophys Chem. 1998;75:163–76.
37. Silin VI, Wieder H, Woodward JT, Valincius G, Offenhausser A, Plant AL. The role of surface free energy on the formation of hybrid bilayer membranes. J Am Chem Soc. 2002;124:14676–83.

38. Tawa K, Morigaki K. Substrate-supported phospholipid membranes studied by surface plasmon resonance and surface plasmon fluorescence spectroscopy. Biophys J. 2005;89:2750–8.
39. Agmo Hernández V, Scholz F. The electrochemistry of liposomes. Isr J Chem. 2008;48:169–84.
40. Radler J, Strey H, Sackmann E. Phenomenology and kinetics of lipid bilayer spreading on hydrophilic surfaces. Langmuir. 1995;11:4539–48.
41. Williams LM, Evans SD, Flynn TM, Marsh A, Knowles PF, Bushby RJ, et al. Kinetics of the unrolling of small unilamellar phospholipid vesicles onto self-assembled monolayers. Langmuir. 1997;13:751–7.
42. Seifert U, Lipowsky R. Adhesion of vesicles. Phys Rev A. 1990;42:4768–71.
43. Castellana ET, Cremer PS. Solid supported lipid bilayers: from biophysical studies to sensor design. Surf Sci Rep. 2006;61:429–44.
44. Johnson JM, Ha T, Chu S, Boxer SG. Early steps of supported bilayer formation probed by single vesicle fluorescence assays. Biophys J. 2002;83: 3371–9.
45. Lipkowski J. Building biomimetic membrane at a gold electrode surface. Phys Chem Chem Phys. 2010;12:13874–87.
46. Li M, Chen M, Sheepwash E, Brosseau CL, Li H, Pettinger B, et al. AFM studies of solid-supported lipid bilayers formed at a Au(111) electrode surface using vesicle fusion and a combination of Langmuir-Blodgett and Langmuir-Schaefer techniques. Langmuir. 2008;24:10313–23.
47. Agmo Hernández V, Scholz F. Kinetics of the adhesion of DMPC liposomes on a mercury electrode. Effect of lamellarity, phase composition, size and curvature of liposomes, and presence of the pore forming peptide mastoparan X. Langmuir. 2006;22:10723–31.
48. Agmo Hernández V, Scholz F. The lipid composition determines the kinetics of adhesion and spreading of liposomes on mercury electrodes. Bioelectrochemistry. 2008;74:149–56.
49. Sek S, Xu S, Chen M, Szymanski G, Lipkowski J. STM studies of fusion of cholesterol suspensions and mixed 1,2-dimyritoyl-sn-glycero-3-phosphocholine (DMPC)/cholesterol vesicles onto a Au(111) electrode surface. J Am Chem Soc. 2008;130:5736–43.
50. Hellberg D, Scholz F, Schubert F, Lovric M, Omanovic D, Agmo Hernández V, et al. Kinetics of liposome adhesion on a mercury electrode. J Phys Chem B. 2005;109:14715–26.
51. Agmo Hernández V, Milchev A, Scholz F. Study of the temporal distribution of the adhesion-spreading events of liposomes on a mercury electrode. J Solid State Electr. 2009;13:1111–4.
52. Agmo Hernández V, Hermes M, Milchev A, Scholz F. The overall adhesion-spreading process of liposomes on a mercury electrode is controlled by a mixed diffusion and reaction kinetics mechanism. J Solid State Electr. 2009;13:639–49.
53. Agmo Hernández V, Karlsson G, Edwards K. Intrinsic heterogeneity in liposome suspensions caused by the dynamic spontaneous formation of hydrophobic active sites in lipid membranes. Langmuir. 2011;27:4873–83.
54. Burgess I, Li M, Horswell SL, Szymanski G, Lipkowski J, Majewski J, et al. Electric field-driven transformations of a supported model biological

membrane—an electrochemical and neutron reflectivity study. Biophys J. 2004;86:1763–76.
55. Whitehouse C, O'Flanagan R, Lindholm-Sethson B, Movaghar B, Nelson A. Application of electrochemical impedance spectroscopy to the study of dioleoyl phosphatidylcholine monolayers on mercury. Langmuir. 2004;20:136–44.
56. Nelson A. Electrochemistry of mercury supported phospholipid monolayers and bilayers. Curr Opin Colloid Interface Sci. 2010;15:455–66.
57. Valincius G, Meškauskas T, Ivanauskas F. Electrochemical impedance spectroscopy of tethered bilayer membranes. Langmuir. 2011;28:977–90.
58. Jeuken LJC, Connell SD, Nurnabi M, O'Reilly J, Henderson PJF, Evans SD, et al. Direct electrochemical interaction between a modified gold electrode and a bacterial membrane extract. Langmuir. 2005;21:1481–8.
59. Jeuken LJC. AFM study on the electric-field effects on supported bilayer lipid membranes. Biophys J. 2008;94:4711–7.
60. Du L, Liu X, Huang W, Wang E. A study on the interaction between ibuprofen and bilayer lipid membrane. Electrochim Acta. 2006;51:5754–60.
61. Vakurov A, Brydson R, Nelson A. Electrochemical modeling of the silica nanoparticle–biomembrane interaction. Langmuir. 2011;28:1246–55.
62. Shirai O, Yamana H, Ohnuki T, Yoshida Y, Kihara S. Ion transport across a bilayer lipid membrane facilitated by valinomycin. J Electroanal Chem. 2004;570:219–26.
63. Shirai O, Yoshida Y, Kihara S, Ohnuki T, Uehara A, Yamana H. Ion transport across a bilayer lipid membrane facilitated by gramicidin A—effect of counter anions on the cation transport. J Electroanal Chem. 2006;595:53–9.
64. Lundgren A, Hedlund J, Andersson O, Brändén M, Kunze A, Elwing H, et al. Resonance-mode electrochemical impedance measurements of silicon dioxide supported lipid bilayer formation and ion channel mediated charge transport. Anal Chem. 2011;83:7800–6.
65. Laredo T, Dutcher JR, Lipkowski J. Electric field driven changes of a gramicidin containing lipid bilayer supported on a Au(111) surface. Langmuir. 2011;27:10072–87.
66. Haller M, Heinemann C, Chow RH, Heidelberger R, Neher E. Comparison of secretory responses as measured by membrane capacitance and by amperometry. Biophys J. 1998;74:2100–13.
67. Neher E, Marty A. Discrete changes of cell-membrane capacitance observed under conditions of enhanced secretion in bovine adrenal chromaffin cells. Proc Natl Acad Sci USA. 1982;79:6712–6.
68. Neher E. Ion channels for communication between and within cells. Science. 1992;256:498–502.
69. Dernick G, de Toledo GA, Lindau M. Exocytosis of single chromaffin granules in cell-free inside-out membrane patches. Nat Cell Biol. 2003;5:358–62.
70. Sakmann B. Elementary steps in synaptic transmission revealed by currents through single ion channels. Science. 1992;256:503–12.
71. Amatore C, Arbault S, Bonifas I, Bouret Y, Erard M, Ewing AG, et al. Correlation between vesicle quantal size and fusion pore release in chromaffin cell exocytosis. Biophys J. 2005;88:4411–20.

72. Hafez I, Kisler K, Berberian K, Dernick G, Valero V, Yong MG, et al. Electrochemical imaging of fusion pore openings by electrochemical detector arrays. Proc Natl Acad Sci USA. 2005;102:13879–84.
73. Leszczyszyn DJ, Jankowski JA, Viveros OH, Diliberto EJ, Near JA, Wightman RM. Secretion of catecholamines from individual adrenal-medullary chromaffin cells. J Neurochem. 1991;56:1855–63.
74. Wightman RM, Jankowski JA, Kennedy RT, Kawagoe KT, Schroeder TJ, Leszczyszyn DJ, et al. Temporally resolved catecholamine spikes correspond to single vesicle release from individual chromaffin cells. Proc Natl Acad Sci USA. 1991;88:10754–8.
75. Chow RH, Vonruden L, Neher E. Delay in vesicle fusion revealed by electrochemical monitoring of single secretory events in adrenal chromaffin cells. Nature. 1992;356:60–3.
76. Marchal D, Boireau W, Laval JM, Moiroux J, Bourdillon C. An electrochemical approach of the redox behavior of water insoluble ubiquinones or plastiquinones incorporated in supported phospholipid layers. Biophys J. 1997;72:2679–87.
77. Gordillo GJ, Schiffrin DJ. The electrochemistry of ubiquinone-10 in a phospholipid model membrane. Faraday Discuss. 2000;116:89–107.
78. Laval JM, Majda M. Electrochemical investigations of the structure and electron-transfer properties of phospholipid-bilayers incorporating ubiquinone. Thin Solid Films. 1994;244:836–40.
79. Moncelli MR, Herrero R, Becucci L, Guidelli R. Kinetics of electron and proton transfer to ubiquinone-10 and from ubiquinol-10 in a self-assembled phophatidylcholine monolayer. Biochim Biophys Acta. 1998;1364:373–84.
80. Mårtensson C, Agmo Hernández V. Ubiquinone-10 in gold-immobilized lipid membrane structures acts as a sensor for acetylcholine and other tetraalkylammonium cations. Bioelectrochemistry. 2012;88:171–80.
81. Largueze JB, El Kirat K, Morandat S. Preparation of an electrochemical biosensor based on lipid membranes in nanoporous alumina. Colloids Surf B. 2010;79:33–40.
82. Yao WW, Lau C, Hui YL, Poh HL, Webster RD. Electrode-supported biomembrane for examining electron-transfer and ion-transfer reactions of encapsulated low molecular weight biological molecules. J Phys Chem C. 2011;115:2100–13.
83. Hosseini A, Collman JP, Devadoss A, Williams GY, Barile CJ, Eberspacher TA. Ferrocene embedded in an electrode-supported hybrid lipid bilayer membrane: a model system for electrocatalysis in a biomimetic environment. Langmuir. 2010;26:17674–8.
84. Correia-Ledo D, Arnold AA, Mauzeroll J. Synthesis of redox active ferrocene-modified phospholipids by transphosphatidylation reaction and chronoamperometry study of the corresponding redox sensitive liposome. J Am Chem Soc. 2010;132:15120–3.
85. Lingler S, Rubinstein I, Knoll W, Offenhäusser A. Fusion of small unilamellar lipid vesicles to alkanethiol and thiolipid self-assembled monolazers on gold. Langmuir. 1997;13:7085–91.

86. Horswell SL, Zamlynny V, Li H, Merrill AR, Lipkowski J. Electrochemical and PM-IRRAS studies of potential controlled transformations of phospholipid layers on Au(111) electrodes. Faraday Discuss. 2002;121:405–22.
87. Agmo Hernández V, Scholz F. Reply to the comment on kinetics of the adhesion of DMPC liposomes on a mercury electrode. Effect of lamellarity, phase composition, size and curvature of liposomes, and presence of the pore forming peptide mastoparan X. Langmuir. 2007;23:8650.
88. Scholz F, Hellberg D, Harnisch F, Hummel A, Hasse U. Detection of the adhesion events of dispersed single montmorillonite particles at a static mercury drop electrode. Electrochem Commun. 2004;6:929–33.
89. Ivosevic N, Zutic V. Spreading and detachment of organic droplets at an electrified interface. Langmuir. 1998;14:231–4.
90. Tsekov R, Kovac S, Zutic V. Attachment of oil droplets and cells on dropping mercury electrode. Langmuir. 1999;15:5649–53.
91. Zutic V, Kovac S, Tomaic J, Svetlicic V. Heterocoalescence between dispersed organic microdroplets and a charged conductive interface. J Electroanal Chem. 1993;349:173–86.
92. Banks CE, Rees NV, Compton RG. Sonoelectrochemistry understood via nanosecond voltammetry: Sono-emulsions and the measurement of the potential of zero charge of a solid electrode. J Phys Chem B. 2002;106:5810–3.
93. Kovac S, Svetlicic V, Zutic V. Molecular adsorption vs. cell adhesion at an electrified aqueous interface. Colloids Surf A. 1999;149:481–9.
94. Zutic V, Ivosevic N, Svetlicic V, Long RA, Azam F. Film formation by marine bacteria at a model fluid interface. Aquat Microb Ecol. 1999;17:231–8.
95. Hellberg D (2002) Untersuchungen von mikroorganismen an elektrodenoberflächen. Diplom, Universität Greifswald, Greifswald
96. Agmo Hernández V, Niessen J, Harnisch F, Block S, Greinacher A, Kroemer HK, et al. The adhesion and spreading of thrombocyte vesicles on electrode surfaces. Bioelectrochemistry. 2008;74:210–6.
97. Cutress IJ, Rees NV, Zhou Y-G, Compton RG. Nanoparticle-electrode collision processes: investigating the contact time required for the diffusion-controlled monolayer underpotential deposition on impacting nanoparticles. Chem Phys Lett. 2011;514:58–61.
98. Maisonhaute E, White PC, Compton RG. Surface acoustic cavitation understood via nanosecond electrochemistry. J Phys Chem B. 2001;105:12087–91.
99. Maisonhaute E, Brookes BA, Compton RG. Surface acoustic cavitation understood via nanosecond electrochemistry. 2. The motion of acoustic bubbles. J Phys Chem B. 2002;106:3166–72.
100. Rees NV, Banks CE, Compton RG. Ultrafast chronoamperometry of acoustically agitated solid particulate suspensions: nonfaradaic and faradaic processes at a polycrystalline gold electrode. J Phys Chem B. 2004;108:18391–4.
101. Rees NV, Zhou Y-G, Compton RG. The aggregation of silver nanoparticles in aqueous solution investigated via anodic particle coulometry. Chemphyschem. 2011;12:1645–7.

102. Rees NV, Zhou Y-G, Compton RG. Making contact: charge transfer during particle-electrode collisions. RSC Advances. 2012;2:379–84.
103. Zhou Y-G, Rees NV, Compton RG. Electrode-nanoparticle collisions: the measurement of the sticking coefficient of silver nanoparticles on a glassy carbon electrode. Chem Phys Lett. 2011;514:291–3.
104. Zhou Y-G, Rees NV, Compton RG. Nanoparticle-electrode collision processes: the underpotential deposition of thallium on silver nanoparticles in aqueous solution. Chemphyschem. 2011;12:2085–7.
105. Zhou Y-G, Rees NV, Compton RG. The electrochemical detection and characterization of silver nanoparticles in aqueous solution. Angew Chem Int Ed. 2011;50:4219–21.
106. Hellberg D, Scholz F, Schauer F, Weitschies W. Bursting and spreading of liposomes on the surface of a static mercury drop electrode. Electrochem Commun. 2002;4:305–9.
107. Hellberg D (2006) Elektrochemische charakterisierung von liposomen. Dissertation, Universität Greifswald, Greifswald
108. Moscho A, Orwar O, Chiu DT, Modi BP, Zare RN. Rapid preparation of giant unilamellar vesicles. Proc Natl Acad Sci USA. 1996;93:11443–7.
109. Saff EB, Kuijlaars ABJ. Distributing many points on a sphere. Math Intell. 1997;19:5–11.
110. Agmo Hernández V (2008) The mechanism of adhesion-spreading of liposomes on a mercury electrode. Dissertation, Greifswald Universität, Greifswald
111. Milchev A. Electrocrystallization: nucleation and growth of nano-clusters on solid surfaces. Russ J Electrochem. 2008;44:619–45.
112. Milchev A. Electrocrystallization. Fundamentals of nucleation and growth. Boston: Kluwer Academic; 2002.
113. Sens P, Safran SA. Pore formation and area exchange in tense membranes. Europhys Lett. 1998;43:95–100.
114. Kornberg RD, McConnell HM. Inside-outside transitions of phospholipids in vesicle membranes. Biochemistry. 1971;10:1111–20.
115. Abreu MSC, Moreno MJ, Vaz WLC. Kinetics and thermodynamics of association of a phospholipid derivative with lipid bilayers in liquid-disordered and liquid-ordered phases. Biophys J. 2004;87:353–65.
116. Rivas L, Luque-Ortega J, Fernández-Reyes M, Andreu D. Membrane-active peptides as anti-infectious agents. J Appl Biomed. 2010;8:159–67.
117. Zetterberg MM, Reijmar K, Pränting M, Engström Å, Andersson DI, Edwards K. PEG-stabilized lipid disks as carriers for amphiphilic antimicrobial peptides. J Control Release. 2011;156:323–8.
118. Hancock REW, Sahl H-G. Antimicrobial and host-defense peptides as new anti-infective therapeutic strategies. Nat Biotechnol. 2006;24:1551–7.
119. Lien S, Lowman HB. Therapeutic peptides. Trends Biotechnol. 2003;21:556–62.
120. Nomura F, Nagata M, Inaba T, Hiramatsu H, Hotani H, Takiguchi K. Capabilities of liposomes for topological transformation. Proc Natl Acad Sci USA. 2001;98:2340–5.
121. Vist MR, Davis JH. Phase-equilibria of cholesterol dipalmitoylphosphatidylcholine mixtures—H-2 nuclear magnetic-resonance and differential scanning calorimetry. Biochemistry. 1990;29:451–64.

122. Gaber BP, Peticolas WL. Quantitative interpretation of biomembrane structure by Raman-spectroscopy. Biochim Biophys Acta. 1977;465: 260–74.
123. Daly TA, Wang M, Regen SL. The origin of cholesterol's condensing effect. Langmuir. 2011;27:2159–61.
124. Liu J, Conboy JC. 1,2-diacyl-phosphatidylcholine flip-flop measured directly by sum-frequency vibrational spectroscopy. Biophys J. 2005;89:2522–32.
125. Hermes M, Czesnick C, Stremlau S, Stöhr C, Scholz F. Effect of NO on the adhesion–spreading of DMPC and DOPC liposomes on electrodes, and the partition of NO between an aqueous phase and DMPC liposomes. J Electroanal Chem. 2012;671:33–7.
126. Zander S, Hermes M, Scholz F, Gröning A, Helm CA, Vollmer D, et al. Membrane fluidity of tetramyristoyl cardiolipin (TMCL) liposomes studied by chronoamperometric monitoring of their adhesion and spreading at the surface of a mercury electrode. J Solid State Electr. 2012;16:2391–7.
127. Tsuda K, Kimura K, Nishio I. Leptin improves membrane fluidity of erythrocytes in humans via a nitric oxide-dependent mechanism—an electron paramagnetic resonance investigation. Biochem Biophys Res Commun. 2002;297:672–81.
128. Tsuda K. Association between homocysteine and membrane fluidity of red blood cells in hypertensive and normotensive men. CVD Prev Control. 2009;4:S153.
129. Tsuda K. Benidipine, a long-acting calcium channel blocker, improves membrane fluidity of erythrocytes in essential hypertension. CVD Prev Control. 2009;4:S160.
130. Mouritsen OG, Jorgensen K. Dynamical order and disorder in lipid bilayers. Chem Phys Lipids. 1994;73:3–25.
131. Loura LMS, de Almeida RFM, Silva LC, Prieto M. FRET analysis of domain formation and properties in complex membrane systems. Biochim Biophys Acta. 2009;1788:209–24.
132. Niemela PS, Hyvonen MT, Vattulainen I. Atom-scale molecular interactions in lipid raft mixtures. Biochim Biophys Acta. 2009;1788:122–35.
133. Somerharju P, Virtanen JA, Cheng KH, Hermansson M. The superlattice model of lateral organization of membranes and its implications on membrane lipid homeostasis. Biochim Biophys Acta. 2009;1788:12–23.
134. Scott HL. Modeling the lipid component of membranes. Curr Opin Struct Biol. 2002;12:495–502.
135. Bloom M, Thewalt J. Spectroscopic determination of lipid dynamics in membranes. Chem Phys Lipids. 1994;73:27–38.
136. Chiantia S, Ries J, Schwille P. Fluorescence correlation spectroscopy in membrane structure elucidation. Biochim Biophys Acta. 2009;1788: 225–33.
137. Day CA, Kenworthy AK. Fluorescence correlation spectroscopy in membrane structure elucidation. Biochim Biophys Acta. 2009;1788:245–53.
138. Perez-Rojas JM, Muriel P. Inhibition of mitochondrial respiration by nitric oxide is independent of membrane fluidity modulation or oxidation of sulfhydryl groups. J Appl Toxicol. 2005;25:522–6.

139. Mesquita R, Picarra B, Saldanha C, Silva JME. Nitric oxide effects on human erythrocytes structural and functional properties—an in vitro study. Clin Hemorheol Microcirc. 2002;27:137–47.
140. Tsyganov I, Maitz MF, Wieser E, Richter E, Reuther H. Correlation between blood compatibility and physical surface properties of titanium-based coatings. Surf Coat Tech. 2005;200:1041–4.
141. Vasilets VN, Kuznetsov AV, Sevastyanov VI. Regulation of the biological properties of medical polymer materials with the use of a gas-discharge plasma and vacuum ultraviolet radiation. High Energ Chem. 2006;40:79–85.
142. Taylor RG, Lewis JC. Microfilament reorganization in normal and cytochalasin-b treated adherent thrombocytes. J Supramol Struct Cell Biochem. 1981;16:209–20.
143. Mikhalovska LI, Santin M, Denyer SP, Lloyd AW, Teer DG, Field S, et al. Fibrinogen adsorption and platelet adhesion to metal and carbon coatings. Thromb Haemost. 2004;92:1032–9.
144. Larsson N, Linder LE, Curelaru I, Buscemi P, Sherman R, Eriksson E. Initial platelet-adhesion and platelet shape on polymer surfaces with different carbon bonding characteristics (an in vitro study of teflon, pellethane and xlon intravenous cannulae). J Mater Sci Mater Med. 1990;1:157–62.
145. Enyedi A, Sarkadi B, Foldespapp Z, Monostory S, Gardos G. Demonstration of 2 distinct calcium pumps in human-platelet membrane-vesicles. J Biol Chem. 1986;261:9558–63.
146. Barber AJ, Jamieson GA. Isolation and characterization of plasma membranes from human blood platelets. J Biol Chem. 1970;245:6357–65.
147. Slayman CL. Electrical properties of neurospora crassa respiration and intracellular potential. J Gen Physiol. 1965;49:93–116.
148. Katyare SS, Satav JG. Altered kinetic-properties of liver mitochondrial membrane-bound enzyme-activities following paracetamol hepatotoxicity in the rat. J Biosci. 1991;16:71–9.
149. Corkery RW. The anti-parallel, extended or splayed-chain conformation of amphiphilic lipids. Colloids Surf B. 2002;26:3–20.
150. Smirnova YG, Marrink S-J, Lipowsky R, Knecht V. Solvent-exposed tails as prestalk transition states for membrane fusion at low hydration. J Am Chem Soc. 2010;132:6710–8.
151. Kasson PM, Lindahl E, Pande VS. Water ordering at membrane interfaces controls fusion dynamics. J Am Chem Soc. 2011;133:3812–5.
152. Smeijers AF, Markvoort AJ, Pieterse K, Hilbers PAJ. A detailed look at vesicle fusion. J Phys Chem B. 2006;110:13212–9.
153. Lee J, Lentz BR. Secretory and viral fusion may share mechanistic events with fusion between curved lipid bilayers. Proc Natl Acad Sci USA. 1998;95:9274–9.
154. Scholz F. Mercury electrodes are indispensable tools for membrane research. Rev Polarogr. 2010;56:63–5.

Chapter 7
Bio-Electrochemistry and Chalcogens

Enrique Domínguez Álvarez, Uma M. Viswanathan,
Torsten Burkholz, Khairan Khairan, and Claus Jacob

7.1 Introduction

The last couple of decades have witnessed the emergence of the wide and diverse field of bio-electrochemistry which nowadays provides enough research to fill several international meetings per year. As part of this research, topics such as the electrochemical analysis of biological samples, including electrochemical biosensors, and the characterization of redox properties of proteins and enzymes first come to mind. Indeed, these areas of biological electrochemistry have blossomed ever since the first pioneering studies on electrochemical biosensors in the 1980s [1]. The field has moved on considerably since then, of course, and various aspects of modern biological electrochemistry have recently formed part of a special issue of *ChemPhysChem* [2].

Today, bio-electrochemistry is moving into various directions, as illustrated in Fig. 7.1, including the development of electrochemical biosensors which do no longer require the "help" of enzymes to detect specific molecules in complex solutions [3]. Related bioanalytical

---

E.D. Álvarez
Universidad de Navarra, Faculty of Pharmacy,
Irunlarrea 1, Pamplona 31008, Spain

U.M. Viswanathan • T. Burkholz • K. Khairan • C. Jacob (✉)
Division of Bioorganic Chemistry, School of Pharmacy, Saarland University,
Campus B 2.1, Saarbruecken 66123, Germany
e-mail: c.jacob@mx.uni-saarland.de

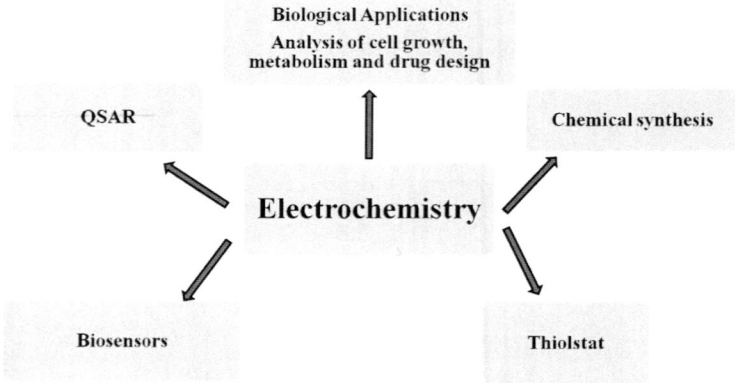

Fig. 7.1 The figure illustrates the widespread applications of traditional and modern electrochemical methods in key areas of biochemical and biological research. Some of these applications are discussed as part of this chapter. Please note that the figure as well as the text can only provide a—necessarily incomplete—selection of applications and do not reflect the full breadth of modern-day bio-electrochemistry

systems have also been developed which enable the electrochemical detection of specific analytes in (partially purified) biological samples, including N-acetylcysteine, cysteine, and disulfides in urine (see also Sect. 5) [4, 5]. The development of such devices has recently been the subject of several expert reviews [6, 7]. Sensors also form the topic of some of the other chapters of this book.

Besides biosensors, bio-electrochemistry also has a long tradition among bioinorganic chemists involved in the study of (redox-active) metalloproteins and enzymes [8]. Here, direct electrochemistry has been at the forefront of developments, often in conjunction with "intelligent" electrodes, such as (modified) film electrodes [9]. While most of these investigations have focused on metal-based biomolecules, more recently, the electrochemistry of non-metal, cysteine-containing proteins and enzymes has also been considered, in part for the characterization of these proteins, but also in part for rapid electrochemical detection [10–13].

Nonetheless, there are also some lesser known and often still emerging fields of bio-electrochemistry which catch the interest of researchers active in the field of biological chemistry or biochemistry. As part of this chapter, we will therefore consider a selection of electrochemical methods, which have already proven to be rather useful as part of biological chemical research, yet often still require further development to become widely accessible and applicable. Here, we will take the perspective of a biological redox chemist, with her/his own specific research interests, topics, and also problems.

Before we start, however, we first need to consider why electrochemical methods are particularly interesting and at the same time also well suited for the analysis of biological materials, from unusual natural products to complex biological systems, such as single cells. This question is not entirely unreasonable, since at first sight, electrochemistry, apart from a few basic applications, is hardly an integral part of Biology, Pharmacy, or Medicine. Nonetheless, we need to realize that electrochemistry is rather diverse and provides a whole arsenal of different methods and techniques which in one aspect or another may be very useful for the rapid and reliable analysis of biological materials. Such electrochemical methods are particularly valuable in bioanalysis because they are often readily *available*, widely *applicable*, extraordinarily *informative* and *sensitive*, yet at the same time also highly *reliable* and *reproducible*, and hence often considerably more *robust* when compared to alternative approaches, such as spectroscopy. Importantly, electrochemical methods, such as (Cyclic) Voltammetry (CV) or differential pulse polarography (DPP), are also highly *selective* and therefore enable the analysis of specific substances and processes in complex biological mixtures.

Within this context, the electrochemical glucose sensor may come to mind: This device is more or less selective for glucose, can be operated in whole blood, and is able to reliably and reproducibly sense even small changes in blood glucose concentrations despite the presence of a wealth of other blood components. Importantly, such a sensor is simple, small, and also rather cost-effective when compared to possible alternatives [14–16].

In the following sections of this chapter, we will focus on four rather interesting and powerful applications of electrochemistry in Biology and drug development: (1) electrochemical methods to monitor the proliferation of cells in real time and without the need for aliquots and staining; (2) electrochemical techniques to analyze cellular processes, such as metabolism, intra- and extracellular levels of reactive oxygen species (ROS), reactive nitrogen species (RNS), and oxidative stress (OS) at the single-cell level; (3) electrochemistry of sulfur and selenium proteins, which are at the center of important cellular regulatory and signaling processes (such as the cellular "thiolstat"); and last but not the least (4) electrochemistry of chalcogen-containing natural products and synthetic substances, with the aim to characterize such compounds and to derive at certain "chemical structure-electrochemical potential-biological activity relationships".

Together, these four applications will highlight the considerable practical potential of electrochemical methods in biochemical and pharmaceutical research, especially in the context of research problems which are difficult to address or to resolve by other, more conventional methods. Nonetheless, these examples represent just a small section of a wider, thriving, and continuously growing area of bioelectrochemical research and applications.

7.2 The Chalcogen-Specific Electrochemical Tool Kit

Several of the techniques considered as part of this chapter are linked to the redox chemistry of chalcogens, i.e., oxygen, sulfur, selenium, and tellurium. Since the methods available for the analysis of such redox systems are often not widely known or straightforward, we will briefly consider them as part of the emerging electrochemical "toolkit" which can be employed for the detection and characterization/analysis of these systems *in vitro* and possibly also *in vivo*.

In stark contrast to the well-established electrochemistry of metal-based redox systems, which include various metalloproteins and enzymes, the analysis of chalcogen-based redox systems is still far from trivial and sometimes even controversial [17–19]. This is rather disappointing, as there is a strong interest among biological chemists to identify and subsequently quantify ROS and RNS inside and outside living cells [20]. Such ROS include, for instance, the superoxide radical anion ($O_2^{\cdot-}$), hydrogen peroxide (H_2O_2), and the hydroxyl radical ($^{\cdot}OH$); nitric oxide ($^{\cdot}NO$) and peroxynitrite ($ONOO^-$) are among the most prominent RNS [21]. Table 7.1 provides a brief overview of the most common ROS and RNS, some of which will be discussed later on in the context of electrochemical detection. From a biochemical point of view, these reactive species are rather important as they play a pivotal role in cell signaling as well as host defense against bacteria and microbes. At the same time, the intracellular concentrations of various ROS and RNS increase during OS and may

Table 7.1 This table summarizes some of the most common ROS (*white background*) and RNS (*light grey*) found in mammalian cells under normal conditions and increased during OS

Reactive species	Formula	Formal oxidation state	Occurrence
Singlet oxygen	$^1O_2$	0	UV-radiation combined with endogenous photosensitizers
Ozone	O_3	0	Air pollution, possibly formation in neutrophils
Superoxide radical	$HO_2^\bullet$	−0.5	Mitochondria, NADPH oxidases, dual oxidases, xanthine oxidase (XO)
	$O_2^{\bullet-}$	−0.5	
Dihydrogen trioxide	H_2O_3	−2/3 (−1, 0, −1)	O_3 chemistry in phagocytes, formation still debatable
Hydrogen peroxide	H_2O_2	−1	Mitochondria (formed from $O_2^{\bullet-}$ in the presence of SOD), immune cells (formed from O_2 by various oxidases)
Aliphatic peroxides	ROOH (R=alkyl)	−1	Lipid damage, e.g., by $^1O_2$, lipid, or DNA damage by O_3
Hydroxyl radical	$HO^\bullet$	−1	Fenton reaction in the presence of H_2O_2 and Cu^+, Fe^{2+}
Nitrogen monoxide	$^\bullet NO$	+2	Nitric oxide synthases (NOS), reduction of NO_2^-
Peroxynitrite	$ONOO^-$	−2, +3, −1, −1	$O_2^{\bullet-} + {}^\bullet NO \rightarrow ONOO^-$
Hypochlorous acid	HOCl	+1, −2, +1	Myeloperoxidase in neutrophils, monocytes, macrophages

It provides oxidation states for the reactive atoms and basic information regarding formation pathways. Most of these species are formed inside the human body as part of metabolism or host defense. Others, such as ozone, are (mostly) due to exogenous sources, such as radiation or environmental pollution. Hypochlorous acid (de facto "bleach", *dark grey*) is an example of a reactive halogen species, which is also formed in the human body by immune cells as part of host defense against invading organisms. Some of these species can be detected rather accurately employing electroanalytical methods. Table adapted from [23]

subsequently cause cell death via apoptosis. Modulating levels of ROS and RNS can therefore also serve a therapeutic purpose, for instance in the context of treating cancer. This has led to the emerging strategy of redox modulation which may also benefit from the input of electrochemistry and will be discussed in Sect. 7 [22].

Not surprisingly, methods enabling the determination of ROS and RNS concentrations in cells have received considerable attention. These detection techniques also include electrochemical methods. ROS, such as H_2O_2, peroxynitrite (ONOO$^-$), nitric oxide ·NO, and nitrite NO_2^-, can be detected fairly easily by a coulometric method using a platinized carbon-based electrode [20, 24].

While most ROS and ·NO therefore can be detected, characterized, and quantified by fairly standard and straightforward electrochemical methods, the electrochemistry of their cellular targets/reaction partners, e.g., sulfur- or selenium-based redox systems, is considerably more complex [17]. This certainly applies to most sulfur- (i.e., cysteine- and methionine-)-based proteins and enzymes, whose redox properties have only sporadically been analyzed by electrochemical methods (see Sect. 5). Yet it also applies to rather simple, small sulfur-containing molecules, such as disulfides and polysulfanes (see Sect. 6). Here, the "electrochemical tool kit" available to date is rather limited, crude, and prone to error: Traditionally, such sulfur-containing substances have been analyzed with the assistance of a metal-based electrode, such as a gold or mercury electrode [25–27]. These electrodes adsorb the sulfur-containing materials rather well, and therefore allow detection and basic characterization. Indeed, some flow-through detectors used in conjunction with chromatography (often FPLC and HPLC) are based on electrochemical detection of the thiol/disulfide redox couple, either in small molecules or in native, modified, or tagged proteins, such as metallothionein (MT) [4, 5, 11]. A schematic view of such an analytical setup is shown in Fig. 7.2.

Unfortunately, the use of such electrodes for analytical purposes has several drawbacks. First of all, it is obvious that the adsorption process used to "capture" the sulfur species on the electrode surface severely affects its "chemistry", including its redox properties. Rather than measuring, for instance, the "true" thiol/disulfide redox potential or monitoring the reduction of a polysulfane (RS_xR, $R \neq H$, $x \geq 3$) to perthiol (RS_xH, $R \neq H$, $x \geq 2$), one is de facto investigating surface-bound sulfur-gold or sulfur-mercury species [1, 27]. The latter obviously

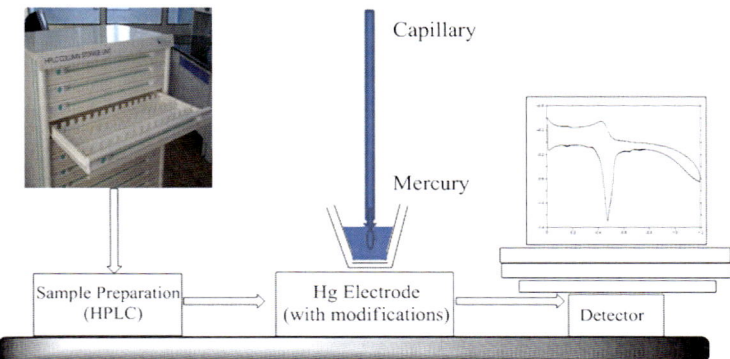

Fig. 7.2 Schematic view of an electrochemical method to monitor the presence of thiol-containing materials (including proteins). The example provided illustrates the use of CV in conjunction with a dropping mercury electrode to monitor the elution of the cysteine-rich protein metallothionein (MT). This method has been used, for instance, for small thiol-containing compounds by Stenken et al. (using a gold-amalgam instead of a "pure" Hg electrode) and for MT proteins by Adam et al. (using a more complex catalytic system also containing cobalt ions). One should emphasize that electrochemical detection (and characterization) has many advantages as far as cysteine-rich proteins such as the MT proteins are concerned, since many of them do not show a signal in traditional UV detectors (as they do not contain any aromatic residues) and are also difficult to analyze otherwise [5]

exhibit their very own redox behavior, which is likely to differ considerably from one of the "free" sulfur species in solution.

Not surprisingly, there have been some attempts in the past to minimize the influence of adsorption phenomena and to optimize the electrochemistry of sulfur- (and selenium-, tellurium-) containing substances. On the one hand, alternative, coupled, or indirect methods have been employed which measure redox behavior by different means. Willem Koppenol and colleagues, for instance, have applied CV to equilibria of chalcogen species with a redox dye and combined this setup with pulse radiolysis to inch closer to the various oxidation and reduction potentials of selenocysteine [18]. Other groups have considered alternative electrodes, including amalgamated copper electrodes and carbon electrodes [4, 28]. While amalgamated electrodes still suffer from adsorption phenomena, as solid rather than liquid electrodes, they nonetheless provide more flexibility when it comes to practical applications. Carbon electrodes may also be used, as they are also easy to handle and are less prone to adsorption phenomena, yet these electrodes do not monitor the classical, biologically

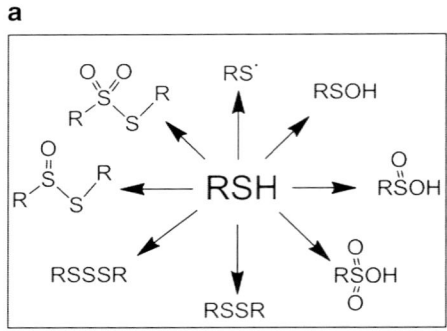

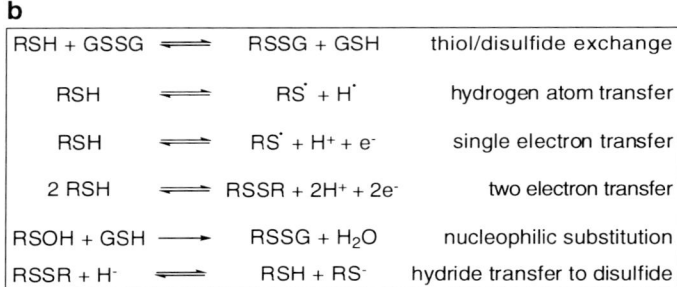

Fig. 7.3 Selection of biochemically relevant Reactive Sulfur Species formally derived from the thiol (RSH) chemotype (*center*). Clockwise from the top: thiyl radical, sulfenic acid, sulfinic acid, sulfonic acid, disulfide, trisulfide, thiosulfinate, thiosulfonate (**a**); sulfur-based redox reactions may proceed via different redox mechanisms, some of which do not involve direct electron transfer and hence are difficult to analyze using electrochemical techniques (**b**)

relevant thiol/disulfide redox behavior, but rather a kind of radical chemistry which is not directly relevant in biological systems [26].

While the thiol/disulfide redox pair - despite some drawbacks - is in principle accessible electrochemically, other sulfur modifications and transformations are more difficult to monitor. Here, one must bear in mind that sulfur is a true redox chameleon in Biology, where it occurs in more than ten different oxidation states (including fractional ones) and in various sulfur "chemotypes". Each of these chemotypes exhibits its own chemical properties. While most of them are also redox-active, the mechanisms at the center of their respective redox chemistries may vary widely.

Figure 7.3 summarizes the most common sulfur chemotypes (Fig. 7.3a) and some of the underlying reaction mechanisms

(Fig. 7.3b) [29]. It is fairly obvious from this figure that despite the fact that there are plenty of redox processes involved in sulfur (bio-) chemistry, few of them involve electron transfer and hence are directly accessible by electrochemical methods. Indeed, nucleophilic exchange and atom transfer mechanisms devoid of direct electron transfer dominate the redox behavior of most sulfur species. It is therefore doubtful that an electrochemical method would easily be able to reveal the redox behavior of sulfur species such as sulfenic and sulfinic acids, thiosulfinates, thiosulfonates, sulfoxides, and sulfones. Nonetheless, these sulfur species should not be ignored outrightly, as they occur inside most living cells and play a pivotal role in intracellular redox signal, response, and control.

Electrochemistry becomes even more difficult when moving from simple sulfur-containing compounds to sulfur proteins and enzymes. Here, the last decade has revealed a central role of cysteine (and, to a lesser extent, methionine) redox chemistry in cell signaling and control. We are now aware of numerous proteins and enzymes which contain redox-active cysteine residues, whose thiol groups can be modified posttranslationally to disulfides, sulfenic, sulfinic, and sulfonic acids as well as to S-nitrosothiols [30]. Such modifications can have a significant impact on protein function and enzyme activity and may ultimately result in a decisive cellular response, such as an antioxidant response or cell death via apoptosis. Unfortunately, electrochemical methods are still not specific enough to determine the redox properties of individual cysteine residues in larger proteins and also fall short of measuring redox properties of proteins which experience sulfenic or sulfinic acid modification, or S-nitrosation (see Sects. 5 and 6).

Electrochemistry of chalcogen-containing compounds and proteins becomes even more complicated when considering selenium- and tellurium-based redox systems. This is rather unfortunate since the last couple of decades has revealed the existence and paramount importance of quite a few (mammalian) selenoproteins and enzymes, including glutathione peroxidase (GPx), the human form of thioredoxin reductase (TrxR), thyroxine deiodinase (T(4)-5′-deiodinase), selenoprotein P (SelP), and selenoprotein W (SelW). While the selenocysteine (SeCys) residues in these proteins and enzymes exhibit a distinct redox chemistry, very little is known so far regarding their oxidation and reduction potentials. To the best of our knowledge, electrochemical techniques such as CV or DPP have not yet been applied in the context of these selenoproteins, despite the fact that the selenol/diselenide redox couple provides a fairly specific and easily distinguishable electrochemical response [17].

While research so far seems to have shied away from employing electrochemical methods to study the redox properties of selenoproteins, various methods have been developed to investigate selenium- and tellurium-containing compounds. During the last two decades, there has been considerable interest in the development of small-molecule synthetic mimics of GPx and of related enzymes [26, 31–36]. These agents, which often exhibit pronounced antioxidant activity *in vitro*, have also been studied by electrochemistry (see Sect. 7). Methods such as CV have been at the forefront of these investigations, and electrodes employed include various types of platinum, carbon, and mercury electrodes [25, 26, 32]. Platinum working electrodes have been particularly useful (and easy to use) in organic solvents [25]. In order to investigate such compounds in a more realistic, physiologically relevant scenario, aqueous, buffered systems whose exact composition may in part depend on the solubility properties of the substance under investigation have also been used. In this case, carbon and mercury working electrodes have been most effective [26, 37–39].

After this brief introduction to the "electrochemical tool kit" to analyze chalcogen-centered redox behavior, we will now turn our attention to individual applications which currently are of particular interest to biological redox chemists. Here, we will take a "top-down" approach, starting with whole cell cultures, then moving to the electrochemistry at or near single cells and then considering specific intracellular target molecules, such as sulfur- and selenoproteins and enzymes. The latter also form a prime target for chalcogen-based redox agents, which will be discussed briefly, with a focus on employing electrochemical methods as part of rational drug design.

7.3 Watching Cells Grow

Cell culture studies form a central part of modern Life Sciences, from microbiology to the development of drugs and pesticides. Here, the growth rate and metabolism of the cell/organism is often one of the most significant determinants. Using such studies, one may, for instance, establish the survival and "fitness" of a particular mutant or screen for the activity of a particular drug prototype. Unfortunately, traditional methods to determine the rate of cell proliferation, such as absorbance readings (OD_{600}) and various staining techniques (e.g., sulforhodamine B, 3-(4,5-dimethylthiazol-2-yl)-2,5-diphenyltetrazolium bromide (MTT) staining, Neutral Red uptake), are rather time consuming and are often marred by artifacts and poor reproducibility. Not surprisingly, alternative methods based on the measurement of cell growth, such as the colony forming unit (CFU) or colony forming assay, are also commonly used in cell culture studies [40]. Unfortunately, all of these methods are based on discontinuous monitoring, i.e., the analysis after certain fixed periods of time, and generally provide only a handful of data points. As time-resolution is low, such studies may miss subtle yet important changes in growth rate, such as initial growth phases and short rest phases. Indeed, cell growth and the effects of agents on cell growth are *not* continuous and linear. The biological activity of a given compound may, for instance, differ at different time points due to complicated metabolic processes. Arsenite, for example, does not show the same toxicity over a prolonged period of time [41]. This circumstance is not necessarily a drawback but may result in additional, extremely valuable data: Amazingly, if recorded and analyzed properly, the time-dependent profile of cell growth in response to a drug can even be used to compare different drugs with each other and ultimately to derive some information regarding drug action. While kinetic information on the dynamics of cell growth clearly is of paramount importance, it cannot be obtained easily by using classical absorbance methods [42].

Within this context, electrochemical methods offer several interesting alternatives. Electrical impedance measurements, for instance, are non-invasive and enable continuous monitoring of cell growth and subsequent (bioinformatic) profiling. This method has been pioneered by the studies of Giaever and Keese [43]. Using impedance measurements to study the spreading, motion, and cell density of mammalian fibroblasts on evaporated gold electrodes, the authors could show that the impedance signals measured were a direct consequence of cell growth, spreading, and motility, as signals disappeared when cytochalasin (a drug that prevents motion) was added to the medium [43].

This method has been developed into several commercial systems which employ multi-well plates, which at first sight resemble classical plastic 96-well plates yet feature integrated (micro-) electrodes at

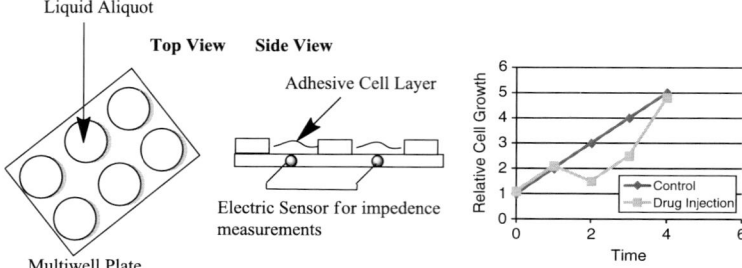

Fig. 7.4 Monitoring cell growth in real time by non-invasive impedance measurements. Several cell samples can be monitored in parallel and the effects of drugs can be measured continuously. This technique has many advantages compared to traditional staining methods. As the "growth" curves for the control (in *black*) and the drug-injected sample (in *grey*) show, there is a clear effect due to the injection of the drug. This effect, however, would be entirely missed by a traditional method which would perform an aliquot-based reading after 1 and 4 h. Furthermore, the growth curves can be profiled and compared to the curves of other drugs in order to pin down similarities and hence help to identify possible (related/different) cellular modes of action and targets. Please note that it is also possible to work with non-adhesive cells as long as there is a specific "anchor" present at the electrode surface

the bottom of each well (see Fig. 7.4). As cells growing on the bottom of these wells increase the electrical resistance of the system, these multi-well plates can be used to monitor the growth of cultured cells in real time and under normal growth conditions (i.e., in the incubator). As long as the cells remain adsorbed to the bottom of the well, it is possible to measure how the cell layer grows and if cells die and detach from the electrode surface, for instance in response to a cytotoxic compound. In this case, the resulting decrease in impedance [44] can be monitored quite precisely and in real time [45]. Although the original method - which is still widely used - is limited to adhesive cell lines, it has provided a major step forward and has already revealed several "secrets" in cell growth behavior which otherwise may have escaped attention (see Fig. 7.4).

Since then, impedance has been applied by several groups with different aims and experimental designs. With impedance sensors it is possible to determine whether a drug causes short-term or long-term cellular responses [46, 47]. In the same context, such methods have proved to be very fast and powerful in establishing the cytotoxicity of test compounds in cancer cells [40, 48–51]. Here, their ability to monitor continuously how a treatment with a known or supposed cancer agent affects to cell population and cell growth has turned out to be of considerable advantage, especially in the context of profiling and comparing effects on cell growth induced by different drugs (for specific examples of applications see below). Presently, attempts are underway to expand the method and its capabilities even further, for instance to answer questions related to cell metabolism or modes of cell death (such as apoptosis) [51]. Some of these additional features will be discussed below.

The application of impedance measurements is not restricted to mammalian cells. Several studies have used impedance to detect fairly selectively pathogenic bacterial cells, using biorecognition elements such as antibodies attached on the electrode surface [52–54]. In this particular case, the cells in question do not grow on the electrode area (as in the original method), but are "fished" and subsequently anchored to the surface by specific recognition antibodies. Once captured at the electrode surface, these bacteria also increase electric resistance which can be measured. Related electric methods, such as the determination of dielectric properties inside a bioreactor through capacitance and conductance measurements, have been used in microbial, fungal, and yeast fermentations in reactors to monitor growth with satisfactory results [55].

Today it is possible to reduce the size of the impedance sensors and to combine them with other electrochemical, microelectrode-based sensors, such as an ion-sensitive field-effect transistor (ISFET) for pH measurement and conductometric sensors for carbon dioxide in well plates of different sizes and geometries [56]. For instance, pH measurements are often crucial in tissue culture, and online, continuous monitoring is desired. Here, novel types of electrodes have recently been tested, including metal (antimony) oxide microelectrodes [57]. It is even possible to interdigital a set of sensors, such as an amperometric sensor (for oxygen consumption) with ISFET and interdigital electrode structure (IDES) sensors in a silicon chip. The values subsequently obtained by such a device are reproducible and reasonable when compared with established methods [58].

While initial studies had to be performed with "home made" electrodes freshly prepared before each experiment [43], several commercial systems are nowadays available which can be employed routinely in daily research. Here, the Bionas 2500 analysis system with IDES and other sensors (ISFET and Clark electrodes) is used widely. Abarzua et al., for instance, have used this system to study the anticancer activity of plant extracts [49]. Similarly, Ceriotti et al. have determined the cytotoxicity of known anticancer molecules and observed that values were comparable with the results obtained by classical methods [50].

A second commercial system, the Real-Time Cell Electronic Sensing System (RT-CES), has been developed by ACEA Biosciences. This device measures the impedance in all wells of a 16- or 96-well microplate in parallel, therefore accelerating data collection enormously. Applications of this advanced, automated technology are very broad and include proliferation studies in different cell lines, drug cytotoxicity determination in cancer cells, induction of apoptosis, receptor tyrosine kinase activity, and protective effects of antagonists of drugs or biological molecules [51, 59].

Besides the two systems mentioned here, there are numerous others which provide the user with a wealth of applications (e.g., the systems developed by MDS Analytical Technologies and Roche Applied Science). Currently, 96-well and even 384-well plates are available for high-throughput assays which ultimately combine these electrochemical sensing techniques with bioinformatics tools. Abassi and collaborators, for instance, have been able to perform a time-resolved screening of 2,000 compounds with known anticancer potential in two cancer cell lines to ascertain which of the compounds present short-term or long-term responses (or both). After determining the impedance-based time-dependent cell response profiles (TCRPs), they clustered the structures into families according to the TCRPs associated with each response. Ultimately, the analysis of similar activities enables the prediction of applications of compounds which may not have been realized previously [60].

7.4 Monitoring Redox Processes at the Cellular Level

As mentioned in the previous section, most traditional cell culture methods used in drug discovery involve the selection and subsequent modification of a target and hence may not be sufficiently predictive because of the interferences/modifications required. As for cell proliferation, label-free technologies able to use native systems instead of the modified ones to monitor (bio-) chemical processes in, at, or near cells are particularly interesting. Such techniques are nondestructive and allow the continuous monitoring of certain processes. They also do not require expensive markers, enzymes, or antibodies, which is a major issue in biochemical research.

While there are various "sensors" available to monitor cellular processes, such as metabolism, at microscopic scale (e.g., pH, dioxygen, hydrogen sulfide), we will focus on electrochemical approaches which may be used to monitor *redox changes* close to or even inside the intact cell. From the perspective of biological redox processes, the detection and quantification of physiologically relevant redox-active species is of particular interest. Some of these chemical species have already been summarized in Table 7.1, which is clearly selective and far from complete. As already mentioned, ROS, RNS, and sulfur-containing species play a major role in normal metabolism, the maintenance of the cellular redox homeostasis, intracellular redox signaling, and host defense [61–64].

In the past, the analysis of intracellular levels of ROS and RNS, but also of GSH and cellular thiol content, in essence has relied on two methods: (a) a rather crude method which involves breaking down the cell and analyzing ROS, RNS, GSH, and thiol levels in the homogenate; (b) various staining methods which involve the application of specific stains to "visualize" and quantify the intracellular level of oxidative stress (OS), individual ROS, and GSH [65]. Both methods have their merits. The first method can be employed rather successfully, for instance, to indirectly determine the formation of ·NO by quantifying its follow-on product nitrite via the Griess reaction [66] or to measure the total intracellular thiol content with 5,5′-dithiobis-(2-nitrobenzoic acid) (DTNB, Ellman's reagent) [67, 68]. This method can also be combined with HPLC, which enables the quantitative determination of several reactive species in the same sample [69, 70]. Staining techniques are somewhat more sophisticated and often rely on expensive equipment, such as fluorescent microplate readers and confocal fluorescence microscopy. These methods can be combined to provide an "intracellular diagnostic platform" which in essence allows the monitoring of processes inside intact cells or cellular organelles and over time [71, 72].

Nonetheless, these methods also suffer from certain drawbacks. Both methods still cannot reliably distinguish between different ROS or different thiols. For instance, 2′,7′-dichlorodihydrofluorescein acetate, which is widely used to establish intracellular levels of OS, is fairly unspecific and detects various oxidizing species, including H_2O_2 and ˙OH radicals [69]. While there are numerous attempts to develop more specific stains [70], the stains available so far are still rather unspecific. At the same time, various ROS are chemically too reactive and/or unstable and hence escape analysis in cell extracts after lysis.

Here, electrochemistry may provide a reliable and in many ways straightforward alternative in the detection of certain ROS and RNS. Indeed, there have been numerous attempts to measure different ROS in biological samples, often employing modified electrodes which provide some specificity for individual ROS or RNS. We will consider these reactive species in the increasing order of their respective cytotoxicity.

The superoxide radical anion ($O_2^{˙-}$) is a byproduct of aerobic mitochondrial respiration which is turned into hydrogen peroxide and dioxygen by one of the various superoxide dismutase enzymes [73–76]. Allen Hill and his colleagues at Oxford were the first to detect $O_2^{˙-}$ electrochemically in a cell-based system employing opsonized microelectrodes of graphite or gold [77, 78]. This pioneering work has subsequently been followed up by studies employing more "sophisticated" electrodes, such as modern carbon fiber microelectrodes and gold electrodes modified with cytochrome *c* [79].

Hydrogen peroxide (H_2O_2) is formed as part of mitochondrial respiration (see above) and also by immune cells as part of host defense. Its particular toxicity mostly results from its participation in the Fenton reaction, whereby H_2O_2 is reduced to the highly aggressive hydroxyl (˙OH) radical [75, 76]. Its stability and appropriate half-life allows electrochemical detection, for instance on platinized surfaces of carbon electrodes and platinized carbon fiber microelectrodes [80–83].

The most reactive ROS is OH˙, which is highly damaging for the cells - it indiscriminately reacts with proteins, DNA, and membrane lipids. As the lifetime of ˙OH in a biological system is in the order of a few nanoseconds, diffusion across membranes and/or toward more remote electrodes (e.g., placed outside the cell) is not feasible [81]. Nonetheless, Zhu et al. have recently described a rather innovative method to indirectly detect and quantify ˙OH radicals. This method exploits the highly aggressive behavior of ˙OH. It measures the impedance at a coated electrode, which decreases once ˙OH radicals gradually damage the coating and expose the electrode surface [84].

The RNS nitric oxide (•NO) [76, 85] and peroxynitrite (ONOO⁻) can be detected directly by electrochemical methods due to their specific redox characteristics. Nitric oxide is determined with the help of polymer-coated carbon microelectrodes [86, 87]. In contrast, peroxynitrite can be detected rather selectively electrochemically using a carbon fiber microelectrode modified by a film of Mn(III)-[2, 2]paracyclophenylporphyrin [88, 89]. Other reactive RNS are very difficult to detect or to distinguish electrochemically due to their high reactivity and short half-life time.

Some of the detection methods for the more common reactive species, however, have already been turned into rather sophisticated sensing devices, such as a biocompatible sensor for •NO, which employs carbon-based screen-printed electrodes [90]. Frequently, these systems are not only specific and sensitive but also suitable to analyze rather low sample volumes ("lab on a chip"), such as a microfluidic device designed to measure OS generated by macrophages [91] or a device to measure •NO release from cells grown directly on the surface of a sensor [92].

Indeed, Christian Amatore and colleagues at the University Paris Descurtes, France have recently developed an electrochemical method which allows researchers to measure various reactive species at the level of a single cell [93]. As most (but not all) ROS and RNS diffuse out of the cell, it is possible to quantify these species outside the cell using specially designed (ultra-) microelectrodes and a specific set of potentials as indicative measuring points - and to subsequently calculate back to intracellular concentrations (see Fig. 7.5a). Most ROS possess a characteristic electrochemical potential, and such methods are able to capture otherwise elusive ROS and to distinguish between them (e.g., H_2O_2, ONOO⁻, •NO, and NO_2^- can be detected and quantified by measuring at 300, 450, 650, and 850 mV vs. the Ag/AgCl electrode (SSCE), respectively) [20]. These methods are particularly suitable for investigating ROS generating cells, such as macrophages, and may even be employed to monitor events associated with a single cell. The recent study by Hu et al. on a single MG63 osteosarcoma cell submitted to mechanical stress, for instance, reflects the enormous potential of this method [20]. Additional publications that describe simultaneous detections of different ROS and RNS have been reviewed by Borgmann [88].

Nonetheless, one should emphasize that such an approach also has its limitations. It is not possible to measure all ROS or RNS outside

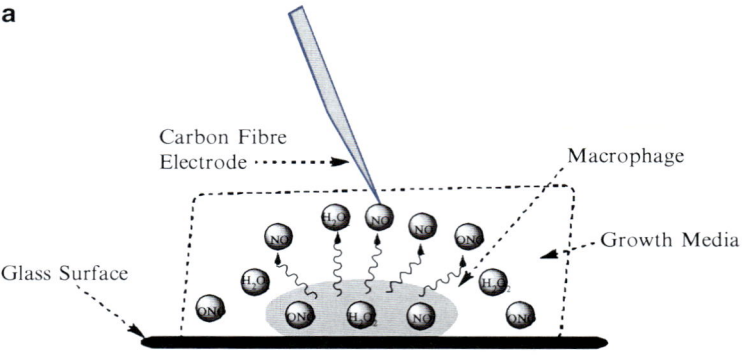

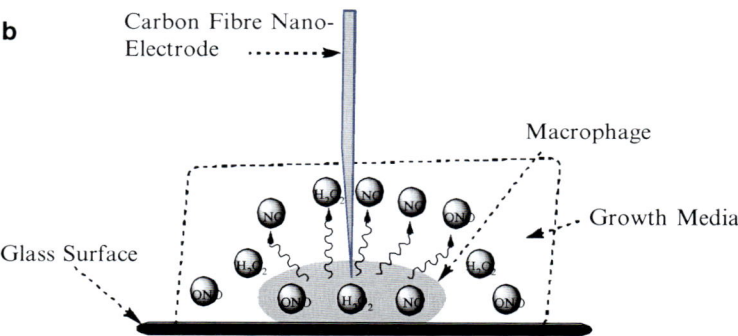

Fig. 7.5 (**a**) Single cell electrochemical measurements to monitor ROS/RNS production at/near a single MG53 macrophage. As measurements are taken in the growth medium close to the macrophage, ROS/RNS need to diffuse out of the macrophage in order to be detected. (**b**) Prospective future measurements of ROS/RNS inside cells. In contrast to the current state-of-the-art setup depicted in part a, this proposed method measures ROS/RNS inside the cell using a tiny electrode able to penetrate the cell membrane. This method does no longer require the diffusion of reactive species out of the cells, yet relies on very small electrodes and hence minute currents. Please note that this kind of intracellular electrochemical monitoring is still in its early developmental phase and not yet available for routine measurements. It may also encounter additional complications due to the presence of numerous (redox-active) biomolecules inside the cell

the cell because of the short lifetimes of the most reactive ones, such as the ˙OH radical. The latter is unable to diffuse out of the cell, where the microelectrode is placed, as it is too reactive and will be "lost" before it can reach the electrode. One alternative would involve the application of nanoscopic electrodes which may be inserted into intact cells (proposed in Fig. 7.5b). Unfortunately, electrodes which would be small enough to operate within cells without interfering with normal cell metabolism and, at the same time, would also be large enough to provide a measurable current, are not yet available.

7.5 Characterization of Cysteine-Containing Proteins and Enzymes

While the determination and in some cases also the quantification of ROS and RNS at or near living cells has witnessed considerable progress during the last decade, electrochemical methods to characterize the prime cellular targets of these reactive species, i.e., redox-sensitive cysteine- and selenocysteine-containing proteins and enzymes, are still in their infancy. This lack of appropriate methods to describe the redox behavior of cysteine- and selenocysteine proteins is out of step with the recent, dramatic progress in the field of sulfur-based intracellular redox signaling and regulatory systems. During the last decade, numerous studies have provided a rather detailed insight into the cellular "thiolstat", a pivotal, overarching regulatory system involved in cellular life and death decisions, including proliferation, differentiation, and apoptosis [30, 94].

Today, there are just a few methods available to appropriately characterize such regulatory processes at the level of proteins. Frequently, sophisticated staining, antibody, and proteomic techniques are employed to identify posttranslational cysteine modifications in the cell, which include disulfide bond formation (intra- and intermolecular disulfides, S-thiolation, S-glutathiolation), S-nitrosation (leading to RSNO species), and sulfenic acid (RSOH) formation (for a review see [94]). Nonetheless, these detection methods still provide a rather static picture. They may answer the question *which* (cysteine) residues are modified in proteins and enzymes, yet do not refer to the *why*. To answer the question why specific residues are prone to modification, one may need to consider the reactivity or oxidation potential of the (cysteine, selenocysteine) residues in question and to compare them to other residues. Such Epa values of thiols may explain and in the future even predict which particular cysteine residues in which particular proteins become oxidized/modified first and under which conditions, assuming the residues in question are accessible to oxidation and also come into contact with oxidants.

It is rather disappointing, and in stark contrast to the thriving area of electrochemistry of metalloproteins, that few electrochemical investigations have been performed so far on cysteine- and selenocysteine-containing proteins and enzymes. Indeed, the zinc/sulfur protein metallothionein (MT-1 and MT-2), which contains seven zinc ions bound to 20 cysteine residues in two zinc/sulfur clusters (Zn_4Cys_{11} and Zn_3Cys_9), seems to be the only protein which has been detected and whose redox properties have been analyzed so far using electrochemical techniques [12, 13, 95]. Here, the cysteine residues (rather than the metal ions) are investigated. The same also applies to related cysteine-based peptides [4, 5, 10, 96]. As these techniques rely on mercury-based electrodes, they are obviously also complicated by adsorption phenomena as discussed in Sect. 2. Nonetheless, these (and related) studies provide the first examples of *electrochemical* investigations of cysteine-redox behavior in proteins. Apart from these examples, and to the best of our knowledge, electrochemistry so far has not been employed to study the redox behavior of cysteine (or selenocysteine) residues in proteins. As a consequence, a reliable redox scale of sulfur and selenium proteins is not available yet, despite the fact that the thiol/disulfide and the selenol/diselenide redox pairs can both be detected at metal (mercury) electrodes (see Sect. 2).

Indeed, only a handful of redox potentials are available for cysteine residues in peptides, proteins, and enzymes so far, for instance for glutathione and the active site cysteine residues in human thioredoxin (Trx), human glutaredoxin (Grx), human protein disulfide isomerase (PDI), and TrxR (see Table 7.2) [97]. These values have been obtained by indirect methods, such as equilibrium redox titrations using the GSH/GSSG couple in conjunction with the Nernst equation.

Apart from oxidation and reduction potentials estimated for these usual suspects (i.e., proteins and enzymes which are well-known to be involved in the maintenance of the intracellular redox balance), our knowledge regarding Epa values of cysteine residues is extremely limited. Since such E values are currently unavailable, most biochemists exploring the cellular thiolstat use pKa values instead. While there is some merit in using acidity of a given cysteine thiol as a measure for its reactivity (the thiolate form is considerably more reactive than the thiol), the pKa value is a poor and too simplistic substitute for true redox parameters.

Table 7.2 Selection of well-known or widespread thiol/disulfide-based redox proteins and enzymes (oxidoreductases)

Protein	Motif in active site	Redox potential (mV vs. NHE)	Biological function
Thioredoxin (12 kDa)	-Cys-Gly-Pro-Cys-	−270	Reduction of disulfide bonds in proteins and enzymes
Glutaredoxin (9 kDa)	-Cys-Pro-Tyr-Cys-	−233(Grx-1); −198 (Grx-3)	Catalyzes reduction of ribonucleotide reductase by GSH
Tryparedoxin (16 kDa)	-Cys-Pro-Pro-Cys-	−249	Leishmania/Trypanosoma oxidoreductase which catalyzes reduction of disulfides and utilizes trypanothione in place of GSH
Protein disulfide isomerases (57 kDa)	-Cys-Gly-His-Cys-	−127	Catalyzes reduction and reformation of (misformed) disulfide bonds
DsbA (21 kDa)	-Cys-Pro-His-Cys-	−125	Catalyzes formation/rearrangement of protein disulfide bonds from thiols/misformed disulfide bonds

The cysteine-redox potential found in these proteins is fine-tuned by the amino acid sequence and microenvironment found in these proteins. As shown in the table, this potential may differ significantly between individual proteins. Not surprisingly, potential differences also affect the biological function and activity of these proteins/enzymes. While some of these proteins, for instance, primarily act as reductases, others also exert a more oxidizing function

7.6 Electrochemistry of Sulfur-Containing Natural Products

While the electrochemistry of cysteine proteins and enzymes is far from trivial, sulfur-containing natural products, such as thiols, disulfides, and polysulfanes, can be characterized electrochemically with comparable ease. Such studies have been used rather extensively in the context of redox-active antioxidants, chemopreventive agents, and pro-oxidant redox modulators, some of which interfere with cellular signaling [4, 5, 96].

Table 7.3 Selection of naturally occurring reactive sulfur species known to occur in (mammalian) biochemistry

Name	Formula	Sulfur oxidation state ($R=+1$)	Occurrence (in nature)
Thiol	RSH	-2	Allyl mercaptan, cysteine, GSH
Disulfide	RSSR	-1	Dialkyl disulfide (DADS), cystine
3-ethenyl-3,4-dihydro-1,2-dithiine	(structure)	-1	Garlic Diallyl-trisulfide (DATS), calicheamicin γ1 Diallyl-tetrasulfide (DATTS)
Trisulfide	RSSSR	$-1, 0, -1$	
Tetrasulfide	RSSSSR	$-1, 0, 0, -1$	
Pentasulfide	RSSSSSR	$-1, 0, 0, 0, -1$	Varacin
Perthiol	RSSH	-1	Allyl perthiol, thiocysteine
Hydropolysulfide	RSxH ($x \geq 3$)	$-1, 0(x-2), -1$	Allyl hydrosulfides
Sulfide anion	S^{2-}	-2	Metal/sulfur clusters, H_2S, HS^-
Polysulfide anion	S_x^{2-} ($x \geq 2$)	$-1, 0(x-2), -1$	Possibly released from organic sulfanes

Chemical formulas, sulfur oxidation states, and basic information regarding their formation and occurrence are provided. This selection of electrochemically accessible sulfur species is neither complete nor final

Table 7.3 provides a brief and necessarily incomplete overview of such - electrochemically accessible - sulfur species which play a role in Biology and on occasion also form part of pharmaceutical research. In most instances, such products can be studied using CV and DPP in conjunction with a metal electrode (often mercury, see Sect. 2).

Examples of particular interest include the various natural sulfur compounds found in garlic and related *Allium* species, such as diallyl disulfide (DADS), diallyl trisulfide (DATS), and diallyl tetrasulfide (DATTS) [27, 29, 98]. While the quality of data obtained by such methods is limited because of adsorption phenomena, CV still provides some crucial information which is lacking otherwise, including aspects of the reversibility of reduction/oxidation and electrochemical potentials. Indeed, CV of disulfides and polysulfanes is superior to alternative, non-electrochemical methods, such as redox titrations with redox dyes, as CV is a direct method and provides fairly precise, extensive, and reproducible information. In the case of the diallyl polysulfanes, for instance, the electrochemical studies reported by our group have shed some light on the redox properties of these unusual sulfur compounds [27, 29, 30, 98, 99].

They have also questioned the notion that polysulfanes are strong oxidants and ultimately have supported the idea that such compounds - once activated - act via a reducing mechanism. Here, electrochemical data supports the concept of Munday and colleagues, which explains the exceptional biological activity of diallyl polysulfanes (and related polysulfanes) with the ability of their reduced forms to reduce dioxygen to $O_2^{\bullet-}$ and the subsequent occurrence of an oxidative insult [100]. Similarly, the electrochemical studies of naturally occurring 3-ethenyl-3,4-dihydro-1,2-dithiine (1,2-DT) have underlined the rather distinct redox properties of this and related α,β-unsaturated disulfide compounds, which sets them apart from "normal" disulfides (see Fig. 7.6) [101].

Similar electrochemical studies are certainly possible for related disulfide-based compounds, including, for instance, thiuram disulfides (such as the drug disulfiram) and 3-dithiolethiones (such as the drug Oltipraz). At the same time, the release of hydrogen sulfide from certain natural products or drug molecules, the biological effects of inorganic polysulfides (S_x^{2-}), and sulfur-metal interactions can also be investigated electrochemically.

In contrast, some of the higher sulfur oxidation states, such as sulfenic, sulfinic, and sulfonic acids, as well as thiosulfinates and thiosulfonates, may not be accessible directly using electrochemical techniques. As already mentioned, such sulfur species undergo redox transformations via substitution mechanisms. Since these reactions do not involve electron transfer, it may be difficult to access such redox systems electrochemically, even if a particular "indicator" reaction is employed (e.g., the electrochemical potential of one-electron abstraction as possible indicator of nucleophilicity or electrophilicity, see Sect. 2). Ultimately, biologically interesting sulfur species, such as sulfenic and sulfinic acids as well as thiosulfinates, may escape electrochemical analysis. Their redox behavior may have to be measured and described by other means, for instance, by equilibrium reactions with GSH or via the kinetics of the underlying exchange reactions.

7.7 Electrochemistry as Part of QSAR

Despite the fact that many sulfur-containing natural products escape a thorough electrochemical analysis, the research performed with disulfides (and related structures) has shown that electrochemistry

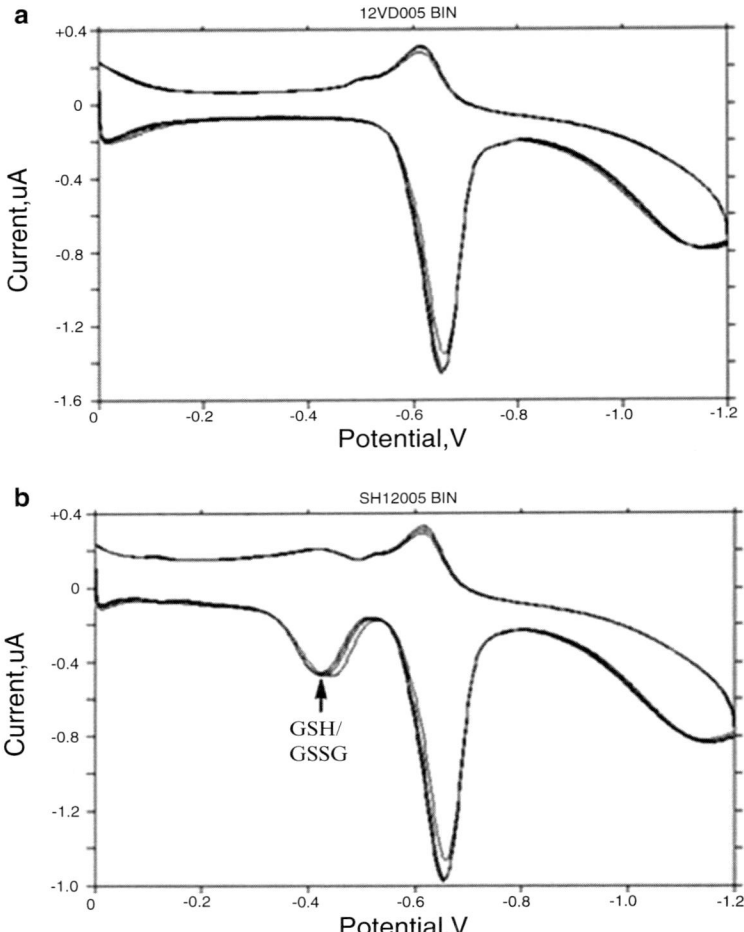

Fig. 7.6 Characterization of sulfur-containing natural products using CV in conjunction with a dropping mercury electrode, as exemplified by the natural compound 3-ethenyl-3,4-dihydro-1,2-dithiine (1,2-DT). This kind of analysis provides valuable information regarding the redox behavior of such compounds, including oxidation and reduction potentials and reversibility (**a**). Such measurements are also capable of distinguishing between different sulfur compounds in the same sample and can be run in the presence of a reference compound, such as GSH (**b**). Voltammograms were recorded for 1,2-DT in the abscence and presence of GSH in phosphate buffer of pH 7.4 at a scan rate of 250 mV/s using a Hg working, Pt wire counter, and Ag/AgCl reference electrode (Khairan et al., unpublished results)

may provide a fast and reliable tool to describe the redox properties of such compounds and subsequently also explain, rationalize, and even predict biological activity (or the lack thereof). Not surprisingly, methods such as CV and DPP have also been used to characterize the redox properties of redox-active selenium- and tellurium agents, especially those used in the context of intracellular redox modulation [25, 26, 32, 102–104]. If redox processes are able to control cell proliferation, differentiation, and apoptosis, so the idea, then agents able to modulate such processes should qualify as highly effective and maybe even selective drugs for a wide range of human diseases [105, 106]. Indeed, chemoprevention by antioxidants, aging and oxidative stress, inflammation and antioxidants, host defense against invading microorganisms via an oxidative burst, and cancer cells going into apoptosis due to crossing the internal redox threshold are all important and current topics related—in one way or another—to redox control [23].

It is therefore hardly surprising that research into redox modulating agents is currently booming and methods such as CV and DPP provide a wide range of opportunities to analyze and describe such compounds (Fig. 7.7). During the 1990s, Ian Cotgreave, Lars Engman, and colleagues at the Karolinska Institute, Stockholm, Sweden have employed CV for the characterization of various GPx mimics, i.e., structurally related selenium and tellurium compounds [25, 102]. These studies, conducted at platinum-button working electrodes in dichloromethane (Bu_4NClO_4 as electrolyte), have, for instance, revealed a certain correlation between the oxidation potentials of structurally related chalcogen compounds and their antioxidant capacity in biological test systems [25]. Indeed, "oxidisability" can be related to the structure of such compounds and also has a profound effect on the compound's antioxidant capacity [102]. A similar link between the molecular structure of redox-active chalcogen compounds, their oxidation potential, and subsequent biological activity has also been observed by Giles et al. using a set of structurally related selenium and tellurium compounds and an *in vitro* (catalytic) antioxidant assay [26, 32, 107]. In these studies, the first oxidation potential (Epa_1) again seems to be indicative of the "oxidisability", and hence chemical redox (re-)activity, which in turn seems to be responsible for biological activity.

Many aspects of these early, pioneering electrochemical investigations into the emerging "chemical structure-electrochemical potential-biological activity relationship" could since be confirmed in organic solvents as well

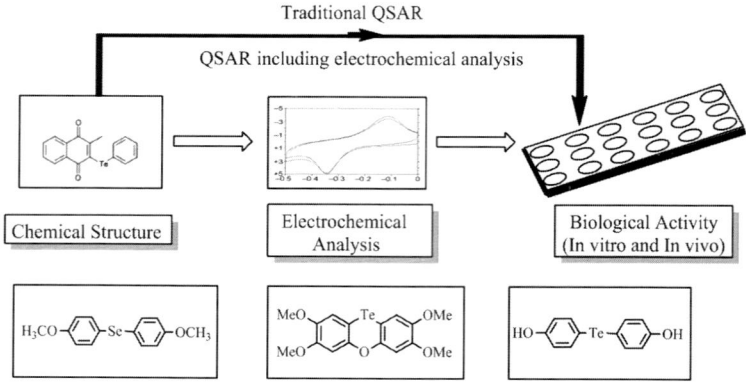

Fig. 7.7 Prospective role of electrochemical analysis in modern QSAR. Methods such as CV and polarography can be used to obtain valuable information regarding the redox properties of prospective drug molecules which exert their biological effects via a redox-based reaction (e.g., oxidative modification of target proteins). The chemical structure determines electrochemical properties (e.g., E_{pa} values), which in turn may influence chemical reactivity, control the biochemical mode of action, and subsequently may explain aspects of biological activity

as in aqueous solutions. The electrochemical parameters clearly occupy a prominent position in this three-way relationship: On one hand, they reflect structural features of the compounds, such as a particular redox mechanism or the presence of electron donating or withdrawing substituents. On the other hand, they are indicative of a particular redox behavior which plays a central role in the compound's biological activity (see Fig. 7.7). While it is hardly surprising that agents acting as redox modulators also exhibit a redox chemistry which in some ways is the cause and hence also predictive of the biological activity, it is indeed surprising to notice that many of these compounds have not yet been studied properly by electrochemical methods. As the therapeutic interest in redox modulation continues to grow, one may anticipate additional, maybe even pioneering electrochemical work in this area of pharmaceutical research.

7.8 Outlook and Conclusions

The previous sections have shown that electrochemical methods already play a significant role in the analysis of chalcogen-centered biological redox processes, from monitoring the formation of ROS and RNS at the single cell level and the characterization of the redox properties of cysteine proteins and enzymes to the elucidation of structure-potential-activity relationships for diverse natural sulfur compounds and synthetic selenium- and tellurium-based redox modulators. Nonetheless, our discussion of these examples has also demonstrated that there are many questions that still need to be addressed and subsequently answered. Here, we would like to point toward a few areas of biological redox chemistry which are still in desperate need of proper analysis, and where electrochemical methods may play a significant role in the future.

First of all, the appropriate description of the complex redox behavior of cysteine proteins and enzymes comes to mind. As an increasing number of redox-active, regulatory, and signaling cysteine proteins of the cellular thiolstat is emerging, there is an urgent need to determine the precise redox properties of these proteins, including their susceptibility toward oxidation and S-thiolation as well as reversibility of such processes. These posttranslational modifications often control protein function and enzyme activity. They do not occur randomly, but rather seem to target particular cysteine (and selenocysteine) residues in specific proteins, probably on the basis of low oxidation potential of the given residue and/or its accessibility for the oxidizing agents. Not surprisingly, electrochemical studies would assist enormously in determining which of these proteins are especially prone toward oxidation/modification, which residues are primarily affected and if such modifications are reversible or not.

Another area of interest is the characterization of sulfur-, selenium-, and tellurium-containing agents which may be beneficial as redox modulating drugs (e.g., as antioxidants, cytostatic/cytotoxic drugs). The last two decades have witnessed considerable progress in this area, yet considerably more studies with more substances need to be performed to establish appropriate structure-potential-activity relationships. Ultimately, electrochemistry may play a central role in this field of research, as the electrochemical potentials reflect the structural properties and also form the basis for biological activity. In the end, Epa and Epc values may not only be considerably more precise, reliable, and reproducible when compared to "activities" in the various antioxidant and redox assays (such as the ferric reducing antioxidant power (FRAP) assay, the total antioxidant capacity (TAC) assay, or the oxygen radical absorption capacity (ORAC) assay) they may also be much easier to obtain and faster to measure.

In any case, the biological chemistry of sulfur, selenium, and tellurium provides ample opportunities for innovative electrochemical investigations in the near and medium future.

References

1. Armstrong FA, Hill HAO, Walton NJ. Direct electrochemistry of redox proteins. Acc Chem Res. 1988;21:407–13.
2. Kolb DM, Amatore C, Compton RG. Modern electrochemistry: interdisciplinary research at the forefront of science. Chemphyschem. 2010;11:2655–6.
3. Park S, Boo H, Chung TD. Electrochemical non-enzymatic glucose sensors. Anal Chim Acta. 2006;556:46–57.
4. Stenken JA, Puckett DL, Lunte SM, et al. Detection of N-acetylcysteine, cysteine and their disulfides in urine by liquid chromatography with a dual-electrode amperometric detector. J Pharm Biomed Anal. 1990;8:85–9.
5. Adam V, Fabrik I, Kohoutkova V, et al. Automated electrochemical analyzer as a new tool for detection of thiols. Int J Electrochem Sci. 2010;5:429–47.
6. Stetter JR, Penrose WR, Yao S. Sensors, chemical sensors, electrochemical sensors, and ECS. J Electrochem Soc. 2003;150:11–6.
7. Pohanka M, Skladai P. Electrochemical biosensors—principles and applications. J Appl Biomed. 2008;6:57–64.
8. Feinberg BA, Lau YK. The electrochemistry of high-potential iron-sulfur proteins and their novel brdicka waves. Bioelectrochem Bioenerg. 1980;7:187–94.
9. Hu NF. Direct electrochemistry of redox proteins or enzymes at various film electrodes and their possible applications in monitoring some pollutants. Pure Appl Chem. 2001;73:1979–91.
10. Mendieta J, Rodriguez AR. Electrochemical study of the binding properties of a metallothionein I related peptide with cadmium or/and zinc. Electroanalysis. 1996;8:473–9.
11. Seiwert B, Karst U. Analysis of cysteine-containing proteins using precolumn derivatization with N-(2-ferroceneethyl)maleimide and liquid chromatography/electrochemistry/mass spectrometry. Anal Bioanal Chem. 2007;388:1633–42.
12. Krizkova S, Zitka O, Adam V, et al. Possibilities of electrochemical techniques in metallothionein and lead detection in fish tissues. Czech J Anim Sci. 2007;52:143–8.
13. Adam V, Baloun J, Fabrik I, et al. An electrochemical detection of metallothioneins at the zeptomole level in nanolitre volumes. Sensors. 2008;8:2293–305.
14. Gerritsen M, Jansen JA, Kros A, et al. Influence of inflammatory cells and serum on the performance of implantable glucose sensors. J Biomed Mater Res. 2001;54:69–75.

15. Mang A, Pill J, Gretz N, et al. Biocompatibility of an electrochemical sensor for continuous glucose monitoring in subcutaneous tissue. J Diabetes Sci Technol. 2005;7:163–73.
16. Keenan DB, Mastrototaro JJ, Voskanyan G, et al. Delays in minimally invasive continuous glucose monitoring devices: a review of current technology. J Diabetes Sci Technol. 2009;3:1207–14.
17. Jacob C, Giles GL, Giles NM, et al. Sulfur and selenium: the role of oxidation state in protein structure and function. Angew Chem Int Ed Engl. 2003;42:4742–58.
18. Nauser T, Dockheer S, Kissner R, et al. Catalysis of electron transfer by selenocysteine. Biochemistry. 2006;45:6038–43.
19. Koppenol WH, Stanbury DM, Bounds PL. Electrode potentials of partially reduced oxygen species, from dioxygen to water. Free Radic Biol Med. 2010;49:317–22.
20. Hu R, Guille M, Arbault S, et al. In situ electrochemical monitoring of reactive oxygen and nitrogen species released by single MG63 osteosarcoma cell submitted to a mechanical stress. Phys Chem Chem Phys. 2010;12:10048–54.
21. Jacob C, Doering M, Burkholz T. The chemical basis of biological redox-control. In: Jacob C, Winyard PG, editors. Redox signaling and regulation in biology and medicine. Weinheim: Wiley-VCH; 2009. p. 63–122.
22. Doering M, Ba LA, Lilienthal N, et al. Synthesis and selective anticancer activity of organochalcogen based redox catalysts. J Med Chem. 2010; 53:6954–63.
23. Jacob C, Winyard P. Redox signalling and regulation in biology and medicine. Weinheim: Wiley-VCH; 2009.
24. Amatore C, Arbault S, Bruce D, et al. Characterization of the electrochemical oxidation of peroxynitrite: relevance to oxidative stress bursts measured at the single cell level. Chemistry. 2001;7:4171–9.
25. Cotgreave IA, Moldeus P, Engman L, et al. The correlation of the oxidation potentials of structurally related dibenzo[1,4]dichalcogenines to their antioxidance capacity in biological systems undergoing free radical-induced lipid peroxidation. Biochem Pharmacol. 1991;42:1481–5.
26. Giles GI, Tasker KM, Johnson RJK et al. (2001) Electrochemistry of chalcogen compounds: prediction of antioxidant activity. Chem Commun. 2490–2491
27. Anwar A, Burkholz T, Scherer C, et al. Naturally occurring reactive sulfur species, their activity against Caco-2 cells, and possible modes of biochemical action. J Sulfur Chem. 2008;29:251–68.
28. Hason S, Simonaho SP, Silvennoinen R, et al. Detection of phase transients in two-dimensional adlayers of adenosine at the solid amalgam electrode surfaces. J Electroanal Chem. 2004;568:65–77.
29. Jacob C, Anwar A. The chemistry behind redox regulation with a focus on sulphur redox systems. Physiol Plant. 2008;133:469–80.
30. Jacob C. Redox signalling via the cellular thiolstat. Biochem Soc Trans. 2011;39:1247–53.
31. Mugesh G, Panda A, Singh HB, et al. Glutathione peroxidase-like antioxidant activity of diaryl diselenides: a mechanistic study. J Am Chem Soc. 2001;123:839–50.

32. Giles GI, Giles NM, Collins CA et al. (2003) Electrochemical, in vitro and cell culture analysis of integrated redox catalysts: implications for cancer therapy. Chem Commun. 2030–2031
33. Jacob C, Knight I, Winyard PG. Aspects of the biological redox chemistry of cysteine: from simple redox responses to sophisticated signalling pathways. Biol Chem. 2006;387:1385–97.
34. Engman L, McNaughton M, Gajewska M, et al. Thioredoxin reductase and cancer cell growth inhibition by organogold(III) compounds. Anticancer Drugs. 2006;17:539–44.
35. Kumar S, Engman L. Microwave-assisted copper-catalyzed preparation of diaryl chalcogenides. J Org Chem. 2006;71:5400–3.
36. Engman L, Al-Maharik N, McNaughton M, et al. Thioredoxin reductase and cancer cell growth inhibition by organotellurium compounds that could be selectively incorporated into tumor cells. Bioorg Med Chem. 2003;11: 5091–100.
37. Collins CA, Fry FH, Holme AL, et al. Towards multifunctional antioxidants: synthesis, electrochemistry, in vitro and cell culture evaluation of compounds with ligand/catalytic properties. Org Biomol Chem. 2005; 3: 1541–6.
38. Mecklenburg S, Collins CA, Doring M, et al. The design of multifunctional antioxidants against the damaging ingredients of oxidative stress. Phosphorus Sulfur Silicon Relat Elem. 2008;183:863–88.
39. Schneider T, Baldauf A, Ba LA, et al. Selective antimicrobial activity associated with sulfur nanoparticles. J Biomed Nanotechnol. 2011;7:395–405.
40. Ceriotti L, Ponti J, Colpo P, et al. Assessment of cytotoxicity by impedance spectroscopy. Biosens Bioelectron. 2007;22:3057–63.
41. Komissarova EV, Saha SK, Rossman TG. Dead or dying: the importance of time in cytotoxicity assays using arsenite as an example. Toxicol Appl Pharmacol. 2005;202:99–107.
42. Repetto G, del Peso A, Zurita JL. Neutral red uptake assay for the estimation of cell viability/cytotoxicity. Nat Protoc. 2008;3:1125–31.
43. Giaever I, Keese CR. Monitoring fibroblast behavior in tissue culture with an applied electric field. Proc Natl Acad Sci USA. 1984;81:3761–4.
44. Du D, Cai J, Ju H, et al. Construction of a biomimetic zwitterionic interface for monitoring cell proliferation and apoptosis. Langmuir. 2005;21:8394–9.
45. Asphahani F, Zhang M. Cellular impedance biosensors for drug screening and toxin detection. Analyst. 2007;132:835–41.
46. Zhu J, Wang X, Xu X, et al. Dynamic and label-free monitoring of natural killer cell cytotoxic activity using electronic cell sensor arrays. J Immunol Methods. 2006;309:25–33.
47. Atienza JM, Zhu J, Wang X, et al. Dynamic monitoring of cell adhesion and spreading on microelectronic sensor arrays. J Biomol Screen. 2005; 10:795–805.
48. Asphahani F, Thein M, Veiseh O, et al. Influence of cell adhesion and spreading on impedance characteristics of cell-based sensors. Biosens Bioelectron. 2008;23:1307–13.
49. Abarzua S, Drechsler S, Fischer K, et al. Online monitoring of cellular metabolism in the MCF-7 carcinoma cell line treated with phytoestrogen extracts. Anticancer Res. 2010;30:1587–92.

50. Ceriotti L, Kob A, Drechsler S, et al. Online monitoring of BALB/3T3 metabolism and adhesion with multiparametric chip-based system. Anal Biochem. 2007;371:92–104.
51. Solly K, Wang X, Xu X, et al. Application of real-time cell electronic sensing (RT-CES) technology to cell-based assays. Assay Drug Dev Technol. 2004;2:363–72.
52. Radke SM, Alocilja EC. A high density microelectrode array biosensor for detection of E. coli O157:H7. Biosens Bioelectron. 2005;20:1662–7.
53. Varshney M, Li Y. Interdigitated array microelectrode based impedance biosensor coupled with magnetic nanoparticle-antibody conjugates for detection of Escherichia coli O157:H7 in food samples. Biosens Bioelectron. 2007;22:2408–14.
54. Yang L, Li Y, Griffis CL, et al. Interdigitated microelectrode (IME) impedance sensor for the detection of viable Salmonella typhimurium. Biosens Bioelectron. 2004;19:1139–47.
55. Olsson L, Nielsen J. On-line and in situ monitoring of biomass in submerged cultivations. Trends Biotechnol. 1997;15:517–22.
56. van Leeuwen M, Li X, Krommenhoek EE, et al. Quantitative determination of glucose transfer between concurrent laminar water streams in a H-shaped microchannel. Biotechnol Prog. 2009;25:1826–32.
57. Wang M, Ha Y. An electrochemical approach to monitor pH change in agar media during plant tissue culture. Biosens Bioelectron. 2007; 22: 2718–23.
58. Brischwein M, Motrescu ER, Cabala E, et al. Functional cellular assays with multiparametric silicon sensor chips. Lab Chip. 2003;3:234–40.
59. Atienza JM, Yu NC, Kirstein SL, et al. Dynamic and label-free cell-based assays using the real-time cell electronic sensing system. Assay Drug Dev Technol. 2006;4:597–607.
60. Abassi YA, Xi B, Zhang WF, et al. Kinetic cell-based morphological screening: prediction of mechanism of compound action and off-target effects. Chem Biol. 2009;16:712–23.
61. Halliwell B. Oxygen and nitrogen are pro-carcinogens. Damage to DNA by reactive oxygen, chlorine and nitrogen species: measurement, mechanism and the effects of nutrition. Mutat Res. 1999;443:37–52.
62. Szabo KE, Gutowski N, Holley JE, et al. Redox control in human disease with a special emphasis on the peroxiredoxin-based antioxidant system. In: Jacob C, Winyard PG, editors. Redox signaling and regulation in biology and medicine. Weinheim: Wiley-VCH; 2009. p. 409.
63. den Hertog J. Protein tyrosine phosphatases as mediators of redox signaling. In: Jacob C, Winyard PG, editors. Redox signalling and regulation in biology and medicine. Weinheim: Wiley-VCH; 2009. p. 197.
64. Charlier E, Piette J, Gloire G. Redox regulation of apoptosis in immune cells. In: Jacob C, Winyard PG, editors. Redox signalling and regulation in biology and medicine. Weinheim: Wiley-VCH; 2009. p. 385.
65. Hansen RE, Winther JR. An introduction to methods for analyzing thiols and disulfides: reactions, reagents, and practical considerations. Anal Biochem. 2009;394:147–58.
66. Fox JB. Kinetics and mechanisms of the griess reaction. Anal Chem. 1979;51:1493–502.

67. Hamilton CJ, Saravanamuthu A, Eggleston IM, et al. Ellman's-reagent-mediated regeneration of trypanothione in situ: substrate economical microplate and time-dependent inhibition assays for trypanothione reductase. Biochem J. 2003;369:529–37.
68. Woodward JJ. The effects of thiol reduction and oxidation on the inhibition of NMDA-stimulated neurotransmitter release by ethanol. Neuropharmacology. 1994;33:635–40.
69. Murrant CL, Reid MB. Detection of reactive oxygen and reactive nitrogen species in skeletal muscle. Microsc Res Tech. 2001;55:236–48.
70. Tarpey MM, Wink DA, Grisham MB. Methods for detection of reactive metabolites of oxygen and nitrogen: in vitro and in vivo considerations. Am J Physiol Regul Integr Comp Physiol. 2004;286:R431–44.
71. Li L, Chen CY, Chun HK, et al. A fluorometric assay to determine antioxidant activity of both hydrophilic and lipophilic components in plant foods. J Nutr Biochem. 2009;20:219–26.
72. Kalivendi SV, Kotamraju S, Zhao H, et al. Doxorubicin-induced apoptosis is associated with increased transcription of endothelial nitric-oxide synthase. Effect of antiapoptotic antioxidants and calcium. J Biol Chem. 2001;276:47266–76.
73. Boveris A, Chance B, Oshino N. The cellular production of hydrogen-peroxide. Biochem J. 1972;128(3):617–30.
74. Gough DR, Cotter TG. Hydrogen peroxide: a Jekyll and Hyde signalling molecule. Cell Death Dis. 2011;2:e213.
75. Thomas C, Mackey MM, Diaz AA, et al. Hydroxyl radical is produced via the Fenton reaction in submitochondrial particles under oxidative stress: implications for diseases associated with iron accumulation. Redox Rep. 2009;14:102–8.
76. Gutteridge JMC, Halliwell B. Free radicals and antioxidants in the year 2000—A historical look to the future. Ann N Y Acad Sci. 2000;899:136–47.
77. Green MJ, Hill HAO, Tew DG, et al. An opsonized electrode—The direct electrochemical detection of superoxide generated by human-neutrophils. FEBS Lett. 1984;170:69–72.
78. Hill HAO, Tew DG, Walton NJ. An opsonized microelectrode—Observation of the respiratory burst of a single human neutrophil. FEBS Lett. 1985;191:257–63.
79. Tammeveski K, Tenno TT, Mashirin AA, et al. Superoxide electrode based on covalently immobilized cytochrome c: modelling studies. Free Radic Biol Med. 1998;25:973–8.
80. Roberts JG, Hamilton KL, Sombers LA. Comparison of electrode materials for the detection of rapid hydrogen peroxide fluctuations using background-subtracted fast scan cyclic voltammetry. Analyst. 2011;136:3550–6.
81. Amatore C, Arbault S, Guille M, et al. Electrochemical monitoring of single cell secretion: vesicular exocytosis and oxidative stress. Chem Rev. 2008;108:2585–621.
82. Cho SH, Jang A, Bishop PL, et al. Kinetics determination of electrogenerated hydrogen peroxide (H_2O_2) using carbon fiber microelectrode in electroenzymatic degradation of phenolic compounds. J Hazard Mater. 2010;175:253–7.

83. Arbault S, Pantano P, Jankowski JA, et al. Monitoring an oxidative stress mechanism at a single human fibroblast. Anal Chem. 1995;67:3382–90.
84. Zhu AW, Liu Y, Rui Q, et al. Selective and sensitive determination of hydroxyl radicals generated from living cells through an electrochemical impedance method. Chem Commun. 2011;47:4279–81.
85. Ferrer-Sueta G, Radi R. Chemical biology of peroxynitrite: kinetics, diffusion, and radicals. ACS Chem Biol. 2009;4:161–77.
86. Katrlik J, Zalesakova P. Nitric oxide determination by amperometric carbon fiber microelectrode. Bioelectrochemistry. 2002;56:73–6.
87. Santos RM, Lourenco CF, Piedade AP, et al. A comparative study of carbon fiber-based microelectrodes for the measurement of nitric oxide in brain tissue. Biosens Bioelectron. 2008;24:704–9.
88. Borgmann S. Electrochemical quantification of reactive oxygen and nitrogen: challenges and opportunities. Anal Bioanal Chem. 2009;394:95–105.
89. Kubant R, Malinski C, Burewicz A, et al. Peroxynitrite/nitric oxide balance in ischemia/reperfusion injury-nanomedical approach. Electroanalysis. 2006;18:410–6.
90. Miserere S, Ledru S, Ruille N, et al. Biocompatible carbon-based screen-printed electrodes for the electrochemical detection of nitric oxide. Electrochem Commun. 2006;8:238–44.
91. Amatore C, Arbault S, Chen Y, et al. Electrochemical detection in a microfluidic device of oxidative stress generated by macrophage cells. Lab Chip. 2007;7:233–8.
92. Trouillon R, O'Hare D, Chang SI. An electrochemical functional assay for the sensing of nitric oxide release induced by angiogenic factors. BMB Rep. 2011;44:699–704.
93. Amatore C, Arbault S, Koh AC. Simultaneous detection of reactive oxygen and nitrogen species released by a single macrophage by triple potential-step chronoamperometry. Anal Chem. 2010;82:1411–9.
94. Jacob C, Ba LA. Open season for hunting and trapping post-translational cysteine modifications in proteins and enzymes. Chembiochem. 2011;12:841–4.
95. Hidalgo J, Aschner M, Zatta P, et al. Roles of the metallothionein family of proteins in the central nervous system. Brain Res Bull. 2001;55:133–45.
96. Ralph TR, Hitchman ML, Millington JP, et al. The electrochemistry of L-cystine and L-cysteine.1. Thermodynamic and kinetic-studies. J Electroanal Chem. 1994;375:1–15.
97. Ren XL, Bjornstedt M, Shen B, et al. Mutagenesis of structural half-cystine residues in human thioredoxin and effects on the regulation of activity by selenodiglutathione. Biochemistry. 1993;32:9701–8.
98. Jacob C, Anwar A, Burkholz T. Perspective on recent developments on sulfur-containing agents and hydrogen sulfide signaling. Planta Med. 2008;74:1580–92.
99. Schneider T, Ba LA, Khairan K, et al. Interactions of polysulfanes with components of red blood cells. MedChemComm. 2011;2:196–200.
100. Munday R, Munday JS. Comparative haemolytic activity of bis(phenylmethyl) disulphide, bis(phenylethyl) disulphide and bis(phenylpropyl) disulphide in rats. Food Chem Toxicol. 2003;41:1609–15.

101. Sarakbi M-B. Natural products and related compounds as promising antioxidants and antimicrobial agents. Saarbruecken: University of Saarland; 2009.
102. Andersson CM, Hallberg A, Linden M, et al. Antioxidant activity of some diarylselenides in biological-systems. Free Radic Biol Med. 1994;16:17–28.
103. Jacob C, Lancaster JR, Giles GI. Reactive sulphur species in oxidative signal transduction. Biochem Soc Trans. 2004;32:1015–7.
104. Giles NM, Gutowski NJ, Giles GI, et al. Redox catalysts as sensitisers towards oxidative stress. FEBS Lett. 2003;535:179–82.
105. Fry FH, Jacob C. Sensor/effector drug design with potential relevance to cancer. Curr Pharm Des. 2006;12:4479–99.
106. Jamier V, Ba LA, Jacob C. Selenium- and tellurium-containing multifunctional redox agents as biochemical redox modulators with selective cytotoxicity. Chemistry. 2010;16:10920–8.
107. Giles GI, Collins CA, Stone TW, et al. Electrochemical and in vitro evaluation of the redox-properties of kynurenine species. Biochem Biophys Res Commun. 2003;300:719–24.

Chapter 8
Conductive Polymer-Based Materials for Medical Electroanalytic Applications

Vessela Tsakova

Abbreviations

AA	Ascorbic acid
ABSA	Aminobenzenesulfonic acid
CA	Chronoamperometry
CP	Conducting polymers
CPE	Carbon paste electrode
CNT	Carbon nanotubes
CV	Cyclic voltammetry
DA	Dopamine
DPV	Differential pulse voltammetry
EPI	Epinephrine
FIA	Flow injection analysis
LbL	Layer-by-layer
LOD	Limit of detection
MIP	Molecularly imprinted polymer
MSE	Mercury sulfate electrode
MWCNT	Multiwalled carbon nanotubes
NAD^+	Nicotinamide adenine dinucleotide
NOREPI	Norepinephrine
NPs	Nanoparticles

V. Tsakova (✉)
Institute of Physical Chemistry, Bulgarian Academy of Sciences,
Sofia, Bulgaria
e-mail: tsakova@ipc.bas.bg

P3MT	Poly(3-methylthiophene)
PAA	Polyacrilic acid
PAMPSA	Poly(2-acrylamido-2-methyl-1-propane-sulfonic acid)
PANI	Polyaniline
PEDOT	Poly(3,4-ethylenedioxythiophene)
POMA	Poly(*o*-methoxyaniline)
PPY	Polypyrrole
PSS	Polysterenesulfonate
PTHI	Polythiophene
PVS	Polyvinylsulfonate
RDE	Rotating disc electrode
SCE	Standard calomel electrode
SDS	Sodium dodecylsulfate
SHE	Standard hydrogen electrode
SWCNT	Single-walled carbon nanotubes
SWV	Square wave voltammetry
UA	Uric acid

8.1 Introduction

Since the discovery of high electrical conductivity in doped polyacetylene in 1977 [1], the investigations in the field of conducting polymers (CPs) have expanded rapidly in number, scope of research, and importance. The 2000 Nobel Prize in Chemistry was awarded to A. J. Heeger, A. McDiarmid, and H. Shirakawa for the discovery and development of electrically conductive polymers. The choice of the Nobel Committee was motivated *by the important scientific position that the field has achieved and the consequences in terms of practical applications and of interdisciplinary development between chemistry and physics* [2]. In the last decade the field of CPs and various CP-based materials has advanced further into new areas of research and technological developments. One of these intensively progressing areas is the involvement of CP-based materials in electrocatalytical applications, with a strong emphasis set on chemical and biochemical sensing. The electroanalytical response of CPs and CP-based composites was studied for a great number of compounds that are involved in the human metabolism, present medications, or harmful chemicals for humans. The large amount of publications on medical electroanalytical applications of CP-based materials are scattered over a number of specialized journals with main scopes in electroanalytical and analytical chemistry, sensing and biosensing, polymer science, medical studies, etc. Due to the highly spread and abundant literature it is nowadays difficult to get a general idea in this specific field of research. The present chapter attempts to outline the state of the art and hopefully to provoke further effort in this challenging scientific area with practical importance for medical diagnostics and medical treatments.

Several books [3–8] and a great number of review articles [9–28] on CPs, CP-based materials, and their applications were published in the recent years. Chemical sensing by CPs was addressed in a number of cases [9–18]. The role of CPs in biosensing devices was also extensively overviewed [14, 18–29]. In the present chapter the focus is set on the involvement of CPs or CP-based materials in electroanalytical determination of some metabolites, neurotransmitters, and drugs (Fig. 8.1) that are relevant for clinical analysis. Due to the already existing extensive literature devoted to the immobilization of biospecific recognition components (i.e., enzymes, antibodies, antigens, nucleic acids, cells, etc.) in CPs [22, 25, 29–33], this specific aspect remains out of the scope of this overview.

8.2 Conducting Polymer and Conducting Polymer-Based Composite Materials: General Presentation

Intrinsically conducting polymers are organic materials with conjugated bonds in the polymer skeleton. The specific bond structure is a prerequisite, but not a sufficient condition for the manifestation of high electrical conductivity. The latter is obtained by oxidative (p-doping) or reductive (n-doping) treatment of the CP material which may occur in electrochemical set-ups by applying suitable potentials. The resulting positive or negative charges along the polymer chains become compensated through ingress or expulsion of ions available in the electrolyte solution. These ionic fluxes are accompanied by the transport of solvent molecules that provoke additional mechanical effects such as swelling and de-swelling of the CP materials. Between the various classes of CPs the most suitable for electrochemical applications (including electroanalysis) are polyaniline (PANI), polypyrrole (PPY), polythiophene (PTHI), and some of their

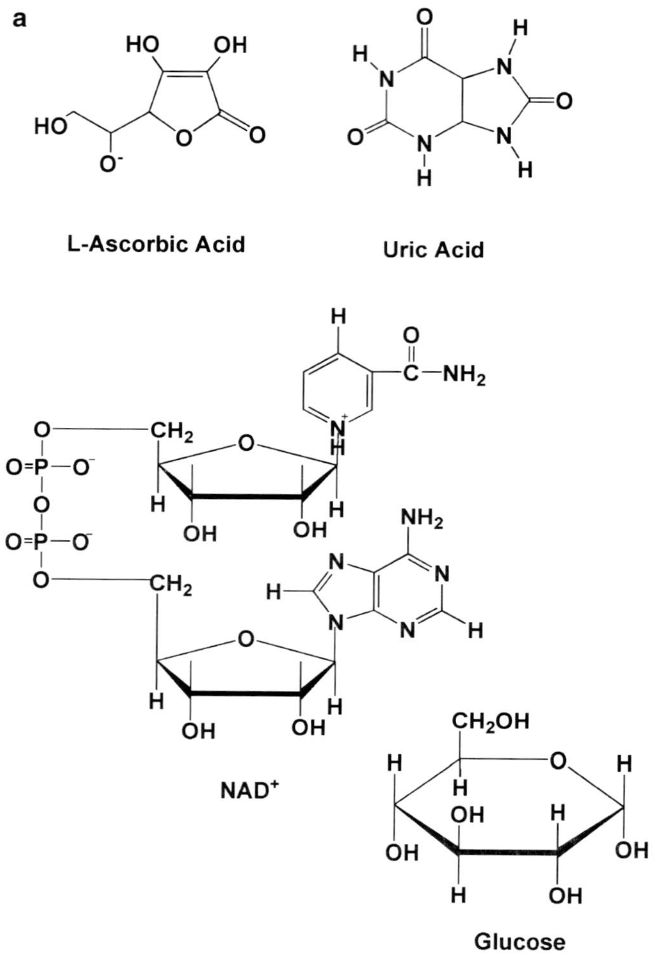

Fig. 8.1 Analytes involved in electroanalytical detection by means of CP-based electrodes

derivatives (Fig. 8.2). They are stable in aqueous solutions and upon multiple electrochemical oxidation/reduction (i.e., doping/dedoping) cycles which provide the possibility for switching between low-conducting and high-conducting states. The redox transitions for some of them (e.g., PANI and PANI derivatives) occur in potential intervals that are suitable for aqueous electrochemistry. The main

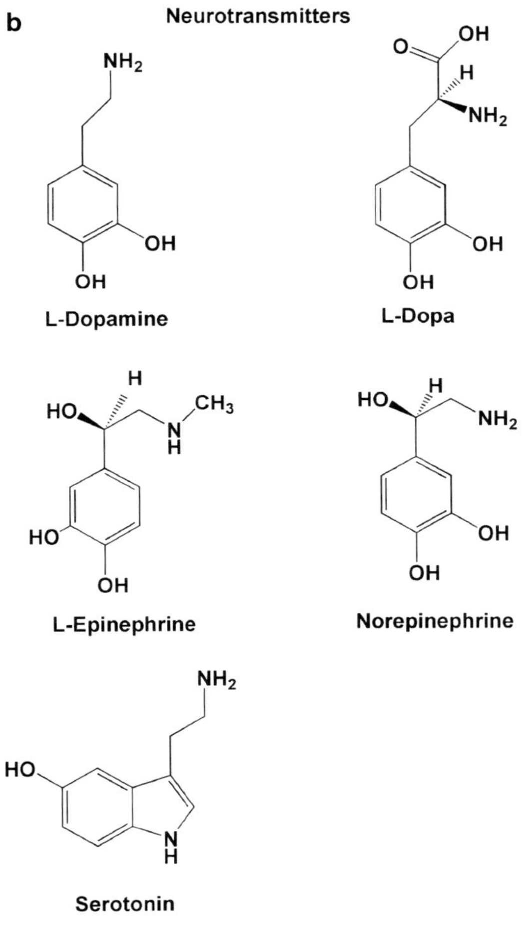

Fig. 8.1 (continued)

redox transition of CPs such as PPY and PTHI occurs at negative potentials below 0 V vs. SHE. In the potential range 0.2–1.5 V vs. SHE, commonly used for electroanalytical studies of biomedical substances, they are in the doped, highly conducting state. For this reason PANI-, PPY-, and PTHI-based materials are good candidates as electrode materials to be used for electrochemical analysis in human fluids such as blood and urine. These polymers along with a great number of their derivatives are extensively studied from an

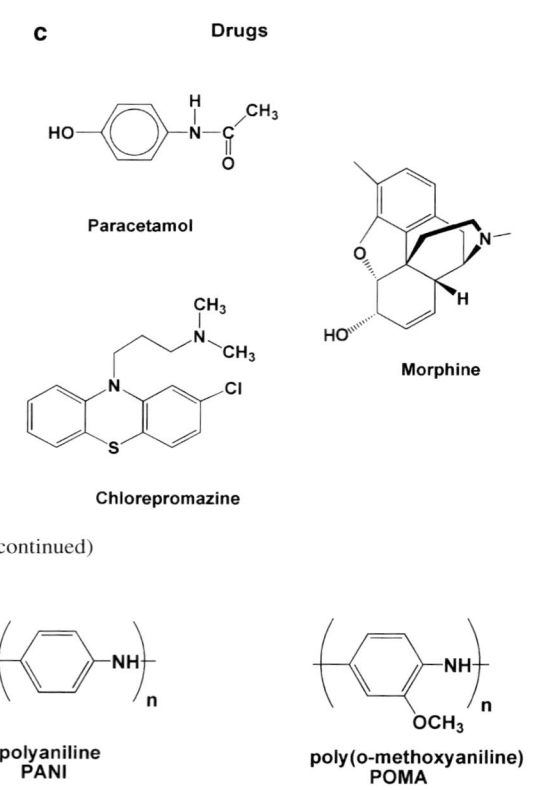

Fig. 8.1 (continued)

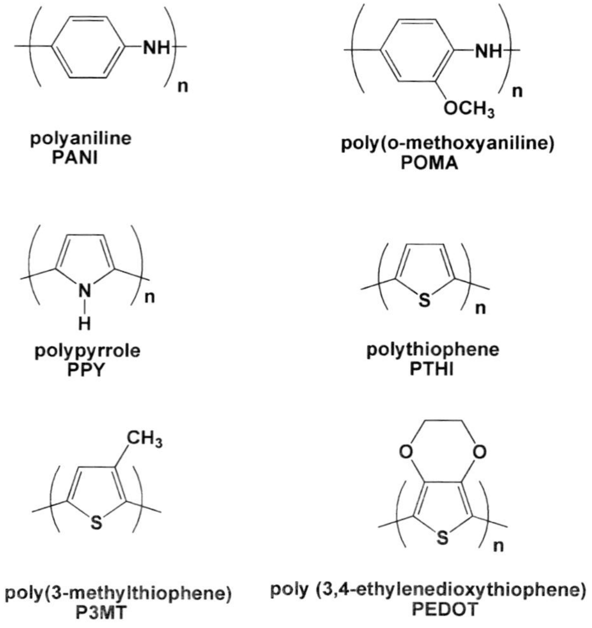

Fig. 8.2 Conducting polymers commonly used for electroanalytical studies

electrochemical point of view [3–7]. It is known that PPY- and PTHI-based CPs are operative in solutions with large range of acidities, including neutral solutions. By contrast, the application of conventional (inorganic anions-doped) PANI is restricted to acidic solutions up to pH 3. Conventional PANI loses its electroactivity and conductivity at higher pH values and is therefore not operative under physiological conditions. Nevertheless there are ways of avoiding this drawback by employing polyanions-doped or self-doped PANIs which preserve their electroactivity in neutral and slightly alkaline solutions [34–37] and therefore present suitable electrode materials for medical applications.

The use of CP-modified electrodes for electroanalytic reactions results usually in highly reduced overpotentials (by several hundreds of mV) in comparison to the non-modified electrodes. This effect is often attributed to the ability of CPs to undergo an intrinsic redox process. It is generally assumed that the internal redox process mediates the electrochemical oxidation reaction of the analyte under study [38] and results in decreased reaction overpotential and increased currents. Moreover, arguments based on presumed electrostatic interactions between the analyte species and charged centers of the CP backbone are often involved in the discussion on the specific role of CPs in electroanalysis. Finally, the possibility for the specific role of hydrophobic (undoped) and hydrophilic (doped) areas of the CP surfaces in the electroanalytical reactions is also addressed [39].

Although CP-modified electrodes are intensively studied for electrooxidation of various analytes, there is often not enough relevant information concerning the role of experimental characteristics of the CP layers, e.g., thickness, type of doping ions, structure and surface morphology, conductivity, etc. for the electroanalytical response. In the following the importance of some of these parameters will be addressed when discussing specific CP-modified electrodes and corresponding electroanalytical applications.

Together with the investigations involving CPs a significant effort is devoted to the development of composite materials containing CPs [6]. Different (wet chemical, electrochemical, adsorption, and physical) synthesis approaches are used to obtain a large variety of CP-based composites, e.g., including metal [40] or metal oxide particles [41]. The aim was initially to use CPs as a conducting medium for dispersing catalytic species (e.g., metallic particles) suitable for the electroanalytical reactions. After realizing the involvement of the CPs in the electrocatalytic reactions, it was further attempted to combine catalytic properties of CPs and of the additional composite components. The final goal in all cases is to obtain materials with large area that are catalytically active and electroanalytically sensitive and selective. The problem of electroanalytical selectivity in the presence of multiple analytes or interfering substances in the analyzed solutions is often pursued by investigating various CP-based composite materials.

The main point to be addressed in such studies concerns the correlation of the electroanalytical response and the properties of the composite (e.g., amount, surface or bulk location of the non-CP component, number, size and size distribution, stabilization shell of the metal nanoparticles, etc.). There are still few studies that succeed in elucidating these aspects and give in the same time reliable description of the electroanalytical response of the studied CP-based modified electrodes. Some of them will be addressed below.

8.3 Metabolites

8.3.1 Ascorbic Acid

l-ascorbic acid (AA) or *vitamin C* (Fig. 8.1) is a naturally occurring organic compound with an important role for various organisms including plants and animals. In many animals it is a metabolite derived from glucose, but human metabolism does not allow for its synthesis and thus ascorbic acid is a necessary component in human nutrition. AA has various functions in the human body [42, 43]. It is known as a powerful reducing agent interacting with free radicals and thus reducing the risk for a number of free radical-induced diseases such as cancer, Parkinson's disease, etc. Apart from its antioxidant properties, AA is involved as an enzyme cofactor in the biosynthesis of important species such as collagen, tyrosine, and neurotransmitters like norepinephrine. Vitamin C is vital to the immune response and wound healing and becomes quickly consumed during infections.

Ascorbic acid is a diprotic acid with pK_a values 4.17 and 11.57 [44]. At physiological conditions AA exists as a monodeprotonated ascorbate ion. The electrooxidation of the ascorbate species involves the transfer of two electrons and one proton and results in the formation of dehydro-L-ascorbic acid. The dehydro-L-ascorbic species undergo

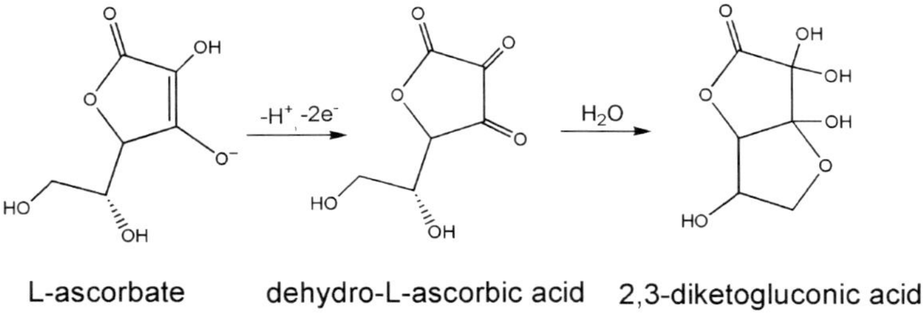

L-ascorbate   dehydro-L-ascorbic acid   2,3-diketogluconic acid

Fig. 8.3 Oxidation of ascorbic acid

hydration to form the non-electroactive product, 2,3-diketogluconic acid (Fig. 8.3) [38]. AA and its two electron oxidation product, dehydroascorbic acid, present a quasireversible redox couple with formal redox potential $E = 0.058$ V vs. RHE [44].

Ascorbic acid is one of the first analytes investigated in view of the possibility for electroanalytical applications of conducting polymer-modified electrodes. The use of conventional, non-modified electrodes results in large overpotentials and fouling of the electrode surface by products of the electrooxidation reaction. PPY [45] and N-substituted PPY with incorporated ferrocyanide ions [46] were used in the early investigations on electrooxidation of AA. The basic idea was that PPY, in contrast to redox polymers, has good electrical conductivity over a broad potential range overlapping with the range of AA oxidation. The fixed cationic charges of oxidized PPY were expected to facilitate the reaction due to the electrostatic interaction with the ascorbate ions. An effective reduction of the overpotential of the AA oxidation reaction was observed. Ferrocyanide ions immobilized within the polymer layer were additionally involved as a redox mediator. In the case of ferrocyanide-modified PPY it was established that both the polymer itself and the additionally bound redox mediator catalyze the reaction [46]. The mediating role of ferrocyanide was found to be important at low pH values where AA is protonated and thus electrostatic interaction with the PPY cationic sites is no longer effective. Detailed electroanalytical studies of AA oxidation on PPY-ferrocyanide films were carried out also at pH 4 [47] and pH 7 [48] (Table 8.1). In the first case the observed electrocatalytical activity for AA oxidation in the 0.5–16 mM concentration interval was assigned mainly to the mediating role of the immobilized ferrocyanide species. The studies carried out at pH 7 have shown a linear

Table 8.1 Electroanalytical detection of ascorbic acid

Conducting polymer, doping, or composite component	Method, pH*	Linear range (mM)	LOD (μM)	Interference studies	Ref.
PPY, Cl-, or dodecylbenzene sulfonate	CA	0.02–1.25			[49]
PPY, dodecylsulfate	FIA	0.0007–0.5	0.4	DA (50 μM)	[50]
PPY, ferrocyanide	CV pH 4	0.5–16.0			[47]
PPY, ferrocyanide, CPE	CV	0.45–9.62	58.2 (2σ)		[48]
PPY	FIA	0.0006–0.006	0.06		[69]
PANI					
PMT					
PANI	FIA pH 5.4	0.001–0.7	1		[52]
PANI	FIA Batch pH 6	0.005–0.1 0.0004–2.0	2.45 0.4	UA, cysteine, acetamidophenol	[53]
PANI	CV pH 5.9	1.5–9.0 (no Cu)	1,000		[51]
POMA with and without Cu particles					
PANI	Conductometry	0.15–1.4			[54]
POMA					
PANI, self-doped (m-amonobenzoic acid)	CA	0.5–4.0		NADH, DA, UA	[55]
PANI, self-doped (3,4-dihydroben-zoic acid)	CA	0.1–10	50		[56]
PANI, PVS	RDE	0.1–25		NADH	[38]
PANI, camphorsulfonate	CV	5–50			[57]
PANI, N-(3-propane-sulfonate)	CV	0.005–0.050	2.2	UA, DA	[58]

PANI, dodecylbenzene sulfonate	CA pH6.8	0.5–8.0	8.3		[59]
PANI, 3-aminobenzenesulfonic acid	CV	0.5–16.5	1.16		[60]
PANI, PVS, PSS	CA	0.014–0.400			[61]
PANI, PVS, PSS	CA	5–35	1	UA, Vitamin E, gluthatione, etc.	[62]
PANI, CuNPs	CA pH 9	0.005–3.5	2	H_2O_2, cysteine, glucoside	[66]
PEDOT	DPV	0.028–0.200 0.028–1.14	0.010		[70]

*In all tables the pH value is explicitly shown only if it differs from the interval 6.8–7.4

range of voltammetric response from 0.45 to 9.62 mM AA, although the exact role of ferrocyanide and PPY was not clarified in this case. Bulky anion (dodecyl sulfate or dodecylbenzene sulfonate)-doped PPY films where electrostatic interaction with the cationic PPY sites is no longer expected were also investigated for AA oxidation [49, 50]. The reaction overpotential was found to be reduced by more than 300 mV in comparison to bare glassy carbon electrode [50]. A direct comparison of Cl-doped and dodecyl benzenesulfonate-doped PPY did not show marked difference in the reaction overpotential. Thus it was concluded that the electrostatic interaction does not play a major role for the electrode process of AA at PPY films [50].

Already in the early investigations involving PPY [49, 50] and later PANI [38] it was shown that the electrocatalytic reaction occurs at the polymer/solution interface and not within the polymer layers. Levich plots with one and the same slope were obtained from RDE experiments irrespective of the CP layer thickness. These experimental observations have shown also that charge transfer across the polymer layer is not the rate limiting step. Mass transport was found to limit the AA oxidation currents [38, 50]. Investigations on the role of the thickness of the CP layer in electroanalytical assays show that thinner layers allow reducing both background capacitive currents and eventual Ohmic drop within the polymer. Therefore thin (and in the same time homogeneous) layers are recommendable for use in electroanalytical devices [50, 51].

The use of inorganic anion-doped PANI or PANI derivatives for electroanalytical determination of AA under physiological conditions encounters difficulties due to the gradual loss of electroactivity and conductivity observed for this polymer at pH >3. Nevertheless investigations carried out at pH values between 5.4 and 6.0 [51–53] demonstrated the possibility for electroanalytical application of this type of PANI layers showing that even at low electrical conductivity PANI acts as a catalytic material for AA oxidation. Structural and/or eventually resistive effects (resulting from different mode of PANI polymerization: potentiodynamic or potentiostatic) were demonstrated to affect the electrocatalytic response [51]. Conventional PANI and POMA layers were also successfully involved in conductometric sensing of AA at pH 7 providing the possibility for AA determination in the sub millimolar concentration range [54].

As already mentioned a way out of the restricted pH range of PANI electroactivity is the use of polyanion-doped or self-sulfonated

PANIs. The involvement of a large variety of sulfonic and polysulfonic acids-doped PANIs in AA electroanalytical assay resulted in some cases in large concentration windows of linear response and good sensitivity (Table 8.1) [55–62]. Investigations on the pH behavior of polysulfonate acid-doped PANI and self-sulfonated PANIs show that in neutral solutions the electroactivity of the polymer layers becomes to a great extent preserved. Nevertheless a change in the mechanism of the intrinsic PANI redox process is observed [36, 37, 42, 63] together with a significant loss in conductivity [64]. In general the electrical conductivity seems not to be crucial for the electrocatalytic process. It is rather the redox mediating ability of the polymer that plays a decisive role. Nevertheless significant differences are observed depending on the dopant used. Plateau oxidation currents twice as large for PVS- in comparison to PSS-doped PANI point to a strong influence of the dopant ion on the polyaniline redox mediating system, possibly through a modification of the conductivity and/or morphology of the conducting polymer in the presence of the various polyanions [61]. A demonstration on the role of the synthetic conditions for AA oxidation on poly(2-acryalamido-2-methyl-1-propane-sulfonic)(PAMPSA)-doped PANI layers is shown in Fig. 8.4 [65].

The electrocatalytic reaction of AA oxidation in the case of PVS-doped PANI was extensively studied by Bartlett et al. [38]. A kinetic model was used to interpret the results obtained in RDE experiments. It was assumed that a thin layer at the polymer/solution interface plays a mediating role (Fig. 8.5) and the surface reaction is of Michaelis–Menten type:

$$CP_{ox} + AA^- \xrightarrow{K_m} \left[CP_{ox}AA^-\right]$$

$$\left[CP_{ox}AA^-\right] \xrightarrow{K_{cat}} CP_{red} + DAA$$

$$CP_{red} \xrightarrow{K_e} CP_{ox} + 2e^- + H^+$$

Here CP_{ox} and CP_{red} denote oxidized and reduced sites at the polymer surface, AA^- is the ascorbate ion and DAA the reaction product—dehydro-L-ascorbic acid.

Only few studies are so far focused on the use of metal particles/CP composites for electroanalytical determination of AA. Copper was chosen as a metal showing intrinsic electrocatalytic activity in neutral and alkaline solutions due to the possible oxidation of the Cu crystalline surface and corresponding transition between several

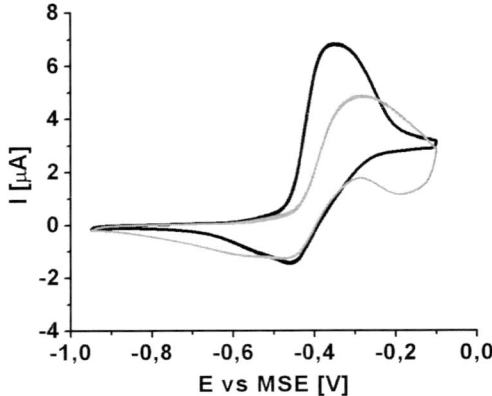

Fig. 8.4 Voltammetric curves measured in the presence of 5 mM ascorbic acid in phosphate buffer solution (pH 7) at PANI/PAMPSA layers obtained through potentiostatic (*black line*) and potentiodynamic (*grey line*) electrodeposition. (Courtesy of V. Lyutov [65])

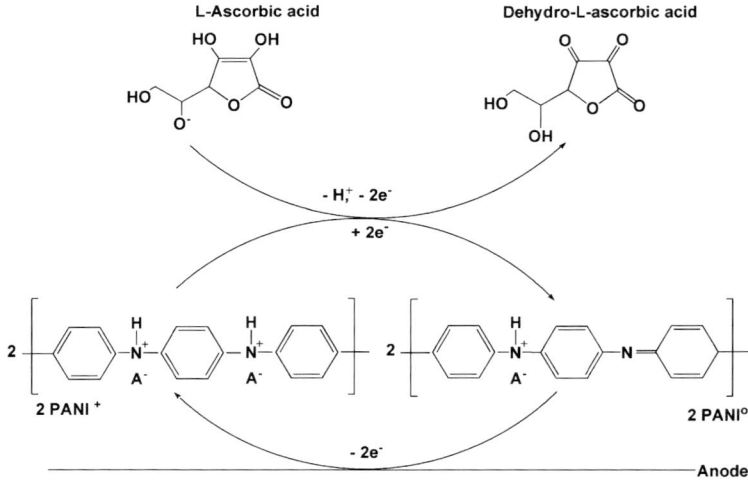

Fig. 8.5 Mediating role of PANI in ascorbic acid oxidation

oxidation states. Cu-modified PANI and POMA electrodes were studied at pH 5.9 [51] and pH 9, respectively [66]. A doubling in the oxidation currents was demonstrated for Cu-modified POMA layers in comparison to the non-modified polymer films. It was found that the two parent polymers (PANI and POMA) behave in a different way, POMA providing larger oxidation currents. The electrocatalytic effect was due to both the polymer and the metal component. In the case CuNPs/PANI-modified electrode investigated at pH 9, the oxidation was mediated only by the CuNPs. The advantage of this system consisted in the negligible background currents observed at this high pH value where the electroactivity of PANI is totally lost.

PTHIs were also involved in studies on the electrooxidation of AA. Basically the electrocatalytic effect was demonstrated in the early investigations of Atta et al. [67, 68] and Erdogdu et al. [69]. A detailed study on AA oxidation on PEDOT-modified electrodes by means of the DPV technique was recently published by Bello et al. [70]. It was demonstrated that the electroanalytical response depends significantly on the parameters (pulse amplitude, step potential, and step time) of the DPV procedure. Careful optimization is needed in all electroanalytical studies involving this technique. By using the optimal DPV parameters three times higher sensitivity of PEDOT in comparison to bare Pt was found and a linear response in the 5–35 and 35–200 ppm concentration intervals was observed (Table 8.1).

CP-modified electrodes reduce in the rule significantly the overpotential for AA oxidation and the voltammetric peak is typically observed at about 0 V vs. SCE. The range of linear response extends usually in the millimolar range which makes the materials suitable for determination in food and beverages. In some cases sensitivity in the µM range that is relevant for biological measurements is also demonstrated. In several cases the modified electrodes were tested in real samples, e.g., for determination of Vitamin C in mono- and multivitamin tablets, juices, etc. [50, 53, 54]. In a recent investigation [62] PANI/PVS films deposited on a needle type microelectrode were successfully applied for AA determination in 100 µL drops of blood plasma by using the standard addition method. The results are encouraging for the development of a suitable PANI-modified electrode for *in vivo* applications. The main problem for biological measurements presents, however, the interfering species (e.g., NADH, dopamine (DA), uric acid (UA), etc.) coexisting in real specimens such as blood or urine. There is a large number of studies addressing this problem that aim basically at the simultaneous determination of more than one analyte, including AA. These studies will be commented in the following parts where the corresponding second analyte becomes addressed.

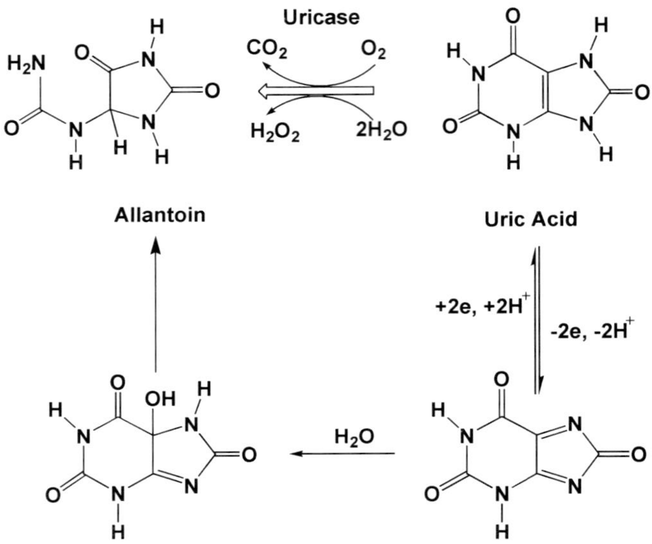

Fig. 8.6 Oxidation of uric acid

8.3.2 Uric Acid

Uric acid (UA) (2,6,8-trihydroxypurine) (Fig. 8.1) is a final product of purine metabolism in the human body. High levels of UA in the blood can lead to various disorders, e.g., gout, kidney and heart diseases, etc. Elevated levels of UA are observed in a wide range of conditions, e.g., leukemia, pneumonia, high blood pressure, ischemia, etc. Together with AA, UA is a strong reducing agents and an effective antioxidant. The clinical significance of urate and the necessity for easy routine screening of this biomarker have been discussed in details in [71, 72].

UA is a diprotic acid with pK_a values 5.4 and 10.3 [73, 74] that forms single charged hydrogen or acid urate ions at biological pH values. The electrooxidation of UA proceeds by the formation of reactive diimine species which are further converted to imine alcohol and finally to allantoin [71] (Fig. 8.6). The involvement of the uricase enzyme provides the possibility for development of biosensors for the electrooxidation of UA (see e.g. [75]). There are few studies involving CPs for electroanalytical *nonenzymatic* detection of UA alone [73, 76].

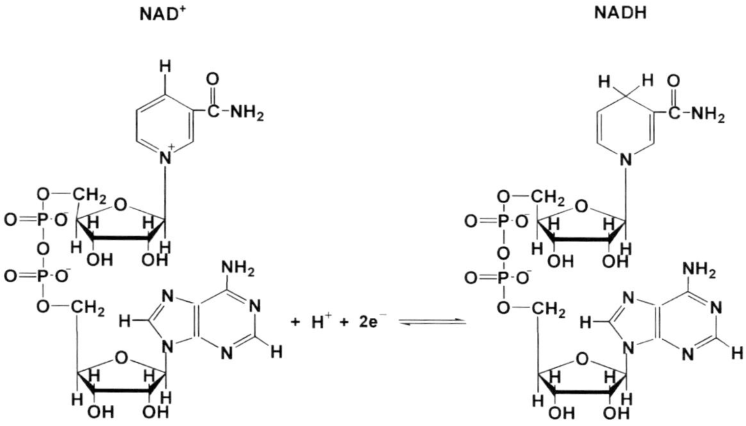

Fig. 8.7 Reduction/oxidation of NAD$^+$/NADH

The use of polyluminol-coated electrodes [76] resulted in relative stability of the UA oxidation peaks in the presence of large amounts of dopamine and ascorbic acid and an extended concentration range (0.03–1 mM) of linear DPV response. Recovery in urine samples was also demonstrated. A large number of investigations address the simultaneous determination of UA and other analytes, most commonly dopamine. They will be commented further in the text.

8.3.3 Nicotinamide Adenine Dinucleotide

Nicotinamide adenine dinucleotide (NAD$^+$) (Fig. 8.1) is an important coenzyme essential for all living cells. It plays a key role as a carrier of electrons involved in metabolic redox reactions as well as in cell signaling [77] (Fig. 8.7). NAD$^+$ functions most often in energy producing reactions involving the degradation (catabolism) of carbohydrates, fats, proteins, and alcohols. The electrochemical oxidation of NADH can be used as transduction reaction for designing amperometric biosensors based on dehydrogenase enzymes. The latter constitute the largest group (over 500) of enzymes known.

Direct electrochemical oxidation of NADH at non-modified electrodes occurs at high overpotentials (typically exceeding 1 V [78]) and results in a rapid fouling of the electrode surface by species produced through the highly reactive one-electron oxidation intermediate of the reaction [79]. The electrochemistry of NADH (and NADP) reactions on mediator-modified electrodes was extensively reviewed by Gorton et al. [80, 81]. The progress in the use of carbon nanotube-based electrodes for NADH oxidation was recently summarized in [13].

The possibility to involve CP-modified electrodes for the oxidation of NADH was identified by exploring the electroanalytical response of P3MT [82], substituted PPY [83], and poly(indole-5carboxylic acid) [84] (Table 8.2). Already in the early investigations an ethanol biosensor was designed based on the enzymatic reaction of alcohol dehydrogenases with NAD^+. In the enzymatic reaction of ethanol oxidation dehydrogenase requires NAD^+ as a cofactor and therefore NADH becomes produced in the course of the process. NADH is subsequently electrooxidized at the CP-modified electrode [83]. A similar approach was used to design an enzymatic L-lactate biosensor involving NADH as a cofactor in the enzymatic reaction [85].

Detailed studies on the mechanism of NADH electrooxidation were carried out for PVS-, PSS-, and PAA-doped PANI [38, 79, 86–88]. It was suggested that in analogy to the AA case PANI plays the role of a redox mediator and a complex is formed between the catalytic site and NADH [88]. This process is followed by two rapid one-electron oxidations of PANI to regenerate the catalytic site [87]. It was argued that the reaction occurs throughout the film and involves the formation of an initial complex between NADH and the reactive sites of PANI within the film. In fact a similar kinetic scheme describes AA oxidation on polyanion-doped PANI with the difference that in the AA case the reaction is confined at the polymer/solution interface. In analogy to the AA oxidation case, a detailed kinetic model was developed [86] providing the possibility to explore four different limiting cases for NADH oxidation depending on the analyte concentration and the thickness of the PSS- and PVS-doped PANI layers.

Polymerized azine dyes were extensively studied for NADH oxidation (Table 8.2) [13, 89–98]. It was expected that the polymerized dyes that have both monomer-based and polymer-based redox electroactivity will facilitate the catalytic reaction. In most cases the polymerized dyes reduce the overpotential for the oxidation reaction by more than 500 mV. A comparative study on the electrocatalytic activity for NADH oxidation of several polyazines (poly(toluidine blue), poly(brilliant cresyl blue), poly(azure A), and poly (methylene

Table 8.2 Electroanalytical detection of NADH

Conducting polymer, doping, or composite component	Method, pH*	Linear range (μM)	LOD (μM)	Interference studies	Ref.
Poly(methylene green)	CA	500–10,000	0.5		[89]
Poly(toluidine blue O)	CA	5–2,000	2.0		[90]
Poly(Meldola Blue)	CA	8–500	20		[91, 92]
		100–1,000	10		
		50–1,700			
Poly(azure A)	FIA	5–100	0.2		[93]
Poly(toluidine blue O)			0.32		
Poly(toluidine blue)	FIA, CV	5–3,200	0.1	AA: interf.	[96]
Poly(toluidine blue O)	CA	2–10	0.5		[94]
CNTs		100–4,500			
Poly(toluidine blue O)	CA	100–1,000		AA, DA	[95]
Poly(methylene green)	CA	5.6–420	3.8		[97]
PPY, methylene blue	FIA	0.1–1,000	0.04		[99]
PMT	CA	1–100	0.76	NAD$^+$, AA	[84]
PANI, PSS, PVS, PAA	CA, RDE	100–2,400		UA, acetoamido-phenol; AA: interf.	[86]
PANI-PSS	CA	1–320	0.3		[109]
On array of Au nanorods					
Poly(p-ABSA), flavins	CV	10–300	1		[100]
	pH 6.4				
PMT, CNTs		0.5–20	0.17		[102]
Poly(thiomine)	FIA	5–100		DA, UA; AA: interf.	[110]

(continued)

Table 8.2 (continued)

Conducting polymer, doping, or composite component	Method, pH*	Linear range (μM)	LOD (μM)	Interference studies	Ref.
PANI, AuNPs (LbL)	CV	10–50 mM			[103, 104]
PEDOT, AuNPs	SWV	10–80		AA, UA (1 mM)	[105]
	CA	0.1–2.2			
PEDOT, AgNPs	CA	10–560	0.1	DA, UA, H_2O_2	[106]
Meldola Blue					
PPY nanocapsules	CA	40–800	0.7		[107]
PtNPs					
PEDOT-PSS, PdNPs	Conductometry	10–1,280 (log dependence)			[108]

blue)) shows that they exhibit redox activity in the same potential range. Their electrocatalytic activity varies depending on the polymerized dye specifics. It was concluded that phenoxazines are better catalysts than phenithiazines. Ring substitution with only tertiary nitrogen atoms as ligands provides higher catalytic activity [98]. Extensive overviews on the involvement of these materials in NADH oxidation were recently published [13, 78].

Combinations of CPs and redox mediating species, e.g., methylene blue [99] or flavins [100], were explored in order to improve the materials' catalytic performance. The incorporation of flavins in self-sulfonated PANI layers [100] resulted in the possibility for both electrocatalytic oxidation of NADH and reduction of NAD^+, the latter mediated by the flavin moieties. The use of poly(neutral red) electrode for NAD^+ reduction was also demonstrated [101].

Carbon nanotubes-based composites of conducting polymers were also used for electroanalytical determination of NADH [13]. Poly(toluidine blue O) [94] and P3MT [102] were combined with CNTs to obtain larger reactive surface. The oxidation currents measured on these electrodes were typically larger than on the polymer-modified electrodes without CNTs.

In the last years metal particles-modified CP electrodes were intensively studied as electrocatalytic materials for NADH oxidation (Table 8.2) [103–108]. Metals such as gold and palladium that are known to catalyze the electrooxidation of NADH were combined with PANI or PEDOT. Different approaches were used to obtain dense populations of AuNPs, e.g., layer-by-layer absorption of pre-synthesized AuNPs and PANI [103, 104] or repetitive electrodeposition in $HAuCl_4$ containing solution on a PEDOT-PSS film [105]. The studies on AuNPs-PEDOT-PSS electrode have shown a decrease in overpotential for NADH oxidation (in comparison to PEDOT-PSS) together with an antifouling effect ascribed to the presence of AuNPs. Typical interferents such as AA and UA did not affect the amperometric response measured at 1 mM concentration levels. This electrocatalytic system was further used to develop an ethanol biosensor based on incorporated alcoholdehydrogenase [105]. It was found that the signal of NADH oxidation can be qualitatively correlated to the amount of ethanol. This is one of the examples for the involvement of NADH oxidation as a transduction reaction in a biosensor set-up.

Finally, a conductometric approach for determination of NADH was recently suggested based on composite materials consisting of PEDOT-PSS and PdNPs. It was suggested that the PdNPs serve as reaction sites for NADH oxidation and the released electrons reduce the PEDOT material thus decreasing its conductance [108].

A further improvement of the electrocatalytic response was sought by incorporating both an electrochemically active mediator (Meldola Blue) and AgNPs in PEDOT [106]. It is suggested that the redox mediator Meldola Blue catalysis the electrochemical reaction whereas AgNPs play the role of electron relays between the redox mediator and the CP. It was established that AA, DA, and H_2O_2 do not affect the amperometric response for NADH oxidation.

8.3.4 Hydrogen Peroxide

Hydrogen peroxide is generated in many enzyme-catalyzed reactions, e.g., of uricase, glucose oxidase, cholesterol oxidase, and choline oxidase with corresponding target molecules, i.e., uric acid, glucose, cholesterol, cholin, etc. Therefore detection of H_2O_2 as a byproduct of enzymatic reactions is one of the strategies to measure these target compounds. Moreover hydrogen peroxide is toxic to "cellular life" and therefore an important analyte in food, pharmaceutical, and environmental analyses.

The oxidation reaction occurs usually at about 0.4–0.7 V vs. SCE [111] and results in the formation of oxygen:

$$H_2O_2 \rightarrow O_2 + 2H^+ + 2e^-$$

The drawback of using oxidation for the detection of hydrogen peroxide is that the potential window for this reaction overlaps with the potential range for oxidation of several organic species (e.g., AA, UA, catechols, etc.) that are coexisting in real samples and interfere with the studied analyte.

The reduction of H_2O_2 occurs usually at about +0.1 to −0.3 V vs. SCE and results in:

$$H_2O_2 + 2H^+ + 2e^- \rightarrow 2H_2O$$

Different approaches are used for the electroanalytical determination of hydrogen peroxide based on the oxidation or reduction reactions both with or without the involvement of enzymatic mediators.

The involvement of enzymes in H_2O_2 detection is based on the possibility for direct electrochemical reduction of enzymes on the CP surface, e.g., horseradish peroxidase on PANI [112]. H_2O_2 is reduced through the enzymatic reaction and subsequently the CP material is used as a redox mediator for recovering the enzyme in its initial state [113].

In the following a short review on the *nonenzymatic* CP-based materials for electrochemical H_2O_2 detection will be presented. The large number of studies devoted to enzymatic catalytic materials for electrochemical sensing of hydrogen peroxide will remain outside the scope of this chapter. Detailed description of enzyme-based electrodes can be found in [12, 22, 28].

There are few studies addressing the possibility to use non-modified CPs for direct electrochemical detection of hydrogen peroxide. The possibility to eliminate the interference of oxygen for reductive sensing of H_2O_2 with a PANI-coated electrode was demonstrated by using dissolved oxygen scavengers (AA or sodium thiosulfate) [114]. Ferrocene sulfonic acid-doped PANI and poly(aniline-co-p-aminophenol) electrodes were tested for oxidative determination of H_2O_2 (Table 8.3) [115, 116]. It was established that the electroanalytical properties are not so good as those obtained with metal particles-modified CPs (see below). An attempt to use polymerized Meldola Blue for reductive electrochemical sensing of hydrogen peroxide [91] has shown that the polymer has inferior electroanalytical performance in comparison to the monomer species with respect to this reaction.

Several studies on electroanalytical determination of H_2O_2 are based on metal particles-modified CP-coated electrodes (Table 8.3) [117–121]. Metals such as Pt, Au, Cu, and Co were used to induce electrocatalytic properties of the modified electrodes. Reduction-based sensing materials were obtained in the Pt/poly(*o*-phenylenediamine) [118], Au/PEDOT [119], and Cu/PPY [120] cases. By using the reduction reaction the interference of many organic compounds could be avoided and in the same time large concentration ranges of linear amperometric response were obtained (Table 8.3). The role of the CP layer for the H_2O_2 reduction was not clarified in all cases. A contribution of both CP layer (PEDOT) and the metal phase (AuNPs) to the reduction current of H_2O_2 was found [119]. On the other hand the electrocatalytic currents in the Cu/PPY system measured in alkaline solution seemed to be due to the metal phase alone [120]. The cobalt oxide particles/poly(*n*-methylaniline) electrode tested for H_2O_2 oxidation in alkaline solutions exhibited electrocatalytic behavior due to the Co species alone [121].

Table 8.3 Electroanalytical detection of hydrogen peroxide

Conducting polymer, dopant, or composite component	Method, pH*	Linear range (μM)	LOD (μM)	Interference studies	Ref.
PANI, ferrocene sulfonic acid	CA, pH 5	4–128		AA, phenol, catechol, formaldehyde: interf.	[115]
Poly(aniline-co-p-aminophenol)	CA	1.25–250	1.25		[116]
PANI	CV, pH 6.2	200–2,500		AA, phenol	[114]
Poly(Meldola Blue)	CA	250–5,000	78		[92]
PANI, Pt	CA, pH 4.6	0.4–240	0.1	RCHO, various inorganic ions	[117]
Poly(o-phenylenediamine), Pt	CA	4–1,000		AA (1 mM), UA (1 mM), L-lactic acid, glutamic acid, glycerol	[118]
PEDOT, AuNPs	RDE	500–5×10^3			[119]
PPY, Cu particles	CA 0.1 M NaOH	7–4,300	2.3	Glucose, glycine, ethanol, acetic acid, L-cysteine (0.4 mM)	[120]
Poly(N-methylaniline), SDS, Co oxide particles	CV	30–12×10^3	18		[121]
	DPV	5–48	3		
	SWV, pH 13	1–12	0.9		

In general the interest towards the development of nonenzymatic catalytic materials for electroanalytical detection of various analytes including H_2O_2 is renewed in the recent years due to the necessity to understand better the mechanism of the reactions occurring on complex composite materials and also to the common problems of immobilized-enzymatic electrodes, e.g., the lack of stability and reproducibility.

8.3.5 *Glucose*

Glucose is a common analyte measured in blood for diagnostics and control of diabetes patients. As a result of the worldwide increasing ratio of the population affected by this condition, there is a continuous effort in the development of glucose sensors, both nonenzymatic and enzyme-based. Overviews on electrochemical nonenzymatic glucose sensors were recently published [122, 123]. The 2010 review article of Toghill and Compton [123] gives a detailed account (with almost 300 references) on the electrode materials involved in nonenzymatic glucose sensing as well as on the mechanism of the reaction and the future of the systems discussed.

The involvement of CPs in nonenzymatic glucose sensing is limited mainly to the role of the polymer as a matrix for dispersing metallic catalytic particles. No intrinsic electrocatalytic effect of the CP materials for glucose oxidation was so far observed. Copper-modified PANI was identified as an amperometric sensor for amino acids and polyhydric compounds in alkaline solutions as early as 1996 [124]. The same system (with rather thin PANI layers offering the possibility to disperse the metal particles) was further investigated [125] to show the possibility for UV photo-induced oxidation of glucose. A study focused on the role of the initial oxidation state of PANI for the copper crystals deposition [126] has revealed that in alkaline solutions the Cu-PANI system becomes operative for glucose oxidation only when copper is electrodeposited in initially reduced PANI layers. At high pH values PANI loses its electroactivity and once thick compact PANI layers become involved in electroanalytic studies, the metal particles to be used for the electrocatalytic reaction should be deposited in a way to have contact with the underlying metal substrate [40, 127]. A linear current response measured at fixed potential was observed in the 2–10 mM glucose concentration range by using compact PANI layers and Cu particles deposited through the polymer porous structure of initially reduced PANI layers [126]. A composite consisting of nickel oxide (NiO_x) and PANI was found to be operative under potentiodynamic conditions in the 1–100 mM concentration range of glucose [128]. Thin PPY layers with dispersed Pt-based alloy particles [129] and PEDOT with incorporated AuNPs [130] were also used for glucose electrooxidation.

CPs play a much more important role in the development of enzymatic-based electrochemical sensors for glucose. Their main purpose is to support the immobilization of enzymes (glucose oxidase or glucose dehydrogenase) on the electrode surface. This specific role of CPs was extensively overviewed [25, 29, 131]. Recent presentations of the enzyme-based electrochemical sensors for glucose are also available [72, 122, 123, 132].

8.4 Neurotransmitters

8.4.1 The Medical Point of View

Neurotransmitters are endogenous chemical compounds that transmit signals from a neuron to a target cell across a synapse. Neurotransmitters are usually synthesized by few biosynthetic steps starting from corresponding precursors, e.g., amino acids. A typical example presents the biosynthesis of epinephrine (adrenaline) [133], based on phenylalanine and tyrosine as precursors (Fig. 8.8). Some of the intermediate products in this biosynthesis path, e.g., dopamine and norepinephrine, play also the role of neurotransmitters (see below).

Studies on neurotransmitter content and release are important for improving the understanding of the role of chemical processes for the central nervous system and thus for the behavioral, cognitive, and emotional state of the human organism. Among other methods, e.g., liquid chromatography, capillary electrophoresis separation, and mass spectrometry, electroanalytical approaches are also used to detect electroactive neurotransmitters such as dopamine, serotonin, epinephrine, etc. (Fig. 8.1b). The main point in electrochemical studies is to find electrode materials that show enough sensitivity and selectivity with respect to the target neurotransmitter. The measurements in real specimens, i.e., blood plasma, urine, or brain tissue, are often compli-

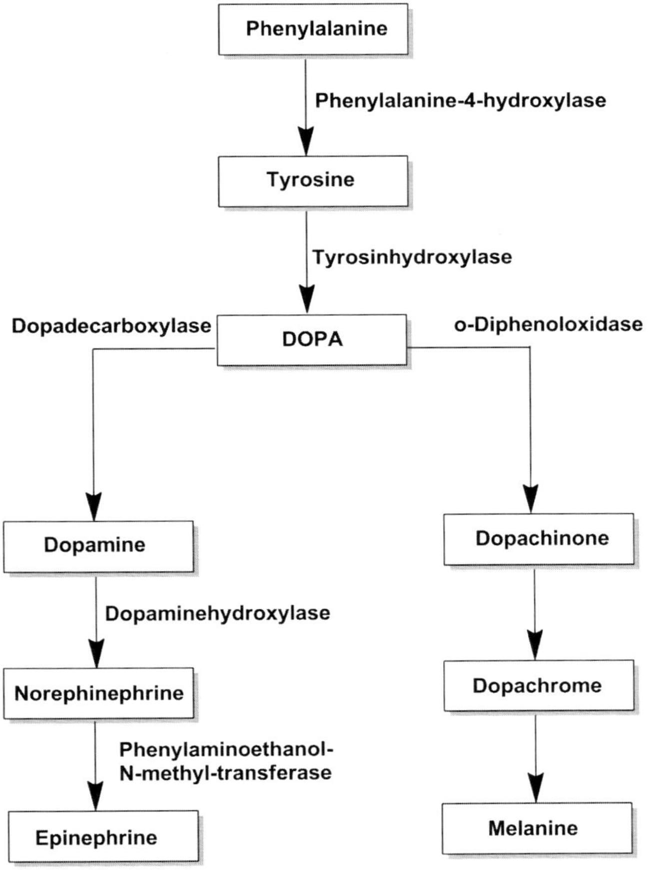

Fig. 8.8 Biosynthesis of epinephrine

cated due to the interference of other electroactive neurotransmitters and metabolites.

A brief description of electroactive neurotransmitters that are the subject of electroanalytical studies involving CP-modified electrodes is given below:

Dopamine (*DA*) (3,4-dihydroxy-β-phenylethylamine or hydroxytyramin) is a neurotransmitter available in the brain, adrenal glands, and sympathetic nerve endings. It is an intermediate product in the biosynthesis of adrenalin (Fig. 8.8). DA plays an important role in the nervous, cardiovascular, renal, and hormonal system of humans [134]. Its abnormal concentration is in the origin of various disorders such as Parkinson's and Huntington's diseases, epillepsy, schizophrenia, Tourette's syndrome, etc. Dopamine helps to control the brain's reward and pleasure centers and to regulate emotional responses. Therefore accurate control on dopamine levels is essential for the brain functions. Moreover, DA is used as a medication that affects the sympathetic nervous system and produces increased heart rate and blood pressure. For these reasons quantitative determination of DA has attracted considerable interest especially in the last years. Due to the fact that DA may undergo electrooxidation electrochemical methods have been widely employed.

The concentration of dopamine in urine is in the 0.46–3.66 µM range [133] whereas *in vivo* measurements in rats' brain show values of 26–40 nM [135].

l-DOPA (3,4-dihydroxy-L-phenylalanine) (Fig. 8.1b) is a precursor in the biosynthetic path of dopamine, epinephrine, and norepinephrine (Fig. 8.8). Since DA cannot cross the blood–brain barrier, L-DOPA is used in medical treatments requiring an increase in DA levels in brain, e.g., in Parkinson's disease. Chronic L-DOPA treatment is sometimes associated with side effects (e.g., nausea, vomiting, etc.) resulting from the increase of L-DOPA levels in blood plasma. Therefore L-DOPA determination is important in medical diagnostics and treatments.

Serotonin (5-hydroxytryptamine) (Fig. 8.1b) is a neurotransmitter available mainly in the gastrointestinal tract and central nervous system. Approximately 80% of serotonin in the human body, located in the gut, is used to regulate the intestinal movements. The remainder is stored in the central nervous system and has various functions, e.g., regulation of mood, appetite, and sleep. Serotonin has also some cognitive functions, including in memory and learning. Modulation of serotonin at synapses is thought to be a major action of several classes of pharmacological antidepressants. Although it is not a hormone, it is usually designed as the "happiness hormone." The concentration of serotonin determined by *in vivo* measurements in rats' brain is 68–70 nM [135].

Epinephrine (EPI) (or *adrenaline*) (Fig. 8.1b) is a hormone and a neurotransmitter. In medical treatments it is used to increase heart rate and to dilate air passages. Epinephrine may be quantitated in blood, plasma, or serum as a diagnostic aid, to monitor therapeutic treatment or to identify the causative agent in poisoning victims. Endogenous plasma epinephrine concentrations in resting adults are normally less than 10 ng/L, but may increase by tenfold during exercise and by 50-fold or more during times of stress. Parenteral administration of epinephrine to acute-care cardiac patients can produce plasma concentrations of 10–100 μg/L.

Norepinephrine (NOREPI) (or noradrenaline) (Fig. 8.1b) is a catecholamine found in the sympathetic nervous system that plays the role of hormone and neurotransmitter. The sympathetic nervous system stimulates the heart, blood vessels, sweat glands, the large internal organs, and the adrenal medulla in the brain. Like dopamine, norepinephrine has a stimulating effect, fosters alertness, and plays an important regulatory role in long-term memory and learning. In clinical medicine norepinephrine is used to treat life-threatening low blood pressure that can occur in certain medical conditions or surgical procedures. Norepinephrine is synthesized from dopamine by dopamine β-hydroxylase (Fig. 8.8).

8.4.2 Electroanalytical Determination of Neurotransmitters Based on CP-Modified Electrodes

The major problem in the electroanalytical determination of electroactive neurotransmitters is selectivity of the electroanalytical response. The involvement of various modified electrodes (CP-based including) in electroanalytical studies aims to widen the potential window in which the oxidation of the target analyte(s) and interfering species occurs and/or to enhance catalytically the target reaction and suppress the interfering ones. For instance, one of the most studied neurotransmitters, dopamine, is coexisting with large excess (about 1,000 times higher concentration) of ascorbic acid. Thus the main goal in electroanalytical studies is to obtain well-resolved peaks of both analytes and/or eventually to suppress the ascorbate oxidation reaction.

Already the first studies on PPY-coated electrodes [45] demonstrated the possibility to resolve voltammetric peaks of dopamine and ascorbic acid which are overlapping at conventional (metal or glassy carbon) electrodes. Extensive studies on P3MT-coated electrodes [67–69, 136] revealed further possibilities to catalyze the oxidation of a number of organic compounds and biological molecules, e.g., DA, AA, EP, NADH, catechol, etc. The comparison of the voltammetric peaks measured at P3MT-coated glassy carbon and platinum electrode has shown significant shift of the anodic oxidation peaks in the negative direction, enhancement of the peak currents, and better resolution of the peaks of analytes that coexist in practically relevant solutions. At the early stage of investigations comparisons of P3MT with other CPs (PANI, PPY, poly-N-methylpyrrole, polyfuran) were carried out and were found to be in favor of P3MT. Later with the development of the knowledge on the synthesis and properties of CP layers the manifold opportunities to modify the catalytic properties of CPs were assessed and employed to study the electrooxidation of neurotransmitters in much more details.

One of the main ideas of the early investigations was to control the size exclusion selectivity of the various analytes by probing different CP-coatings as perm selective membranes. For this reason overoxidized CPs were investigated [137–141] (Table 8.4). Upon irreversible oxidation (overoxidation) at sufficiently positive potentials the electrical conductivity is lost although the overoxidized layer still has ionic conductivity similarly to Nafion. Overoxidized PPY was tested for the oxidation of the neurotransmitters DA, NE, EP, SR, and L-DOPA and the interferents AA and UA [137, 139–141]. The oxidation peaks of AA and UA were found to be suppressed at the expense of an enhancement of the neurotransmitter oxidation peaks. *In vivo* measurements of DA with overoxidized PPY-coated microelectrodes were carried out as early as 1996 [139]. However in all these applications CPs are not used for their intrinsic electronic conductivity, but rather as a permselective membrane with surface charged groups (i.e., carboxylic [139, 142]) introduced as a result of overoxidation and/or specific hydrophobic interactions [140].

Over the last 10 years the investigations on electrooxidation of neurotransmitters involving CP-based materials have rapidly increased. Various strategies, i.e., chemical binding of the target species, electrostatic discrimination, involvement of metallic catalytic centers, combination with CNTs, etc., have been probed in order to achieve good sensitivity, selectivity, and stability of the electroanalytical response. Biosensing based on CP-immobilized enzymes involved in the biosynthetic pathways of the target analytes was also widely studied. In the following electroanalytical investigations employing CP-based electrodes involving also CPs in combination with carbon nanotubes, metal nanoparticles and specifically chosen

Table 8.4 Electroanalytical detection of dopamine

Conducting polymer, dopant, or composite component	Method, pH*	Linear range (μM)	LOD (μM)	Co-analytes (μM)	Interference studies	Ref.
P3MT	SWV	0.5–50	0.2	AA	AA: independent peak	[168]
γ-Cyclodextrine	0.1 M H$_2$SO$_4$					
PTHI (overoxidized)	CV	50–1,000		100–500	AA: no peak	[138]
Poly(4,4'-bis(butylsulfanyl)-2,2'-bithiophene)	DPV	50–10,000	1		AA: separate peak	[181]
POLY(2-amino-1,3,4-thiadiazole)	DPV pH 5	5–50	0.33	AA: 30–300 LOD: 2 UA: 10–100 LOD: 0.19 Xanthine: 10–100 LOD: 0.59		[74]
Polycarbazole	DPV CV	0.002–0.02 50–800			AA	[182]
Poly(1,2-phenylenediamine) (overoxidized)	SWV	0.05–10 (0.3 mM AA)	0.01 (0.002 no AA)	AA: 200–2,000, LOD: 80	AA: 0.3 mM	[160]
Poly(o-phenylenediamine)	DPV 0.1 M Na$_2$SO$_4$	10,000–50,000 (10 mM AA)			AA: 10 mM. no peak	[154]
Poly(o-phenylenediamine)	DPV	20		Serotonin: 20	AA (1 mM), UA (1 mM), no peaks	[155]
Poly(N,N'-dimethylaniline)	SWV	0.2–36 (0.2 mM AA)			AA, non-interf.	[183]
Poly(aniline-boronic acid)	Conductometric	1–100	10			[184]

(continued)

Table 8.4 (continued)

Conducting polymer, dopant, or composite component	Method, pH*	Linear range (µM)	LOD (µM)	Co-analytes (µM)	Interference studies	Ref.
Poly(aniline-boronic acid)	DPV	5–200			AA (0.5 mM) no peak	[157]
Poly(p-aminobenzenesulfonic acid)	DPV	0.1–1 1.0–10 10–100		AA: 1,000–4,000	AA, hipuric acid, citric acid, glucose, cysteine, etc.	[165]
PPY (overoxidized)	CV	1–100	0.8	Epinephrine: 1–100 LOD: 0.6	AA, UA: non-interf.	[140]
PPY (overoxidized)	DPV	2.5–500 µM	0.1		AA (0.1 mM)	[141]
Polyindole (overoxidized)	CV	100–500			AA	[153]
Poly(1-naphtylamine) (overoxidized)	SWV	1–100	0.16		AA (3 times) fully suppressed	[151]
PPY, ferrocyanide	LSV DPV pH 6	100–1,200 200–950	39 15	AA: 100–1,000 100–950		[185]
PPY	CA	8–40	3.2		AA	[159]
Sulfonated β-cyclodextrine PEDOT	SWV	1–30			AA (1 mM)	[186]
PEDOT	DPV	20–80	1.2	AA: 500–3,500 LOD: 7.4 UA: 20–130 LOD: 1.4	AA (up to 1,000 times), glucose, urea, acetaminophen, aspirin, creatinin: non-interf.; L-DOPA, NOREPI, EPI: interf.	[166]

PEDOT β-Cyclodextrine	CV SWV	200–1200 100–500		AA: 100–1,000 (CV) 300–1,500 (SWV)	[167]
PEDOT, Prussian blue	SWV	2–100	24.6	AA (1 mM)	[187]
PEDOT, Prussian blue array	CV	2–5 (no AA) 10–50 (1 mM AA)	2 4.3	AA (1 mM)	[188]
PEDOT-SDS	CV	0.5–25 30–100	0.061 0.086	AA UA	[160]
PEDOT-PANS	LSV pH 5	No data		AA, UA: good selectivity for DA	[144]
Polycarbazole poly(carbazole-co-p-tolylsulfonyl pyrrole)	CA	30–130 μM	0.27 0.5	AA (0.1–0.5 mM) suppressed	[189]
Poly(indole-6-carboxylic acid)/Nafion, tetracyanoquinodimethane	DPV	10–120		AA (1 mM)	[190]
PANI, MWNT, β-cyclodextrine	CV DPV	Log–log dependence 1–10		AA (0.8 mM) suppressed	[143]
P3MT, CNT, Nafion	DPV	0.02–0.1 0.1–1.0 1–6	0.005	AA: suppressed; UA: non-interf.	[158]
PPY (overoxidized), CNT	DPV	0.5–6.8 for PPY 0.04–1.4 for PPY/CNT	0.080 0.0017	AA (0.5 mM) no peak	[152]

(continued)

Table 8.4 (continued)

Conducting polymer, dopant, or composite component	Method, pH*	Linear range (μM)	LOD (μM)	Co-analytes (μM)	Interference studies	Ref.
PPY (overoxidized) SWNT	DPV	1–50	0.38	AA: 20–1,000 UA: 2–100		[161]
Poly(4-aminoiphenol) β-Cyclodextrine, AuNPs	DPV	15–50 recovery of spiked DA in blood serum at the 0.3–0.9 μM level		AA: 20–45 (at 12.5 DA)	UA (25 μM)	[191]
PPY (overoxidized) AuNPs	DPV	0.075–20	0.015	Serotonin: 0.007–2.2 LOD: 0.001	AA (up to 5 mM)	[192]
PEDOT, AuNPs	DPV, CA	0.002–0.022 2–20	0.002		AA (1 mM)	[39]
PEDOT, AuNPs	DPV	20–80		AA: 500–3,500 UA: 20–130		[162]
PEDOT, AuNPs	DPV	10–264 (no AuNPs) 13–268			AA: suppressed for negatively charged shell	[179]
P3MT, AuNPs	DPV	1–35	0.24	UA: 1–32 LOD: 0.17	AA (0.1 mM)	[171]
PANI, polycationic electrolyte, AuNPs	CV	50–1,000				[193]

Material	Method	Linear range	LOD	Interferents	Ref
PANI, AuNPs	DPV	7–600			[194]
PANI, AuNPs (LbL)	DPV	7–148 (UA: 176)	3	UA: 29–720 LOD: 20	No	[172]
PEDOT, PdNPs	DPV	0.5–1.0		UA: 7–12	AA	[174]
Polyfuran, PdNPs	DPV	0.5–100	0.048	AA: 50–1,000 paracetamol: 0.5–100	Glucose (1 mM), UA (1 mM)	[170]
Poly(N-methyl-pyrrole) PdNPs	DPV	0.1–10	0.012	AA: 50–1,000 LOD: 7 UA: 0.5–20 LOD: 0.027	Glucose Tryptophan Serotonin, tyramine, tyrosine, etc.	[145]
P3MT, PdNPs	DPV	0.05–1.00 (0.1 mM AA)	0.008	AA: 10–160 LOD: 7	AA, UA, glucose	[169]
PPY, CuNPs	DPV	0.001–0.1	0.00085	UA: 0.001–10 LOD: 0.0008	AA	[174]
PEDOT, Cu particles	DPV	6–200 (1 mM AA) or 0.006–0.060 0.3–2 (1 mM AA)	0.004	AA: 1,000–5,000	AA	[163, 164]
AgCl/PANI core-shell nanoparticles	SWV	0.2–3.5 (1 mM AA)			AA (1 mM): inhibited reaction	[195]

*The pH value is explicitly shown only if it differs from the pH interval 7.0–7.4

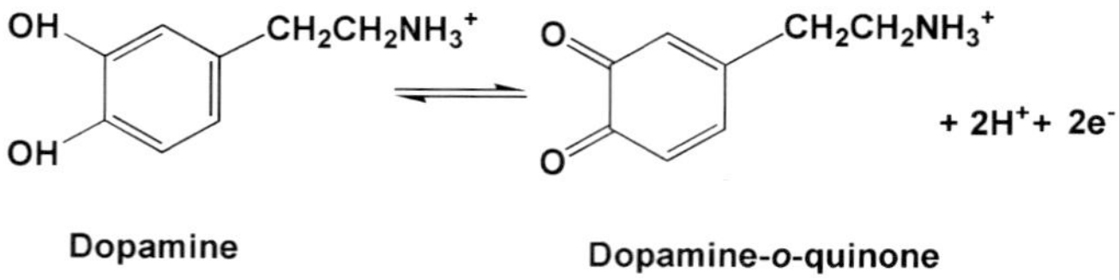

Fig. 8.9 Oxidation of dopamine

dopants are addressed. The emphasis is given to studies which are showing important effects in terms of increased selectivity and sensitivity and which are close to the conditions relevant for measurements in real specimens.

8.4.3 Electroanalytical Determination of Dopamine

DA is a monoamine catechol that undergoes a two electron oxidation resulting in the formation of the o-quinone form of DA (Fig. 8.9). The electrooxidation reaction is pH-dependent. According to voltammetric experiments up to pH 7 the peak potential dependence on pH is linear and the values of the slopes dE_p/dpH confirm a two e$^-$ per two H$^+$ process [143–145]. At higher pH values (about pH 9) a deviation from linearity is observed due to expected deprotonation of DA. The two pK_a values of DA are 8.9 and 10.6 [146]. Electroanalytical studies are usually carried out in PBS at pH 7.0–7.4, i.e., close to physiological conditions.

Ascorbic acid is one of the main interferents for the electroanalytical determination of DA. Moreover the oxidation of AA may become mediated by DA [147–149]. The main problem is that the oxidation product of DA, dopamine-o-quinone, catalytically reacts with AA and is reduced back to DA. Thus the same DA molecule may be involved several times in this process. Therefore, it seems difficult to achieve accurate determination of DA in the presence of ascorbic acid unless AA becomes oxidized well before the oxidation potential of DA is reached [150] or the AA oxidation reaction is fully suppressed.

A complete suppression of the AA oxidation peak was observed when using overoxidized CPs, e.g., PTHI [138], poly(1-naphtylamine) [151], PPY-CNT composite [152], and polyindole [153] as well as poly(o-phenylenediamine)-coated electrodes [154, 155, 158]. For the overoxidized CPs and poly(o-phenylenediamine) operating in neutral solutions, this effect is commented in terms of permselectivity of the cationic dopamine species through the non-electronically conducting polymer membranes. This process is supported by electrostatic repulsion between the surface carboxyl groups and the negatively charged ascorbate anions. Gradual suppression of the AA peak with increasing film thickness was observed for the permselective layers. The dependence of the DA oxidation peaks on film thickness showed a plateau-like curve with two falling side parts for the very thin (not enough pre-concentration) and very thick (not enough permeation) layers [158]. The uric acid oxidation peak was also suppressed in some of these cases [140, 154].

A complete suppression of the AA oxidation peak was also observed when using boronic acid-doped PANI codeposited with Nafion [157], Nafion-CNT-coated P3MT [158], PPY film doped with sulfonated-β-cyclodextrin [159], and PEDOT in the presence of SDS [160]. Nafion is used as a cationic exchange polymer attracting positively charged species. Boronic acid and possibly β-cyclodextrin interact with dopamine and build corresponding complexes whereas SDS and the sulfonated groups in the case of sulfonated-β-cyclodextrin play a repulsive role for the ascorbate ions and thus favor the selective oxidation of dopamine.

Apart from the possibility to eliminate the interfering oxidation signals, many investigations were directed to the opportunities for simultaneous determination of dopamine with other analytes. Ascorbic acid and uric acid are between the most studied species. In several cases [156, 161–164] AA determination was demonstrated in the presence of DA although the level of the DA peak depended on the amount of AA in the analyzed solution. Independently, linearly increasing oxidation peaks with the amounts of DA and AA were demonstrated in few studies involving poly(*p*-aminobenzenesulfonic acid) [165], PEDOT [166], PEDOT (synthesized in the presence of β-cyclodextrine) [167], P3MT (synthesized in the presence of β-cyclodextrine) [168], PdNPs-PMPPY [145], PdNPs-polyfuran [150], and PdNPs-P3MT [169]. The good selectivity and lack of interference of both peaks were attributed to electrostatic repulsion in the case of sulfonated polymers [165], to hydrophobic interactions of the cyclodextrines incorporated in the polymer layers [168], to hydrophobic interaction between the CP (PEDOT) and the aromatic part of DA [167], and finally to specific combinations of catalytic metal NPs and CP [150, 169, 170].

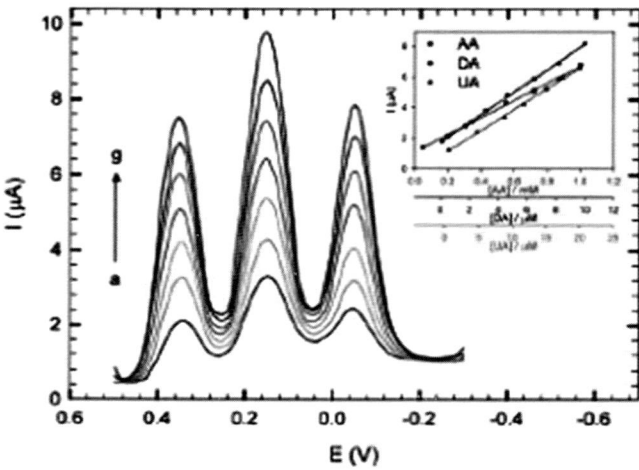

Fig. 8.10 Simultaneous determination of AA, DA, and UA using poly(*N*-methylpyrrole)/PdNPs-modified electrode [145]. (Reprinted from N. Atta, M.F. El-Kady, A. Galal, Simultaneous determination of catecholamines, uric acid, and ascorbic acid at physiological levels using poly(*N*-methylpyrrole)/Pd-nanoclusters sensor, Anal. Biochem. 400(2010) 78–88, With permission from Elsevier)

Simultaneous determination of DA and UA was achieved also in several studies employing overoxidized PPY codeposited with SWNTs [161], AuNPs-PEDOT [162], AuNPs-P3MT [171], AuNPs-PANI [172], PdNPs-PEDOT [173], PdNPs-PMPPY [145], and CuNPs-PPY [176]. Examples demonstrating the possibility for simultaneous determination of all three analytes (AA, DA, and UA) were published by Jeyalakshmi et al. [166] and Atta et al. [145] although better sensitivity was achieved in the latter case (Fig. 8.10).

Most of the materials suitable for simultaneous determination of AA and DA or UA and DA are composites combining metal (Au, Cu, and Pd) NPs and CPs. The role of the metallic catalytic centers is not always well identified although studies on corresponding bulk electrodes give evidence for their role. Gold single crystal electrodes provide the possibility to resolve AA and DA [175]. Copper-plated electrodes were found to be sensitive for DA due to the formation of Cu(II)-*o*-quinolate complexes whereas AA oxidation becomes strongly suppressed [176].

Metal (Au, Cu, and Pd) NPs immobilized on supporting, noncatalytical electrodes were explored for their electroanalytic perfor-

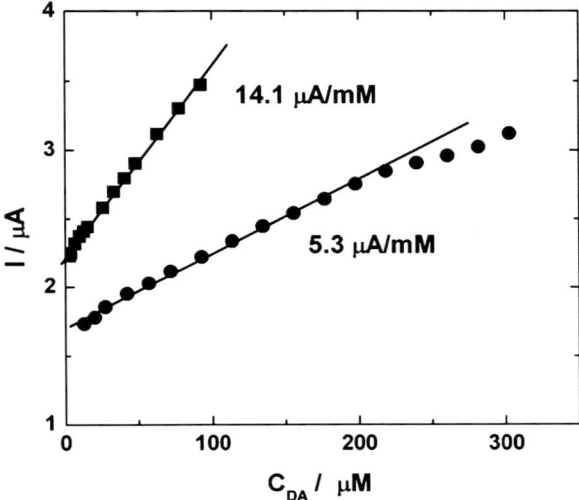

Fig. 8.11 Concentration dependence of DA peak currents obtained in DPV experiments at LbL-deposited PANI/AuNPs-modified electrodes with different amount of AuNPs [172]. (Reprinted from A. Stoyanova, S. Ivanov, V. Tsakova, A. Bund, Au nanoparticle-polyaniline nanocomposite layers obtained through layer-by-layer adsorption for the simultaneous determination of dopamine and uric acid, Electrochim. Acta 56 (2011) 3693–3699, with permission from Elsevier)

mance for DA determination in a number of cases. PdNPs deposited directly on glassy carbon show a good resolution of the DA and AA oxidation peaks [177]. On the other hand, Pd/carbon paste electrode did not show selectivity for the AA and DA oxidation peaks whereas PdNPs-loaded carbon nanofibers were found to resolve well DA, AA, and UA [178]. To compare different studies involving one and the same type of metal NPs is usually a difficult task because of the lack of information concerning the size, number, and distribution of the metal phase. In rare cases the role of the size, number, and loading of the metal particles as well as the type of their stabilization shell in the electroanalytical process is explored.

The role of the metal loading was demonstrated in the case of layer-by-layer (LbL) deposited composite layers consisting of PANI and AuNPs [172]. An increase in the metal particles loading by increasing the duration of the AuNPs adsorption step resulted in a threefold increase in the sensitivity for DA determination (Fig. 8.11).

AuNPs encapsulated by N-dodecyl-N,N-dimethyl-3-ammonium-1-propane-sulfonate (SB12) and tannic acid (Tann) were used to obtain AuNPs-PEDOT layers [179]. It was suggested that the SB12 species build a compact spherical shell around the metal core and a large fraction of the deprotonated sulfonic groups remains located at the outer surface of the NP. On the other hand the Tann molecules provide branched species with high density of negative charge and high permeability after encapsulating the AuNP core. The obtained composite materials were studied for their electroanalytic response with respect to DA, AA, and UA oxidation. The sensitivity for DA oxidation was found to depend significantly on the organic shell of the AuNPs, Tann-stabilized particles exhibiting larger oxidative currents. Moreover, strong suppression of the AA and UA oxidation peaks was observed for Tann-AuNPs. This was not the case for SB12-AuNPs. Although the range of linear response (Table 8.4) was not affected by the different types of AuNPs used to synthesize the PEDOT-composite layers, the sensitivity for DA determination varied by a factor of 2.6 in favor of the Tann-AuNPs-PEDOT material (Fig. 8.12). This result was commented in terms of the formation of a compact negatively charged layer around the SB12-AuNPS impeding the access of DA molecules to the AuNP core [180].

The role of the size and number of copper crystals deposited on PEDOT for the DA oxidation electroanalytical response was investigated by Stoyanova et al. [163, 164]. Copper crystals were deposited from $CuSO_4$ and CuC_2O_4 solutions providing the possibility to obtain different populations of crystals on the PEDOT surface (Fig. 8.13a, b). The electroanalytical performance of Cu-modified PEDOT with about the same metal loading but distributed in different way has shown a significant difference in both range of linear response and sensitivity (Fig. 8.13c, d). The large Cu crystals provided an extended range of linear response (6–200 μM) with a relatively low sensitivity (0.013 μA/μM). In these conditions the reaction was found to occur under diffusion control. On the other hand, the large population of much smaller crystals deposited on PEDOT resulted in a most sensitive material exhibiting two separate ranges of linear response (12–69 nM and 0.3–2.0 μM). The sensitivity in both ranges was 9 and 0.3 μA/μM, respectively [164]. The reaction was found to be surface-limited and adsorption was assumed to play a crucial role. In these studies the thickness of the polymer layer was found to affect both directly (through the PEDOT-based DA oxidation current) and indirectly

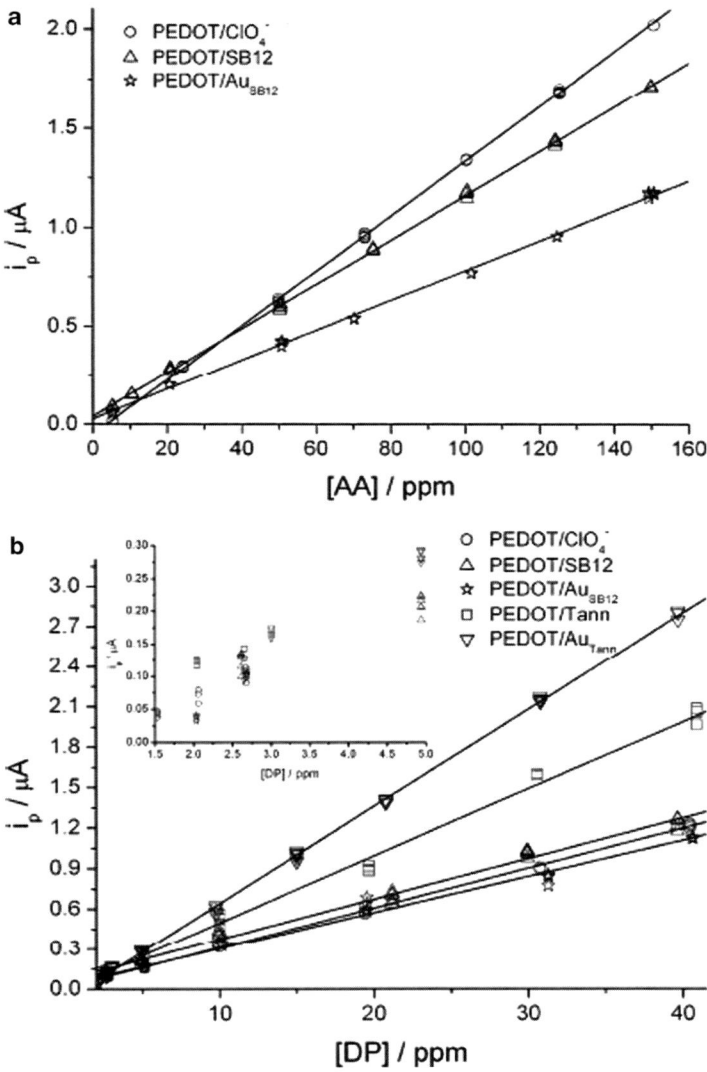

Fig. 8.12 Concentration dependence of AA and DA peak currents obtained at PEDOT and PEDOT/AuNPs electrodes with different stabilization shell of the AuNPs [180]. (Reprinted from C. Zanardi, F. Terzi, R. Seeber, Composite electrode coatings in amperometric sensors. Effect of differently encapsulated gold nanoparticles in poly(3,4-ethylenedioxythiophene) system, Sens. Actuators B 148 (2010) 277–282, with permission from Elsevier)

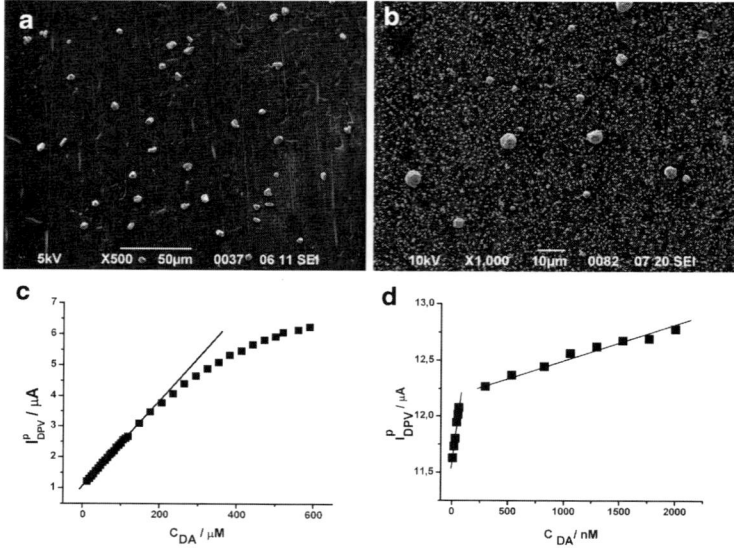

Fig. 8.13 Copper particles deposited from: (**a**) $CuSO_4$ solution at a thin PEDOT layer; (**b**) CuC_2O_4 solution at a thick PEDOT layer. Concentration dependence of the DA peak currents obtained in DPV experiments at Cu-modified PEDOT layers: (**c**) from the type shown in (**a**); (**d**) from the type shown in (**b**) [163, 164]. (Adapted from A. Stoyanova, V. Tsakova, Copper-modified poly(3,4-ethylenedioxythiophene) layers for selective determination of dopamine in the presence of ascorbic acid I. Role of the polymer layer thickness, J. Solid State Electrochem. 14 (2010) 1947–1955 and A. Stoyanova, V. Tsakova, Copper-modified poly(3,4-ethylenedioxythiophene) layers for selective determination of dopamine in the presence of ascorbic acid II. Role of the characteristics of the metal deposit, J. Solid State Electrochem. 14 (2010) 1957–1965. Reprinted with permission of Springer)

(through the type of the metal deposit obtained) the electroanalytical response. Similar studies demonstrate the possibility to obtain very different electroanalytical performance depending on the physical and chemical properties of both the polymer and metallic components of the composite CP-modified electrode and the necessity to look for further optimization in already known metal/polymer systems.

It is interesting to point out that in some cases two or three linear (or pseudo-linear) concentration regions were identified [164, 165]. This becomes evident when detailed measurements with a number of experimental points are carried out in different concentration intervals. In several cases seeming linear dependences are in fact pseudo-linear due to the lack of sufficient measurement points especially in the low concentration range.

8.4.4 Electroanalytical Determination of Other Neurotransmitters

The main challenge in the electroanalytical determination of *l-DOPA* is to obtain good selectivity against interfering substances, e.g., DA, epinephrine, or norepinephrine (Table 8.5). A polycarbazole electrode was first used to obtain a selective electrochemical response of L-DOPA oxidation in the presence of epinephrine [196]. The attempts to combine β-cyclodextrine with P3MT [197] and PPY [198, 199] and to use the obtained functionalized coatings for L-DOPA determination did not result in selectivity with respect to DA, epinephrine, and norepinephrine. In a recent study on a bi-layered PPY/MWCNT electrode [200] a good sensitivity for L-DOPA and selectivity with respect to AA was obtained. Measurements in synthetic human serum demonstrated good recovery. Nevertheless the direct interference of other catecholamines was not studied.

The electrooxidation of *epinephrine* and *norepinephrine* on CP-based electrodes was studied in few cases (Table 8.5) [198, 201–204]. An AuNPs/overoxidized PPY electrode was used for simultaneous electroanalytical determination of epinephrine and UA in the coexistence of large amount of AA [201]. A better sensitivity with respect to epinephrine (in the presence of UA and AA) was obtained with a SWCNT/AuNPs/PPY electrode [202]. A poly(2-amino-1,3,4-thiadiazole)-modified electrode was suggested for sensitive determination of norepinephrine also in the presence of AA and UA [203]. The interference of other catecholamines was not studied in these cases.

Overoxidized PPY-coated electrodes were used for electroanalytical studies on serotonin oxidation [192, 205] (Table 8.5). High sensitivity and selectivity with respect to DA was obtained by using AuNPs/overoxidized PPY [192].

Table 8.5 Electroanalytical detection of L-DOPA, epinephrine, norepinephrine, and serotonin

Conducting polymer, dopant, or composite component	Analyte	Method, pH*	Linear range (μM)	LOD (μM)	Co-analytes Linear range LOD (μM)	Interference studies	Ref.
Polycarbazole	L-DOPA	CV CA	1,000–80,000			EPI suppressed	[196]
P3MT, γ-cyclodextrine	L-DOPA	SWV 0.1 M H$_2$SO$_4$	2–50	1		DA peaks coincide	[197]
PPY, CNT	L-DOPA	DPV	1–100			AA suppressed	[200]
P3MT	EPI		No data				[201]
PPY, β-cyclodextrine	EPI		No data			DA, L-DOPA, NOREPI: no selectivity	[198]
PPY (overoxidized) AuNPs	EPI	DPV	0.3–20 (0.1 mM AA)	0.03	UA LR: 0.05–28 LOD: 0.012	AA	[204]
PPY, AuNPs, CNTs	EPI	DPV	0.004–0.1	0.002	AA (200) UA (400)		[202]
Poly(2-amino-1,3,4-thiadiazole)	NOREPI	CA pH 5	0.04–25	0.00017		AA (0.2 mM) UA (0.2 mM)	[203]
PPY (overoxidized) SDS or PSS	Serotonin	DPV	1–100	1			[205]
PPY (overoxidized) AuNPs	Serotonin	DPV	0.007–2.2	0.001	DA LR: 0.075–20 LOD: 0.015	AA (1,000-fold)	[192]

8.5 Various Analytes Including DNA, Foodborne Pathogens, and Drugs

8.5.1 DNA and Foodborne Pathogens

DNA analysis plays an important role in medical diagnostics and prevention. In contrast to other sophisticated methods for DNA analysis, electrochemical DNA sensing is one of the suitable and fast methods providing the possibility to transform the DNA hybridization effect into electrical signal. CPs have been widely used for the immobilization of DNA probes serving as sensing elements for the corresponding DNA targets. In analogy to the immobilization of enzymes various techniques such as electrochemical entrapment, covalent immobilization, or affinity interactions have been used. The specific applications of CPs in DNA sensing are addressed in the recent extensive review of by Peng et al. [206] with more than 80 cited references relevant for CP and CP-based materials for DNA analysis. Foodborne pathogen detection was investigated by immobilization of corresponding DNA or antibodies in CP materials, e.g., PANI and PPY [207].

8.5.2 Drugs

Paracetamol (*N*-(4-hydroxyphenyl)acetamide) (referred also as acetaminophen or *p*-acetamidophenol) is an antipyretic and an analgesic used for the treatment of fever of bacterial or viral origin and for the relief of headache, arthritis, and postoperative pains.

There are several investigations addressing the possibility for electroanalytical determination of paracetamol by CP-modified electrodes (Table 8.6). An extended linear concentration region of the voltammetric response was found by using a Cu-complex of poly(terthiophene-carboxylic acid) [208]. Recent investigations have demonstrated the possibility for simultaneous determination of paracetamol and one of its products, *p*-aminophenol, by PEDOT[209] and PANI/graphene-modified electrodes [210].

Morphine ((5α,6α)-7,8-didehydro-4,5-epoxy-17-methylmorphinan-3,6-diol), apart from being an opiate, is regarded in clinical medicine as a strong analgesic used to relieve severe pains. PEDOT was successfully applied for the electroanalytical determination of morphine in

Table 8.6 Electroanalytical detection of drugs

Conducting polymer, dopant, or composite component	Substance	Method, pH*	Linear range (μM)	LOD (μM)	Co-analytes	Interference studies	Ref.
Poly(terthiophene-carboxylic acid) Cu-complex	Paracetamol	CV	20–5,000	5		N-acetylcysteine, AA, UA, glucose, p-aminophenol, glutamic acid, methionine, salicylic acid, tyrosine	[208]
PPY, MIP paracetamol	Paracetamol	DPV	5–500 1,250–4,500	0.79		AA, DA, glucose, phenacetin, phenol	[211]
PANI, poly(flavin adenine dinucleotide), MWCNT	Paracetamol	CV, FIA	2–60			None	[212]
PEDOT	Paracetamol	DPV	1–100	0.4	p-Aminophenol 4–320 μM LOD: 1.2 μM		[208]
PANI, graphene	p-Aminophenol (degradation product of paracetamol)	DPV	0.2–20 20–100	0.065	Paracetamol 10–100 μM		[210]
PEDOT	Morphine	CV SWV	0.3–8 10–60	0.050 0.068		AA, UA	[213]
PEDOT, MIP morphine	Morphine	pH 5.3	100–1,000			Codeine	[214]
PEDOT, MIP	Morphine	CA pH 5.6	100–2,000			Codeine	[215]
P3MT, γ-cyclodextrine	Chlorpromazine (neuroleptic)	CV pH 9	2–8	0.1		AA (100 times)	[197]

urine samples [213]. A good discrimination against codeine was achieved by using PEDOT molecularly imprinted with morphine itself [214] or polymer particles imprinting [215].

Finally, electroanalytical detection of *chlopromazine* [197] and *p-aminophenol* (as a degradation product of paracetamol) was also investigated based on P3MT/cyclodextrin [197] and PANI/graphene [210] electrodes.

8.6 Outlook

Conducting polymer-modified electrodes are studied in a variety of applications concerning electroanalytical detection of metabolites, neurotransmitters, and drugs. In most cases CPs play a mediating role in the electrochemical reaction and provide the possibility to reduce the reaction overpotential and enhance the analyte electrooxidation currents. In the quest of highly selective and sensitive electrodes composite materials consisting of conducting polymers and polyanions, cyclodextrines, metal nanoparticles, carbon nanotubes, more recently graphene, etc., are widely studied. The composite components have a specific catalytic role, enhance the real reacting surface, and/or provide better selectivity. As a result in individual studies selective electrodes resolving the signals of several analytes occurring simultaneously in real specimens are obtained. Nevertheless in most cases, especially for complex composite materials, a number of materials characteristics could be further varied in order to obtain the optimal electroanalytical response. A deeper understanding of the role of factors such as thickness, morphology, electrical conductivity, and type of the doping ions of the CP, as well as amount and distribution of the composite component, size, and coordination shell of metal NPs, etc., is necessary for the optimization of these complex systems. The assessment of these factors together with the possibility to develop electrode arrays consisting of different CPs or CP-based composites may lead the way to complex electroanalytical analysis in blood, urine, or brain tissue.

In general, the efforts of the scientific community seem to be spread without a clearly identified aim in terms of combinations of analytes, relevant concentration ranges, and corresponding needs of the clinical analysis. A closer collaboration of materials scientists, analytical chemists, and medical specialists will undoubtedly result in the involvement of CP-based materials in the solution of specific practical electroanalytical problems emerging in clinical analysis and medical research.

Acknowledgments Financial support of project DTK 02/25 with National Science Fund of Bulgaria is gratefully acknowledged.

References

1. Shirakawa H, Louis EJ, Mac Diarmid AG, Chiang CK, Heeger AJ. Synthesis of electrically conducting organic polymers: halogen derivatives of polyacetylene, $(CH)_x$. J Chem Soc Chem Commun. 1977;16:578–80.
2. Norden B, Krutmeijer E. The Nobel Prize in Chemistry, 2000: conductive polymers (advanced information). Stockholm: Information Department, The Royal Swedish Academy of Sciences; 2000. p. 1–16
3. Skotheim TA, Reynolds JR, editors. Handbook of conducting polymers, 3rd edn. Boca Raton: CRC; 2007.
4. Inzelt G. Conductimg polymers. A new era in electrochemistry. Berlin: Springer; 2008.
5. Wallace G, Spinks GM, Kane-Maguire LAP, Teasdale PR. Conductive electroactive polymers: intelligent polymer systems. 3rd ed. Boca Raton: CRC; 2009.
6. Eftekhari A, editor. Nanostructured conductive polymers. Chichester: Wiley; 2010.
7. Cosnier S, Karyakin A, editors. Electropolymerization. Concepts, materials and applications. Weinheim: Wiley; 2010.
8. Elschner A, Kirchmeyer S, Loevenich W, Merker U, Reuter K. PEDOT, Principles and applications of an intrinsically conductive polymer. Boca Raton: CRC; 2011.
9. Dai L, Soundarrajan P, Kim T. Sensors and sensor arrays based on conjugated polymers and carbon nanotubes. Pure Appl Chem. 2002;74:1753–72.
10. Adhikari B, Majumdar S. Polymer in sensor applications. Prog Polym Sci. 2004;29:699–766.
11. Ramanavicius A, Ramanaviciene A, Malinauskas A. Electrochemical sensors based on conducting polymer—polypyrrole. Electrochim Acta. 2006;51:6025–37.
12. Lange U, Roznyatovskaya NV, Mirsky VM. Conducting polymers in chemical sensors and arrays. Anal Chim Acta. 2008;614:1–26.
13. Ashok Kumar S, Chen SM. Electroanalysis of NADH using conducting and redox active polymer/carbon nanotubes modified electrodes—a review. Sensors. 2008;8:739–66.
14. Ates M, Sarac AS. Conducting polymer coated carbon surfaces and biosensor applications. Prog Org Coat. 2009;66:337–58.
15. Rajesh T, Ahuja D. Recent progress in the development of nano-structured conducting polymers/nanocomposites for sensor applications. Sens Actuators B. 2009;136:275–86.

16. Bobacka J, Ivaska A. Chemical sensors based on conducting polymers. In: Cosnier S, Karyakin A, editors. Electropolymerization concepts, materials and applications. Weinheim: Wiley; 2010. p. 173–87.
17. Long YZ, Li MM, Gu C, Wan M, Duvail JL, Liu Z, et al. Recent advances in synthesis, physical properties and applications of conducting polymer nanotubes and nanofibers. Prog Polym Sci. 2011;36:1415–42.
18. Vashist SK, Zheng D, Al-Rubeaan K, Luong JHT, Sheu FS. Advances in carbon nanotube based electrochemical sensors for bioanalytical applications. Biotechnol Adv. 2011;29:169–88.
19. Gerard M, Chaubey A, Malhotra BD. Application of conducting polymers to biosensors. Biosens Bioelectron. 2002;17:345–59.
20. Malhotra BD, Chaubey A. Biosensors for clinical diagnostics industry. Sens Actuators B. 2003;91:117–27.
21. Vidal JC, Garcia-Ruiz E, Castillo JR. Recent advances in electropolymerized conducting polymers in amperometric biosensors. Microchim Acta. 2003;143:93–111.
22. Ahuja T, Mir IA, Kumar D, Rajesh. Biomolecular immobilization on conducting polymers for biosensing applications. Biomaterials. 2007;28:791–805.
23. Guimar NK, Gomez N, Schmidt CE. Conducting polymers in biomedical engineering. Prog Polym Sci. 2007;32:876–921.
24. Xia L, Wei Z, Wan M. Conducting polymer nanostructures and their application in biosensors. J Colloid Interface Sci. 2010;341:1–11.
25. Cosnier S, Holzinger M. Electrosynthesized polymers for biosensing. Chem Soc Rev. 2011;40:2146–56.
26. Mulchandani A, Myung NV. Conducting polymer nanowires-based label-free biosensors. Curr Opin Biotechnol. 2011;22:502–8.
27. Nambiar S, Yeow JTW. Conductive polymer-based sensor for biomedical applications. Biosens Bioelectron. 2011;26:1825–32.
28. Dhand C, Das M, Datta M, Malhotra BD. Recent advances in polyaniline based biosensors. Biosens Bioelectron. 2011;26:2811–21.
29. Cosnier S, Holzinger M. Bisensors based on electropolymerized films. In: Cosnier S, Karyakin A, editors. Electropolymerization. Concepts, materials, applications. Weinheim: Wiley; 2010. p. 189–213.
30. Bartlett PN, Cooper JM. A review on the immobilization of enzymes in electropolymerized films. J Electroanal Chem. 1993;362:1–12.
31. Cosnier S. Biomolecule immobilization on electrode surfaces by entrapment or attachment to electrochemically polymerized films. A review. Biosens Bioelectron. 1999;14:443–56.
32. Cosnier S. Affinity sensors based on electropolymerized films. Electroanalysis. 2005;17:1701–15.
33. Cosnier S. Recent advances in biological sensors based on electrogenerated polymers: a review. Anal Lett. 2007;40:1260–79.
34. Hyodo K, Nozaki M. High ion selective electrochemical synthesis of polyaniline. Electrochim Acta. 1988;33:165–6.
35. Karyakin AA, Maltsev IA, Lukachova LV. The influence of defects in polyaniline structure on its electroactivity: optimization of 'self-doped' polyaniline synthesis. J Electroanal Chem. 1996;402:217–9.
36. Tarver J, Yoo JE, Dennes TJ, Schwatz J, Loo YL. Polymer acid doped polyaniline is electrochemically stable beyond pH 9. Chem Mater. 2009;21:280–6.

37. Lyutov V, Tsakova V, Bund A. Microgravimetric study on the formation and redox behavior of poly(2-acrylamido-2-methyl-1-propanesulfonate)-doped thin polyaniline layers. Electrochim Acta. 2011;56:4803–11.
38. Bartlett PN, Wallace ENK. The oxidation of ascorbate at poly(aniline)-poly(vinylsulfonate) composite coated electrodes. Phys Chem Chem Phys. 2001;3:1491–6.
39. Senthilkumar S, Mathiyarasu J, Phani KLN. Exploration of synergism between a polymer matrix and gold nanoparticles for selective determination of dopamine. J Electroanal Chem. 2005;578:95–103.
40. Tsakova V. Metal-based composites of conducting polymers. In: Eftekhari A, editor. Nanostructured conductive polymers. New York: Wiley; 2010. p. 289–340.
41. Bissessur R. Inorganic-based nanocomposites of conductive polymers. In: Eftekhari A, editor. Nanostructured conductive polymers. Chichester: Wiley; 2010. p. 261–88.
42. Halliwell B. Drug antioxidant effects: a basis for drug selection. Drugs. 1991;42:569–605.
43. Harrison FE, May JM. Vitamin C function in the brain: vital role of the ascorbate transporter SVCT2. Free Radic Biol Med. 2009;46:719–30.
44. Malinauskas A, Garjonyte R, Mazeikiene R, Jureviciute I. Electrochemical response of ascorbic acid at conducting and electrogenerated polymer modified electrodes for electroanalytical applications: a review. Talanta. 2004;64:121–9.
45. Saraceno RA, Pack JG, Ewing AG. Catalysis of slow charge transfer reactions at polypyrrole-coated glassy carbon electrodes. J Electroanal Chem. 1986;197:265–78.
46. Mao H, Pickup P. Electronically conductive anion exchange polymers based on polypyrrole preparation, characterization and electrocatalysis of ascorbic acid oxidation. J Electroanal Chem. 1989;265:127–42.
47. Pournaghi-Azar MH, Ojani R. Electrochemistry and electroactivity of polypyrrole/ferrocyanide films on a glassy carbon electrode. J Solid State Electrochem. 2000;4:75–9.
48. Raoof JB, Ojani R, Rashid-Nadimi S. Preparation of polypyrrole/ferrocianide films modified carbon paste electrode and its application on the electrocatalytic determination of ascorbic acid. Electrochim Acta. 2004;49:271–80.
49. Lyons MEG, Breen W. Ascorbic acid oxidation at polypyrrole-coated electrodes. J Chem Soc Farad Trans. 1991;87:115–23.
50. Gao Z, Chen B, Zi M. Electrochemistry of ascorbic acid at polypyrrole/dodecyl sulphate film-coated electrodes and its application. J Electroanal Chem. 1994;365:197–205.
51. Komsiyska L, Tsakova V. Ascorbic acid oxidation at nonmodified and copper-modified polyaniline and poly-ortho-methoxyaniline coated electrodes. Electroanalysis. 2006;18:807–13.
52. Casella IG, Guascito MR. Electrocatalysis of ascorbic acid on the glassy carbon electrode chemically modified with polyaniline films. Electroanalysis. 1997;9:1381–6.
53. O'Connell P, Gormally C, Pravda M, Guilbault GG. Development of amperometric L-ascorbic acid (Vitamin C) sensor based on electropoly-

merised aniline for pharmaceutical and food analysis. Anal Chim Acta. 2001;431:239–47.
54. Ivanov S, Tsakova V, Mirsky VM. Conductometric transducing in electrocatalytical sensors: detection of ascorbic acid. Electrochem Commun. 2006;8:643–6.
55. Zhou DM, Xu JJ, Chen HY, Fang HQ. Ascorbate sensor based on "self-doped" polyaniline. Electroanalysis. 1997;9:1185–8.
56. Sun JJ, Zhou DM, Fang HQ, Chen HY. The electrochemical copolymerization of 3,4-dihydroxybenzoic acid and aniline at microdisk gold electrode and its amperometric determination for ascorbic acid. Talanta. 1998;45:851–6.
57. Zhang L, Dong S. The electrocatalytic oxidation of ascorbic acid on polyaniline film synthesized in the presence of camphorsulfonic acid. J Electroanal Chem. 2004;568:189–94.
58. Heras JY, Giacobone AFF, Battaglini F. Ascorbate amperometric determination using conducting copolymers from aniline and N-(3-propane sulfonic acid) aniline. Talanta. 2007;71:1684–9.
59. Ambrosi A, Morrin A, Smyth MR, Killard AJ. The application of conducting polymer nanoparticle electrodes to the sensing of ascorbic acid. Anal Chim Acta. 2011;609:37–43.
60. Zhang L, Lang Q, Shi Z. Electrochemical synthesis of three-dimensional polyaniline network on 3-aminobenzenesulfonic acid functionalized glassy carbon electrode and its application. Am J Anal Chem. 2010;1:102–12.
61. Kilmartin PA, Martinez A, Bartlett PN. Polyaniline-based microelectrodes for sensing ascorbic acid in beverages. Curr Appl Phys. 2008;8:320–3.
62. Bonastre AM, Bartlett PN. Electrodeposition of PANI films on platinum needle type microelectrodes. Application to the oxidation of ascorbate in human plasma. Anal Chim Acta. 2010;676:1–8.
63. Ge C, Armstrong NR, Saavedra SS. pH-sensing properties of poly(aniline) ultrathin films self-assembled on indium-tin oxide. Anal Chem. 2007;79:1401–10.
64. Lyutov V, Tsakova V. Silver particles-modified polysulfonic acid-doped polyaniline layers: electroless deposition of silver in slightly acidic and neutral solutions. J Solid State Electrochem. 2011;15:2553–61.
65. Lyutov V. Functionalised conducting polymer layers through incorporation of polyanions and metal particles. PhD Thesis, Institute of Physical Chemistry, Bulgarian Academy of Sciences, Sofia; 2013.
66. Xi L, Ren D, Luo J, Zhu Y. Electrochemical analysis of ascorbic acid using copper nanoparticles/polyaniline modified glassy carbon electrode. J Electroanal Chem. 2010;650:127–34.
67. Atta NF, Galal A, Karagoezler E, Russel GC, Zimmer H, Mark Jr HB. Electrochemical detection of some organic and biological molecules at conducting poly(3-methylthiophene) electrodes. Biosens Bioelectron. 1991;6:333–41.
68. Atta NF, Marawi I, Petticrew KL, Zimmer H, Mark HB, Galal A. Electrochemistry and detection of some organic and biological molecules at conducting polymer electrodes. Part3. Evidence of the electrocatalytic effect of the heteroatom of the poly(hetetroarylene) at the electrode/electrolyte interface. J Electroanal Chem. 1996;408:47–52.

69. Erdogdu G, Karagoezler E. Investigation and comparison of the electrochemical behavior of some organic and biological molecules at various conducting polymer electrodes. Talanta. 1997;44:2011–8.
70. Bello A, Giannetto M, Mori G, Seeber R, Terzi F, Zanardi C. Optimization of the DPV potential waveform for determination of ascorbic acid on PEDOT-modified electrodes. Sens Actuators B Chem. 2007;121:430–5.
71. Dutt JSN, Cardosi MF, Livingstone C, Davis J. Diagnostic implications of uric acid in electroanalytical measurements. Electroanalysis. 2005;17:1233–43.
72. Wang Y, Xu H, Zhang J, Li G. Electrochemical sensors for clinic analysis. Sensors. 2008;8:2043–81.
73. Kumar SS, Mathiyarasu J, Phani KLN, Jain YK, Yegnaraman V. Determination of uric acid in the presence of ascorbic acid using poly(3,4-ethylenedioxythiophene)-modified electrodes. Electroanalysis. 2005;17:2281–6.
74. Kalimuthu P, John SA. Simultaneous determination of ascorbic acid, dopamine, uric acid and xanthine using nanostructured polymer film modified electrode. Talanta. 2010;80:1686–91.
75. Dobay R, Harsanyi G, Visy C. Detection of uric acid with a new type of conducting-polymer based enzymatic sensor by bipotentiostatic technique. Anal Chim Acta. 1999;385:187–94.
76. Kumar SA, Cheng HW, Chen SM. Selective detection of uric acid in the presence of ascorbic acid and dopamine using polymerized luminol film modified glassy carbon electrode. Electroanalysis. 2009;21:2281–6.
77. Belenky P, Bogan KL, Brenner C. NAD+ metabolism on health and disease. Trends Biochem Sci. 2007;32:12–9.
78. Karyakin A. Electropolymerized azines: a new group of electroactive polymers. In: Cosnier S, Karyakin A, editors. Electropolymerization: concepts, materials and applications. Weinheim: Wiley; 2010. p. 93–110.
79. Bartlett PN, Simon E, Toh CS. Modified electrodes for NAD oxidation and dehydrogenase-based biosensors. Bioelectrochemistry. 2002;56:117–22.
80. Gorton L, Dominguez E. Electrochemistry of NAD(P)+/NAD(P)H, encyclopedia of electrochemistry. In: Wilson GS, editor. Bioelectrochemistry. Weinheim: Wiley-VCH; 2002.
81. Gorton L, Dominguez E. Electrocatalytic oxidation of NAD(P)/H at mediator-modified electrodes. Rev Mol Biotechnol. 2002;82:371–92.
82. Atta NF, Galal A, Karagoezler E, Zimmer H, Rubinson JF, Mark Jr HB. Voltammetric studies of the oxidation of reduced nicotinamide adenine dinucleotide at a conducting polymer electrode. Journal of Chemical Society. Chem Commun. 1990;19:1347–9.
83. Schuhmann W, Lammert R, Haemmerle M, Schmidt HL. Electrocatalytic properties of polypyrrole in amperometric electrodes. Biosens Bioelectron. 1991;6:689–97.
84. Jaraba P, Agui L, Yanez-Sedeno P, Pingarron JM. NADH amperometric sensor based on poly(3-methylthiophene)-coated cylindrical carbon fiber microelectrodes: application to the enzymatic determination of L-lactate. Electrochim Acta. 1998;43:3555–65.
85. Somasundrum M, Bannister JV. Mediatorless electrocatalysis at a conducting polymer electrode: application to ascorbate and NADH measurement. J Chem Soc Chem Commun. 1993;1993:1629–31.

86. Bartlett PN, Wallace ENK. The oxidation of β-nicotinamide adenine dinucleotide (NADH) at poly(aniline)-coated electrodes Part II. Kinetics of reaction at poly(aniline)-poly(styrenesulfonate)composites. J Electroanal Chem. 2000;486:23–31.
87. Bartlett PN, Simon E. Measurement of the kinetic isotope effect for the oxidation of NADH at a poly(aniline)-modified electrode. J Am Chem Soc. 2003;125:4014–5.
88. Toh CS, Bartlett PN, Mano N, Aussenac F, Kuhn A, Dufourc EJ. The effect of calcium ions on the electrocatalytic oxidation of NADH by poly(aniline)-poly(vinylsulfonate) and poly(aniline)-poly(sterenesulfonate) modified electrodes. Phys Chem Chem Phys. 2003;5:588–93.
89. Zhou DM, Fang HQ, Chen HY, Ju HX, Wang Y. The electrochemical polymerization of methylene green and its electrocatalysis for the oxidation of NADH. Anal Chim Acta. 1996;329:41–8.
90. Cai CX, Xue KH. Electrochemical polymerization of toluidine blue o and its electrocatalytic activity toward NADH oxidation. Talanta. 1998;47:1107–19.
91. Vasilescu A, Noguer T, Andreescu S, Calas-Blanchard C, Bala C, Marty JL. Strategies for developing NADH detectors based on Meldola Blue and screen-printed electrodes: a comparative study. Talanta. 2003;59:751–65.
92. Vasilescu A, Andreescu S, Bala C, et al. Screen-printed electrodes with electropolymerized Meldola Blue as versatile detectors in biosensors. Biosens Bioelectron. 2003;18:781–90.
93. Gao Q, Wang W, Ma Y, Yang X. Electrooxidative polymerization of phenotiazine derivatives on screen-printed carbon electrode and its application to determine NADH in flow injection system. Talanta. 2004;62:477–82.
94. Zeng J, Wei W, Wu L, Liu X, Li Y. Fabrication of poly(toluidine blue O)/carbon nanotubes composite nanowires and its stable low-potential detection of NADH. J Electroanal Chem. 2006;595:152–60.
95. Dilgin Y, Gorton L, Nisli G. Photoelectrocatalytic oxidation of NADH with electropolymerized toluidine blue O. Electroanalysis. 2007;19:293.
96. Chen Y, Yuan J, Tian C, Wang X. Flow injection analysis and voltammetric detection of NADH with a poly-toluidine blue modified electrode. Anal Sci. 2004;20:507–11.
97. Dai ZH, Liu FX, Lu GF, Bao JC. Electrocatalytic detection of NADH and ethanol at glassy carbon electrode modified with electropolymerized films from methylene green. J Solid State Electrochem. 2008;12:175–80.
98. Karyakin AA, Karyakina EE, Schuhmann W, Schmidt HL. Electropolymerized azines: Part II. In a search of the best electrocatalyst for NADH oxidation. Electroanalysis. 1999;11:553–7.
99. Chi Q, Dong S. Electrocatalytic oxidation and flow injection determination of reduced nicotinamide coenzyme at a glassy carbon electrode modified by a polymer thin film. Analyst. 1994;119:1063–6.
100. Kumar SA, Chen SM. Electrochemically polymerized composites of conducting poly(p-ABSA) and flavins (FAD, FMN, RF) films and their use as electrochemical sensors: a new potent electroanalysis of NADH and NAD^+. Sens Actuators B. 2007;123:964–77.
101. Karyakin AA, Bobrova OA, Karyakina EE. Electroreduction of NAD+ to enzymatically active NADH at poly(neutral red) modified electrodes. J Electroanal Chem. 1995;399:179–84.

102. Agui L, Pena-Farfal C, Yanez-Sedeno P, Pingarron JM. Poly-(3-methylthiophene)/carbon nanotubes hybrid composite-modified electrodes. Electrochim Acta. 2007;52:7946–52.
103. Tian SJ, Liu JY, Zhu T, Knoll W. Polyaniline doped with modified gold nanoparticles and its electrochemical properties in neutral aqueous solutions. Chem Commun. 2003;2003:2738–9.
104. Tian SJ, Liu J, Zhu T, Knoll W. Polyaniline/gold nanoparticle multilayer films: assembly, properties, and biological applications. Chem Mater. 2004;16:4103–8.
105. Manesh KM, Santhosh P, Gopalan AI, Lee KP. Electrocatalytic oxidation of NADH at gold nanoparticles loaded poly(3,4-ethylenedioxythiophene)-poly(sterene sulfonic acid) film modified electrode and integration of alcohol dehydrogenase for alcohol sensing. Talanta. 2008;75:1307–14.
106. Balamurugan A, Ho KC, Chen SM, Huang TY. Electrochemical sensing of NADH based on Meldola Blue immobilized silver nanoparticle-conducting polymer electrode. Colloids Surf A Physicochem Eng Asp. 2010;362:1–7.
107. Mao H, Li Y, Liu X, Zhang W, Wang C, El-Deyab SS, et al. The application of novel spindle-lile polypyrrole hollow nanocapsules containing Pt nanoparticles in electrocatalysis oxidation of nicotinamide adenine dinucleotide (NADH). J Colloid Interface Sci. 2011;356:757–62.
108. Lange U, Mirsky VM. Chemosensitive nanocomposite for conductometric detection of hydrazine and NADH. Electrochim Acta. 2011;56:3679–84.
109. Zhang M, Yamaguchi A, Morita K, Termae N. Electrochemical synthesis of Au/polyaniline-poly(4-sterenesulfonate) hybrid nanoarray for sensitive biosensor design. Electrochem Commun. 2008;10:1090–3.
110. Gao Q, Cui X, Yang F, Ma Y, Yang X. Preparation of poly(thionine) modified screen printed electrode and its application to determine NADH in flow injection analysis system. Biosens Bioelectron. 2003;19:277–82.
111. Malitetsta C, Palmisano F, Torsi L, Zambonin PG. Glucose fast-response amperometric sensor based on glucose oxidase immobilized in an electropolymerized poly(o-phenylenediamine) film. Anal Chem. 1990;62:2735–40.
112. Bartlett PN, Birkin PR, Palmisano F, De Benedetto G. A study on the direct electrochemical communication between horseradish peroxidase and a poly(aniline) modified electrode. J Chem Soc Farad Trans. 1996;92:3123–30.
113. Raffa D, Leung KT, Battaglini F. A microelectrochemical enzyme transistor based on an N-alkylated polyaniline and its application to determine hydrogen peroxide at neutral pH. Anal Chem. 2003;75:4983–7.
114. Gu Y, Chen CC. Eliminating the interference of oxygen for sensing hydrogen peroxide with polyaniline modified electrode. Sensors. 2008;8:8237–47.
115. Yang Y, Mu S. Determination of hydrogen peroxide using amperometric sensor of polyaniline doped with ferrocenesulfonic acid. Biosens Bioelectron. 2005;21:74–8.
116. Chen C, Sun C, Gao Y. Amperometric sensor for hydrogen peroxide based on poly(aniline-co-p-aminophenol). Electrochem Commun. 2009;11:450–3.
117. Wang Y, Huang J, Zhang C, Wei J, Zhou X. Determination of hydrogen peroxide in rainwater by using a polyaniline film and platinum particles co-modified carbon fiber microelectrode. Electroanalysis. 1998;10:776–8.
118. Li Y, Lenigk R, Wu X, Gruendig B, Dong S, Renneberg R. Investigation of oxygen- and hydrogen peroxide reduction on platinum particles dispersed on poly(o-phenylenediamine) film modified glassy carbon electrodes. Electroanalysis. 1998;10:671–6.

119. Kondratiev VV, Pogulaichenko NA, Tolstopjatova EG, Malev VV. Hydrogen peroxide electroreduction on composite PEDOT films with included gold nanoparticles. J Solid State Electrochem. 2011;15:2383–93.
120. Zhang T, Yuan R, Chai Y, Li W, Ling S. A novel nonenzymatic hydrogen peroxide sensor based on a polypyrrole nanowire-copper nanocomposite modified gold electrode. Sensors. 2008;8:5141–52.
121. Ojani R, Raoof JB, Norouzi B. Carbon paste electrode modified by cobalt ions dispersed into poly(N-methylaniline) preparing in the presence of SDS: application in electrocatalytic oxidation of hydrogen peroxide. J Solid State Electrochem. 2010;14:621–31.
122. Park S, Boo H, Chung TD. Electrochemical non-enzymatic glucose sensors. Anal Chim Acta. 2006;556:46–57.
123. Toghill KE, Compton RG. Electrochemical non-enzymatic glucose sensors: a perspective and an evaluation. Int J Electrochem Sci. 2010;5:1246–301.
124. Casella IG, Cataldi TRI, Guerrieri A, Desimoni E. Copper dispersed into polyaniline films as an amperometric sensor in alkaline solutions of amino acids and polyhydric compounds. Anal Chim Acta. 1996;335:217–25.
125. Farrel ST, Breslin CB. Oxidation and photo-induced oxidation of glucose at a polyaniline film modified by copper particles. Electrochim Acta. 2004;49:4497–503.
126. Stoyanova A, Tsakova V. Electrooxidation of glucose on copper-modified polyaniline layers in alkaline solution. Bulg Chem Commun. 2008;40:286–90.
127. Tsakova V. How to affect number, size, and location of metal particles deposited in conducting polymer layers. J Solid State Electrochem. 2008;12:1421–34.
128. Peng XY, Li W, Liu XX, Hua PJ. Electrodeposition of NiOx/PANI composite film and its catalytic properties towards electrooxidations of polyhydroxyl compounds. J Appl Polym Sci. 2007;105:2260–4.
129. Becerik I, Kadirgan F. Glucose sensitivity of platinum-based alloys incorporated in polypyrrole films at neutral media. Synth Met. 2001;124:379–84.
130. Terzi F, Zanfrognini B, Zanardi C, Pigani L, Seeber R. Poly(3,4-ethylenedioxythiophene)/Au-nanoparticles composite as electrode coating suitable for electrocatalytic oxidation. Electrochim Acta. 2011;56:3575–9.
131. Geetha S, Rao CRK, Vijayan M, Trivedi DC. Biosensing and drug delivery by polypyrrole. Anal Chim Acta. 2006;568:119–25.
132. Zhang X, Ju H, Wang J, editors. Electrochemical sensors, bisensors and their biomedical applications. Academic, Boston; 2008.
133. Lexikon R. Medizin. Muenchen: Urban & Schwarzenberg; 1987.
134. Bozzi Y, Borrelli E. Dopamine in neurotoxicity and neuroprotection: what do D2 receptors have to do with it? Trends Neurosci. 2006;29:167–74.
135. Perry M, Li Q, Kennedy RT. Review on recent advances in analytical techniques for the determination of neurotransmitters. Anal Chim Acta. 2009;653:1–22.
136. Mark HB, Atta N, Ma YL, Petticrew KL, Zimmer H, Shi Y, et al. The electrochemistry of neurotransmitters at conducting organic polymer electrodes: electrocatalysis and analytical applications. Bioelectrochem Bioelectron. 1995;38:229–45.
137. Gao Z, Zi M, Chen B. Permeability controllable overoxidised polypyrrole film modified glassy carbon electrodes. Anal Chim Acta. 1994;286:213–8.

138. Gao Z, Yap D, Zhang Y. Voltammetric determination of dopamine in a mixture of dopamine and ascorbic acid at a deactivated polythiophene film modified electrode. Anal Sci. 1998;14:1059–63.
139. Pihel K, Walker QD, Wightman RM. Overoxidized polypyrrole-coated carbon fiber microelectrodes for dopamine measurements with fast-scan cyclic voltammetry. Anal Chem. 1996;68:2084–9.
140. Kang TF, Shen GL, Yu RQ. Permselectivity of neurotransmitters at overoxidized polypyrrole-film-coated glassy carbon electrodes. Talanta. 1996;43:2007–13.
141. Zhang X, Ogorevc B, Tavcar G, Svegl IG. Over-oxidized polypyrrole-modified carbon fibre ultramicroelectrode with an integrated silver/silver chloride reference electrode for the selective voltammetric measurement of dopamine in extremely small sample volumes. Analyst. 1996;121:1817–22.
142. Palmisano F, Malitetsta C, Centonze D, Zambonin PG. Correlation between permselectivity and chemical structure of overoxidized polypyrrole membranes used in electroproduced enzyme biosensors. Anal Chem. 1995;67:2207–11.
143. Yin T, Wei W, Zeng J. Selective detection of dopamine in the presence of ascorbic acid by use of glassy-carbon electrodes modified with both polyaniline film and multi-walled carbon nanotubes with incorporated β-cyclodextrin. Anal Bioanal Chem. 2006;386:2087–94.
144. Balamurugan A, Chen SM. Poly(3,4-ethylenedioxythiophene-co-(5-amino-2-naphthalenesulfonic acid)) (PEDOT-PANS) film modified glassy carbon electrode for selective detection of dopamine in the presence of ascorbic acid and uric acid. Anal Chim Acta. 2007;596:92–8.
145. Atta N, El-Kady MF, Galal A. Simultaneous determination of catecholamines, uric acid and ascorbic acid at physiological levels using poly(N-methylpyrrole)/Pd-nanoclusters sensor. Anal Biochem. 2010;400:78–88.
146. Lide DR, editor. Handbook of chemistry and physics, 84th edn. Boca Raton: CRC; 2004.
147. Tse DCS, McCreey RL, Adams RN. Potential oxidative pathways of brain catecholamines. J Med Chem. 1976;19:37–40.
148. Justice Jr JB, Jaramillo A. Selectivity and kinetics of ctecholamines at modified carbon paste electrodes. J Electrochem Soc. 1984;131:106C.
149. Justice Jr JB. Introduction to in-vivo volatmmetry. In: Adams RN, Justice Jr JB, editors. Voltammetry in the neurosciences. Clifton: Humana; 1987. p. 3–101.
150. Atta NF, El-Kady MF, Galal A. Palladium nanoclusters-coated polyfuran as a novel sensor for catecholamine neurotransmitters and paracetamol. Sens Actuators B. 2009;141:566–74.
151. D'Eramo F, Sereno LE, Arevalo AH. Preparation, characterization and analytical applications of a new and novel electrically conducting polymer. Electroanalysis. 2006;18:1523–30.
152. Tu X, Xie Q, Jiang S, Yao S. Electrochemical quartz crystal impedance study on the overoxidation of polypyrrole-carbon nanotubes composite films for amperometric detection of dopamine. Biosens Bioelectron. 2007;22:2819–26.
153. Ghita M, Arrigan DWM. Dopamine voltammetry at overoxidised polyindole electrodes. Electrochim Acta. 2004;49:4743–51.

154. Ekinci E, Erdogdu G, Karagoezler E. Preparation, optimization and voltammetric characteristics of poly(o-phenylenediamine)film as a dopamine-selective polymeric membrane. J Appl Polym Sci. 2001;79:327–32.
155. Selvaraju T, Ramaraj R. Simultaneous determination of dopamine and serotonin in the presence of ascorbic acid and uric acid at poly(o-phenylenediamine) modified electrode. J Appl Electrochem. 2003;33:759–62.
156. Mo JW, Ogorevc B. Simultaneous measurement of dopamine and ascorbate at their physiological levels using voltammetric microprobe based on overoxidized poly(1,2-phenylenediamine)-coated carbon fiber. Anal Chem. 2001;73:1196–202.
157. Mathiyarasu J, Senthilkumar S, Phani KLN, Yegnaraman V. Selective detection of dopamine using functionalized polyaniline composite electrode. J Appl Electrochem. 2005;35:513–9.
158. Wang HS, Li TH, Jia WL, Xu HY. Highly selective and sensitive determination of dopamine using a Nafion/carbon nanotubes coated poly(3-methylthiophene) modified electrode. Biosens Bioelectron. 2006;22:664–9.
159. Harley CC, Rooney AD, Breslin CB. The selective determination of dopamine at a polypyrrole film doped with sulfonated β-cyclodextrins. Sens Actuators B. 2010;150:498–504.
160. Atta NF, Galal A, Ahmed RA. Poly(3,4-ethylene-dioxythiophene) electrode for the selective determination of dopamine in presence of sodium dodecyl sulfate. Bioelectrochemistry. 2011;80:132–41.
161. Li Y, Wang P, Wang L, Lin X. Overoxidized polypyrrole film directed single-walled carbon nanotubes immobilization on glassy carbon electrode and its sensing applications. Biosens Bioelectron. 2007;22:3120–5.
162. Mathiyarasu J, Senthilkumar S, Phani KLN, Yegnaraman V. PEDOT-Au nanocomposite film for electrochemical sensing. Mater Lett. 2008;62:571–3.
163. Stoyanova A, Tsakova V. Copper-modified poly(3,4-ethylenedioxythiophene) layers for selective determination of dopamine in the presence of ascorbic acid I. Role of the polymer layer thickness. J Solid State Electrochem. 2010;14:1947–55.
164. Stoyanova A, Tsakova V. Copper-modifiedpoly(3,4-ethylenedioxythiophene) layers for selective determination of dopamine in the presence of ascorbic acid II. Role of the characteristics of the metal deposit. J Solid State Electrochem. 2010;14:1957–65.
165. Jin G, Zhang Y, Cheng W. Poly(p-aminobenzene sulfonic acid)-modified glassy carbon electrode for simultaneous detection of dopamine and ascorbic acid. Sens Actuators B. 2005;107:528–34.
166. Jeyalakshmi SR, Senthilkumar S, Mathiyarasu J, Phani KLN, Yegnaraman V. Simultaneous determination of ascorbic acid, dopamine and uric acid using PEDOT polymer modified electrode. Indian J Chem. 2007;46A:957–61.
167. Vasantha VS, Chen SM. Electrocatalysis and simultaneous detection of dopamine and ascorbic acid using poly(3,4-ethylenedioxy) thiophene film modified electrodes. J Electroanal Chem. 2006;592:77–87.
168. Bouchta D, Izaoumen N, Zejli H, El Kaoutit M, Temsamani KL. Electroanalytical properties of a novel PPY/β-cyclodextrin coated electrode. Anal Lett. 2005;38:1019–36.
169. Atta NF, El-Kady MF. Novel poly(3-methylthiophene)/Pd, Pt nanoparticle sensor: synthesis, characterization and its application to the simultaneous

analysis of dopamine and ascorbic acid in biological fluids. Sens Actuators B. 2010;145:299–310.
170. Atta NF, El-Kady MF. Poly(3-methylthiophene)/palladium sub-micro-modified sensor electrode. Part II: voltammetric and EIS studies, and analysis of catecholamine neurotransmitters, ascorbic acid and acetaminophen. Talanta. 2009;79:639–47.
171. Huang X, Li Y, Wang P, Wang L. Sensitive determination of dopamine and uric acid by the use of a glassy carbon electrode modified with poly(3-methylthiophene)/gold nanoparticle composites. Anal Sci. 2008;24:1563–8.
172. Stoyanova A, Ivanov S, Tsakova V, Bund A. Au nanoparticle-polyaniline nanocomposite layers obtained through Layer -by-Layer adsorption for the simultaneous determination of dopamine and uric acid. Electrochim Acta. 2011;56:3693–9.
173. Harish S, Mathiyarasu J, Phani KLN, Yegnaraman V. PEDOT/Palladium composite material: synthesis, characterization and application to simultaneous determination of dopamine and uric acid. J Appl Electrochem. 2008;38:1583–8.
174. Ulubay S, Dursun Z. Cu nanoparticles incorporated polypyrrole modified GCE for sensitive simultaneous determination of dopamine and uric acid. Talanta. 2011;80:1461–6.
175. Dursun Z, Pelit L, Taniguchi I. Voltammetric determination of ascorbic acid and dopamine simultaneously at a single crystal Au(111) electrode. Turk J Chem. 2009;33:223–31.
176. Zen JM, Chung HH, Senthilkumar S. Selective detection of o-diphenols on copper-plated screen-printed electrodes. Anal Chem. 2002;74:1202–6.
177. Thiagarajan S, Yang RF, Chen SM. Palladium nanoparticles modified electrode for the selective detection of catecholamine neurotransmitters in presence of ascorbic acid. Bioelectrochemistry. 2009;75:163–9.
178. Huang J, Liu J, Hou H, You T. Simultaneous electrochemical determination of dopamine, uric acid and ascorbic acid using palladium nanoparticle-loaded carbon nanofibers modified electrode. Biosens Bioelectron. 2008;24:632–7.
179. Terzi F, Zanardi C, Martina V, Pigani L, Seeber R. Electrochemical, spectroscopic and microscopic characterisation of novel poly(3,4-ethylenedioxythiophene)/gold nanoparticles composite materials. J Electroanal Chem. 2008;619–620:75–82.
180. Zanardi C, Terzi F, Seeber R. Composite electrode coatings in amperometric sensors. Effect of differently encapsulated gold nanoparticles in poly(3,4-ethylenedioxythiophene) system. Sens Actuators B. 2010;148:277–82.
181. Lupu S, Parenti F, Pigani L, Seeber R, Zanardi C. Differential pulse techniques on modified conventional-size and microelectrodes. Electroactivity of poly[4,4′-bis(butylsulfanyl)-2,2′-bithiophene] coating towards dopamine and ascorbic acid oxidation. Electroanalysis. 2003;15:715–25.
182. Kawde RB, Santhanam KSV. An in vitro electrochemical sensing of dopamine in the presence of ascorbic acid. Bioelectrochem Bioenerg. 1995;38:405–9.
183. Roy PR, Okajima T, Ohsaka T. Simultaneous analysis of dopamine and ascorbic acid using poly(N, N-dimethylaniline)-modified electrodes. Bioelectrochemistry. 2003;59:11–9.
184. Fabre B, Taillebois L. Poly(aniline boronic acid) based conductimetric sensor of dopamine. Chem Commun. 2003;2003:2982–3.

185. Raoof JB, Ojani R, Rashid-Nadimi S. Voltammetric determination of ascorbic acid and dopamine in the same sample at the surface of a carbon paste electrode modified with polypyrrole/ferrocyanide films. Electrochim Acta. 2005;50:4694–8.
186. Senthilkumar S, Mathiyarasu J, Phani KLN. Simultaneous determination of dopamine and ascorbic acid on poly (3,4–ethylenedioxythiophene) modified glassy carbon electrode. J Solid State Electrochem. 2006;10:905–13.
187. Lupu S, Lete C, Marin M, Totir N, Balaure PC. Electrochemical sensors based on platinum electrodes modified with hybrid inorganic–organic coatings for determination of 4-nitrophenol and dopamine. Electrochim Acta. 2009;54:1932–8.
188. Lupu S, del Campo FJ, Munoz FX. Development of microelectrode arrays modified with inorganic–organic composite materials for dopamine electroanalysis. J Electroanal Chem. 2010;639:147–53.
189. Ates M, Castillo J, Sarac AS, Schuhmann W. Carbon fiber microelectrodes electrocoated with polycarbazole and poly(carbazole-co-p-tolylsulfonyl pyrrole) films for the detection of dopamine in presence of ascorbic acid. Microchim Acta. 2008;160:247–51.
190. Pandey PC, Chauhan DS, Singh V. Poly(indole-6-carboxylic acid) and tetracyanoquinodimethane-modified electrode for selective oxidation of dopamine. Electrochim Acta. 2009;54:2266–70.
191. Gopalan AI, Lee KP, Manesh KM, Santhosh P, Kim JH, Kang JS. Electrochemical determination of dopamine and ascorbic acid at a novel gold nanoparticles distributed poly(4-aminothiophenol) modified electrode. Talanta. 2007;71:1774–81.
192. Li J, Lin X. Simultaneous determination of dopamine and serotonin on gold nanocluster/overoxidized-polypyrrole composite modified glassy carbon electrode. Sens Actuators B. 2007;124:486–93.
193. Prakash S, Rao CRK, Vijayan M. Polyaniline-polyelectrolyte-gold(0) ternary nanocomposites: synthesis and electrochemical properties. Electrochim Acta. 2009;54:5919–27.
194. Tsakova V, Ivanov S, Lange U, Stoyanova A, Lyutov V, Mirsky VM. Electroanalytical applications of nanocomposites from conducting polymers and metallic nanoparticles prepared by layer-by-layer deposition. Pure Appl Chem. 2011;83:345–58.
195. Yan W, Feng X, Chen X, Li X, Zhu JJ. A selective dopamine biosensor based on AgCl@polyaniline core-shell nanocomposites. Bioelectrochemistry. 2008;72:21–7.
196. Kawde RB, Laxmeshwar NB, Santhanam KSV. A selective sensing of dopa in the presence of adrenaline. Sens Actuators B. 1995;23:35–9.
197. Bouchta D, Izaoumen N, Zejli H, Kaoutit ME, Temsamani KR. A novel electrochemical synthesis of poly-3-methylthiophene-γ-cyclodextrin film Application for the analysis of chlorpromazine and some neurotransmitters. Biosens Bioelectron. 2005;20:2228–35.
198. Izaoumen N, Bouchta D, Zejli H, El Kaoutit M, Stalcuup A, Temsamani K. Electrosynthesis and analytical performances of functionalized poly(pyrrole/β-cyclodextrin) films. Talanta. 2005;66:111–7.
199. Izaoumen N, Bouchta D, Zejli H, El Kaoutit M, Temsamani K. The electrochemical behavior of neurotransmitters at a poly(pyrrole-β-cyclodextrin) modified glassy carbon electrode. Anal Lett. 2005;38:1869–85.

200. Shahrokhian S, Asadian E. Electrochemoical determination of L-dopa in the presence of ascorbic acid on the surface of the glassy carbon elecrode modified by a bilayer of multi-walled carbon nanotube and poly-pyrrole doped with tiron. J Electroanal Chem. 2009;636:40–6.
201. Wang HS, Huang DQ, Liu RM. Study on the electrochemical behavior of epinephrine at a poly(3-methylthiophene)-modified glassy carbon electrode. J Electroanal Chem. 2004;570:83–90.
202. Lu X, Li Y, Du J, Zhou X, Xue Z, Liu X, et al. A novel nanocomposites sensor for epinephrine detection in the presence of uric acid and ascorbic acids. Electrochim Acta. 2011;56:7261–6. doi:10.1016/j.electacta.2011.06.056.
203. Kalimuthu P, John SA. Selective determination of norepinephrine in the presence of ascorbic and uric acids using an ultrathin polymer modified electrode. Electrochim Acta. 2011;56:2428–32.
204. Li J, Lin XQ. Electrodeposition of gold nanoparticles on overoxidized polypyrrole film modified glassy carbon electrode and its application for the simultaneous determination of epinephrine and uric acid under coexistence of ascorbic acid. Anal Chim Acta. 2007;596:222–30.
205. Shiigi H, Okamura K, Kijima D, Deore B, Sree U, Nagaoka T. An overoxidized polypyrrole/dodecylsulfonate micelle composite film for amperometric serotonin sensing. J Electrochem Soc. 2003;150:H119–23.
206. Peng H, Zhang L, Soeller C, Travas-Sejdic J. Conducting polymers for electrochemical DNA sensing. Biomaterials. 2009;30:2132–48.
207. Arshak K, Velusamy V, Korostynska O, Oliwa-Stasiak K, Adley C. Conducting polymers and their applications to biosensors: emphasizing on foodborne pathogen detection. IEEE Sens J. 2009;9:1942–51.
208. Boopathi M, Won MS, Shim YB. A sensor for acetaminophen in a blood medium using a Cu(II)-conducting polymer complex modified electrode. Anal Chim Acta. 2004;512:191–7.
209. Mehretie S, Admassie S, Hunde T, Tessema M, Solomon T. Simultaneous determination of N-acetyl-p-aminophenol and p-aminophenol with poly(3,4-ethylenedioxythiophene) modified glassy carbon electrode. Talanta. 2011;85:1376–82.
210. Fan Y, Liu JH, Yang CP, Yu M, Liu P. Graphene-polyaniline composite film modified electrode for voltammetric determination of 4-aminophenol. Sens Actuators B. 2011;157:669–74.
211. Oezcan L, Sahin Y. Determination of paracetamol based on electropolymerized-molecularly imprinted polypyrrole modified pencil graphite electrode. Sens Actuators B. 2007;127:362–9.
212. Li Y, Sankar Y. Polyaniline and poly(flavin adenine dinucleotide) doped multi-walled carbon nanotubes for p-acetamidophenol sensor. Talanta. 2009;79:486–92.
213. Atta N, Galal A, Rasha A. Direct and simple electrochemical determination of morphine at PEDOT modified Pt electrodes. Electroanalysis. 2011;23:737–46.
214. Yeh WM, Ho KC. Amperometric morphine sensing using a molecularly imprinted polymer-modified electrode. Anal Chim Acta. 2005;542:76–82.
215. Ho KC, Yeh WM, Tung TS, Liao JY. Amperometric detection of morphine based on poly(3,4-ethylenedioxythiophene) immobilized molecularly imprinted particles prepared by precipitation polymerization. Anal Chim Acta. 2005;542:90–6.

Chapter 9
Experimental and Theoretical Issues of Nanoplasmonics in Medicine

Daniel A. Travo, Ruby Huang, Taiwang Cheng, Chitra Rangan, Erden Ertorer, and Silvia Mittler

9.1 Introduction

Biosensors are comprised of two components: a biomolecule that binds highly specifically to an analyte and a transducer that converts the biomolecular recognition to signal. Gold nanoparticles (GNPs) exhibit unique optical properties, which make them suitable for bio-sensing applications. The high affinity of gold to thiol (–SH) groups (which is strong but reversible) allows functional group attachments to biomolecules. The attachment of any organic material to nanogold produces a large change in its optical scattering properties. The combination of these two properties has lead to the development of GNP biosensors.

Recently, GNPs have found tremendous use in biological assays, detection, labeling, and sensing. GNP-based methods have been applied to screening for hepatitis A, hepatitis B, HIV, Ebola, smallpox,

---

D.A. Travo • R. Huang • T. Cheng • C. Rangan (✉)
Department of Physics, University of Windsor, Windsor, ON, Canada
e-mail: rangan@uwindsor.ca

E. Ertorer
Biomedical Engineering Program, Western University, London, ON, Canada

Department of Physics and Astronomy, Western University, London, ON, Canada

S. Mittler (✉)
Department of Physics and Astronomy, Western University, London, ON, Canada
e-mail: smittler@uwo.ca

and anthrax without any false negatives or false positives [8]. A GNP's optical spectrum is a probe of the localized surface plasmon resonance (LSPR) phenomenon, a collective electronic excitation that is localized in spatial extent due to the small size of the nanoparticle compared to the wavelength. The optical resonance frequency of the GNPs is a function of size, shape, interparticle distance and surrounding medium [18, 24].

In modern sensing applications, nanoparticles are immobilized on a surface in order to present the maximum detection surface to the analyte. A surface platform also allows for multiplexing—a device that tests multiple analytes in the same sample, which opens doors to the possibility of screening multiple diseases simultaneously. This configuration is well known to researchers in the surface science community as surface quantum dots or supported thin-film islands.

Surface immobilized GNP platforms provide significant advantages over 3D or colloidal gold sensor platforms. In spite of the simplicity of production of colloidal gold, solution-based GNPs need a stabilizing agent to prevent aggregation. This is achieved by adding a citrate layer or a polymer coating to create a core-shell structure [10]. This introduction of additional surface chemistry complicates further functionalization. The same issues are problematic for substrate-immobilized colloidal GNP systems. Colloidal gold requires additional handling considerations for lab-on-a-chip applications. In contrast, surface immobilized particles do not need stabilization against aggregation. Their bare surface allows easy functionalization using thiol chemistry.

This chapter documents the experimental issues concerning the fabrication of surfaces upon which GNPs are immobilized, and the theoretical issues concerning the modeling of such structures. In particular, the challenges and advantages of Organometallic Chemical Vapor Deposition (OMCVD) are presented. We also compare and contrast various theoretical approaches of modeling the optical properties of this system.

9.2 Fabrication of Surface Immobilized GNP Platforms

Conventional fabrication methods to produce surface immobilized nanoparticles such as Focused Ion Beam (FIB) or Electron Beam Lithography not only require high investment for the setups, their operating costs are high too and they suffer from low speed and small coverage area. For research and development purposes these disadvantages can be neglected; however, for industrial or clinical applications where mass production is required, simple, cost-effective, and fast methods are required.

OMCVD is a well-known method to create metallic thin films on substrate surfaces and can also be used for fabricating surface immobilized GNPs. OMCVD grown NP are randomly distributed and chemically attached to the substrate [14, 20, 44]. Although it is a statistical process, it yields narrow GNP size distributions which makes the surfaces suitable for biosensor applications. The hemispherical shape of the GNPs provides more stability [4]. The fabrication process is simple and inexpensive: it involves simple beaker chemistry and is performed at low temperatures (~65 °C). In this section, we will discuss OMCVD grown GNPs for plasmonic sensor applications with all the aspects; from preparation of the substrate surface for OMCVD process to biosensor application.

9.2.1 Substrate Selection

There are two main aspects in substrate selection: chemical and optical properties. Chemical composition and surface chemistry have a huge impact on the quality of the final product: particle size distribution, homogeneity, and stability of the particles. Silicon wafers are a good candidate material; however, transmission-based LSPR sensors require transparency at the range of the LSPR wavelength. This wavelength depends on the size of the GNPs and corresponds typically in the visible to near infrared region of the spectrum. On the other hand, if the sensor is based on an optical waveguide in an optical lab-on-a-chip, the substrate should be suitable for waveguide fabrication procedures, e.g., ion exchange. BK7 (Schott, Germany) is a high quality glass, exhibiting more than 90% transmission between 350 nm and 2,500 nm [37]. Ion exchange and surface functionalization properties are well studied [4, 40, 42].

9.2.2 Surface Preparation with Silane Self-Assembled Monolayers

In contrary to physical vapor deposition (PVD), CVD and OMCVD are selective processes [14]. Selectivity is provided by the surface chemistry; chemical interaction between the substrate and the material to be deposited play a major role. Substrate surfaces need to be adapted and therefore modified to provide chemisorption of the target atoms. The stability of the NPs depends on the chemical with which they bind to the substrate surface. Quality of the surface functionalization is critical. For plasmonic sensor applications, physisorbed particles would be able to move on the surface due to a lack of proper bonding and will aggregate. This aggregation is irreversible and provides false sensor signal due to this undesired clustering effect (see Fig. 9.5b, c). Furthermore, losing NPs during the sensing process due to detachment will decrease the signal and the signal-to-noise ratio. Therefore the substrate surface must be functionalized in such a way that it is ensured to obtain particles that are chemically attached to the surface. Gold has a strong affinity to thiols [15] and amines [6]. Therefore surface functionalization can be carried out employing thiol or amine groups in self-assembly processes. Mercapto-propyl-trimethoxy-silane (MPTMS) is a compound to silanize glass surfaces for a functionalization with thiol (mercapto) groups [41] (Fig. 9.1e). Hexamethyldisilazane (HMDS) is commonly used for adhesion promoter of photoresists for lithography applications [31], with an amine head group (Fig. 9.1f). Surface functionalization to promote GNP growth via OMCVD can be done by either of these methods.

Figure 9.1 illustrates the steps towards silanized surfaces and the necessary silanes. For the silanization the cleaned glass surface (Fig. 9.1a) should be oxidized since the silane groups of MPTMS and HMDS attach to oxidized surfaces (–OH) to form a silane network on the substrate (Fig. 9.1c, d). A piranha solution procedure is a common method to oxidize the substrate surfaces. Two hours of immersion in a freshly prepared 3:1 sulfuric acid:hydrogen peroxide solution forms hydroxyl groups on the substrate surface (Fig. 9.1b). (Caution: Piranha solution is dangerous, extremely corrosive, and violently reacts with organic materials. All safety measures should be taken to handle it). Oxygen plasma treatment is another way to oxidize glass substrate surfaces. Twenty minutes of treatment cleans

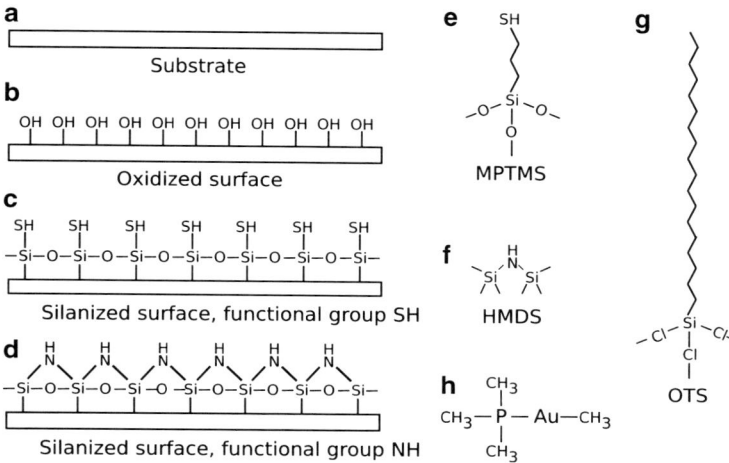

Fig. 9.1 (**a**) BK7 substrate, (**b**) oxidized surface, (**c**) thiol functionalized surface, (**d**) amine functionalized surface, (**e**) MPTMS, (**f**) HMDS, (**g**) OTS, (**h**) gold precursor ($[(CH_3)_3P]AuCH_3$)

the surface and functionalized the surface with hydroxyl groups. Surface modification due to the piranha or plasma treatment can be verified by contact angle measurements. After the oxidation treatment the glass surface turns hydrophilic and the contact angle of water drops below 4°.

After piranha treatment, substrates should be rinsed with abundant amounts of deionized water. Nitrogen-dried samples are placed in a vacuum oven at 95 °C for dehydration for an hour. The removing of possible moisture on the substrates is necessary as moisture causes uncontrolled polymerization of MPTMS and reduces the quality of self-assembled monolayer. For MPTMS silanization samples are immersed in 1:100 (V/V) MPTMS and ethanol (anhydrous) solution under argon environment in a glove box overnight. After rinsing and drying, baking in a vacuum oven at 95 °C for 20 min establishes the silane network on the surface.

HMDS functionalization is carried out in an oven (YES-3TA HDMS Oven, Yield Engineering, CA, USA). Because of the automatization of the HDMS process, more stable and reproducible samples can be achieved with the HMDS and oxygen plasma treatment combination.

9.3 OMCVD Process

OMCVD is a common method to deposit metallic thin film layers. If the interaction energy between the deposited metal atoms is higher than the metal–substrate interaction energy, island growth occurs [31]. This principle can be implemented to produce metallic NP instead of thin films [14]. Conventional CVD setups are usually dynamic which means they have a mass transport system to carry the precursor. The precursor is the chemical compound which contains the metal to be deposited. Previous studies show that for GNPs growth dynamic reactors do not show any significant advantages over static ones [3]. Additionally, it requires complex modules such as a gas flow handling system and consumes higher amounts of precursor. The static setup is simple; the reactor consists only of a vacuum chamber with a valve. The volatile precursor and the samples are placed on the bottom of the reactor (Fig. 9.2). Heating the evacuated reactor evaporates the precursor and heats the samples for the OMCVD process.

GNPs could attach to the reactor surface. That means less gold for the substrates. Silanization of the chamber with octadecyltrichlorosi-

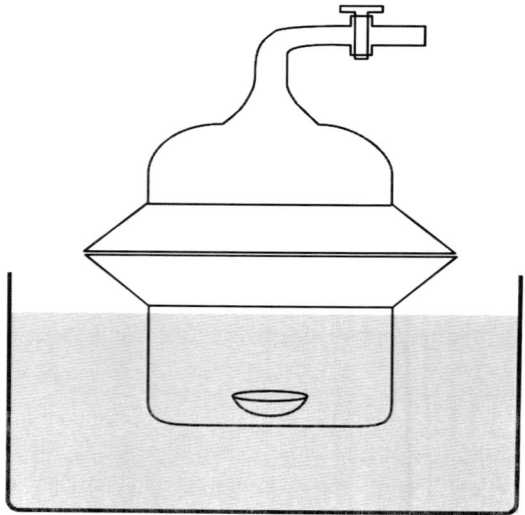

Fig. 9.2 Reactor in a water bath. A watch glass containing the precursor is placed on the reactor bottom with the samples

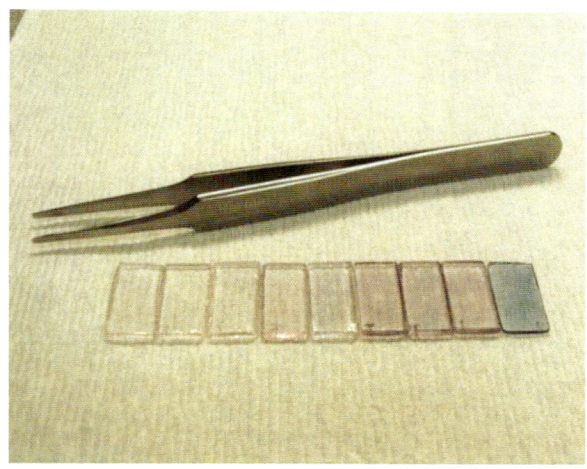

Fig. 9.3 Samples with OMCVD grown gold nanoparticles

lane (OTS) would avoid physisorption of the GNPs on the reactor chamber.

Trimethylphosphinegoldmethyl ($[(CH_3)_3P]AuCH_3$) is an organometallic precursor (Fig. 9.1h) which is known to deliver pure thin films of gold at relatively low temperatures [16]. Working at low temperature makes this precursor compatible with organic compounds such as self-assembled monolayers of MPTMS or HMDS. $[(CH_3)_3P]AuCH_3$ has a very small vapor pressure and evaporates at temperatures as low as room temperature. When the precursor evaporates and touches the functionalized surface, the precursor molecule breaks up and releases the gold atom from the molecule. The gold atom attaches to the functionalized substrate surface or another gold atom previously deposited forming GNPs. The phosphine group and the methyl stay in the vapor phase.

Using an oven as a heat source for the reactor increases the evaporation speed of the precursor. However, this method yields high inhomogeneities; the sample to sample differences are large and the GNP distribution along an individual sample is inhomogeneous. This nonuniformity can be seen qualitatively by the naked eye due to the color the GNPs produce (Fig. 9.3), but can also be detected quantitatively by UV-Vis absorption spectroscopy (Fig. 9.4a). Since there is no external gas flow into the chamber,

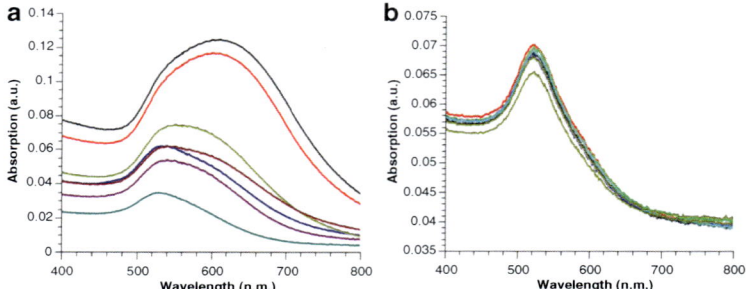

Fig. 9.4 Absorption spectra of two different GNP batches; GNP growth performed (**a**) in an uneven bottom reactor and an oven as the heating source, and (**b**) in a flat bottom reactor in a water bath

mass transport is provided by internal convection. Heating the entire chamber does not create a uniform convection. Placing the chamber partially in a water bath provides a uniform temperature gradient increasing the homogeneity along the sample surface as well as between the samples and increases reproducibility significantly. On the other hand, surface temperature of the samples affects the reactivity, therefore reactor surface roughness adversely affect the quality of the samples and the batch. Flat bottom reactors, allowing the flat samples to have an optimum heat contact to the reactor wall, deliver the best sample homogeneity. Figure 9.4 shows UV-Vis absorption spectra of two different batches; one was oven heated in a rough bottom chamber (Fig. 9.4a) and the other one was fabricated in a flat bottom chamber in the water bath (Fig. 9.4b). OMCVD parameters were 65 °C reaction temperature and 0.050 mbar initial reactor pressure, 20 mg precursor, 17 min reaction time.

Depending on the surface chemistry and the cleanness of the inner reactor walls, undesired physisorbed GNPs are formed on the inner reactor surface. This consumption of the precursor on the reactor surfaces yields fewer nanoparticles on the samples. In order to avoid that, the reactor surface should be non-growth surfaces for the precursor. Non-growth surfaces are iter alia $-CH_3$ functionalized surfaces. These surfaces can be fabricated via silanization with OTS[26]. Following a piranha procedure, the reactor is immersed in 1:500 (v/v) OTS:toluene solution in an argon environment in a glove box overnight. After rinsing the reactor with toluene, a treatment in a vacuum oven at 95 °C for 20 min evaporates the toluene and forms the silane network on the inner surface of the reactor.

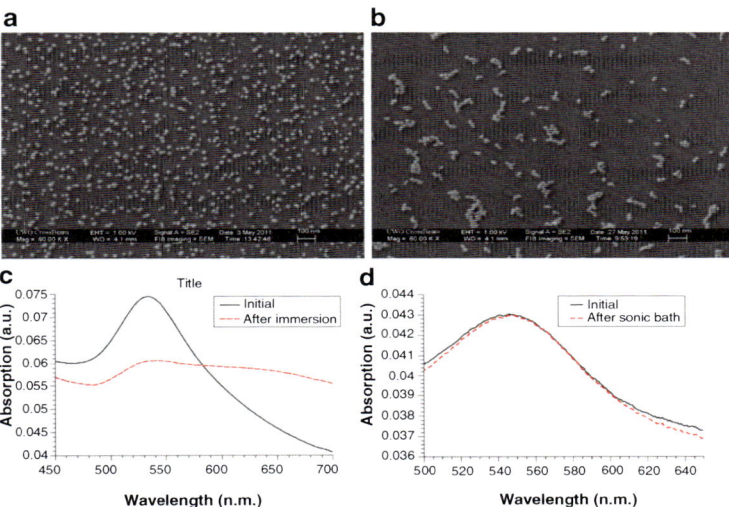

Fig. 9.5 SEM images of a sample with too many physisorbed GNPs; (**a**) before processing, (**b**) after immersing in ethanol and subsequent drying. (**c**) UV-Vis spectrum of the same sample before and after processing. (**d**) UV-Vis spectra of a sample with chemically bonded GNPs, before any treatment and after an immersion/sonication treatment

Due to the vapor process, both sides of the samples are coated. However, the bottoms of the samples (the sides that are in direct contact with the reactor wall) show less homogeneity and more physisorbed particles. This side is gently wiped with a tissue (Kimwipe) to remove the undesired GNPs. The success of all individual steps of the sample preparation procedure allows the formation of the chemical bonds between the substrate and the GNPs. If the quality of the self-assembly layer is poor, the number of physisorbed NP is high. Since they are not strongly attached to the substrate surface they can aggregate or detach from the surface while using the sample in a solution. Figure 9.5a, b shows a sample with too many physisorbed GNPs before and after immersion in ethanol and subsequent drying. Aggregation is clearly observed in the SEM image (Fig. 9.5b) as well as a decrease in the amount of GNPs. This can be observed in the UV-Vis spectra of the samples (Fig. 9.5c). Lost particles cause a decrease in the absorption signal and due to the aggregation a crosstalk shoulder appears. Therefore, the LSPR peak gets significantly wider diminishing the high sensitivity of the method. On the other hand, the treatment of the samples in an ultrasonic bath in ethanol not only removes the physisorbed particles, but also provides a simple quality test. UV-Vis spectra of samples should not change significantly after 5 min of sonication. Figure 9.5d shows absorption spectra of a sample with chemically bonded GNPs. After 5 min of sonication in ethanol and subsequent drying the absorption spectrum does not change significantly. There is only a small amount of physisorbed particles removed.

9.3.1 Characterization

With scanning electron microscopy (SEM) the particles are imaged (Fig. 9.6a). Although glass is an insulator, the MPTMS layer in combination with the GNPs allows imaging at 1 kV gun potential. Samples with HMDS layer required a thin conductive layer for imaging with SEM, such as 1 nm of osmium.

UV-Vis absorption spectroscopy delivers some information the characteristic properties of the nanoparticles (Fig. 9.6b) since the spectra are related to the size, shape, interparticle distance, and surrounding medium of the GNPs [18, 47].

Image processing methods are employed to obtain size and interparticle distance distribution. Eight images at 100k magnification are collected from all over the sample. To yield the size information, the freeware image processing software ImageJ [1] converts the SEM image (Fig. 9.7a) into a black and white mask image (Fig. 9.7b). ImageJ then calculates the area of the particles from the mask image. Areas are converted to diameters assuming a semi-spherical particle and the diameters are displayed in histograms (Fig. 9.7c). The findmax function in ImageJ is used to find the coordinates of all particles in an image. A small program written in ImageJ calculates from these coordinates information of all particles, e.g. the center-to-center nearest neighbor distance between GNP pairs. Results are displayed in histograms as well (Fig. 9.7d).

9.3.2 Sensing

Plasmonic sensors are sensitive to the refractive index changes in their vicinity. Figure 9.8 shows a bulk refractive index experiment.

9 Experimental and Theoretical Issues of Nanoplasmonics in Medicine 353

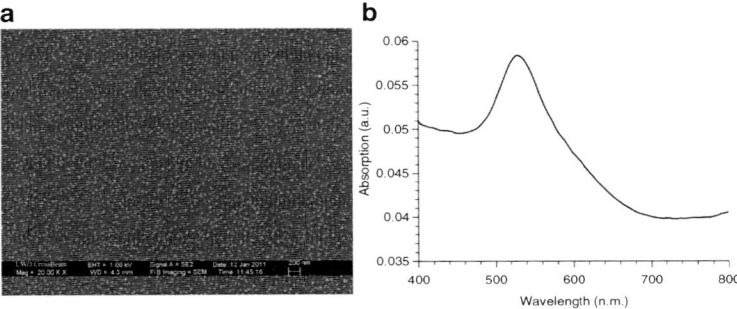

Fig. 9.6 (a) SEM image and (b) UV-Vis absorption spectrum of a sample with only chemically bonded GNPs

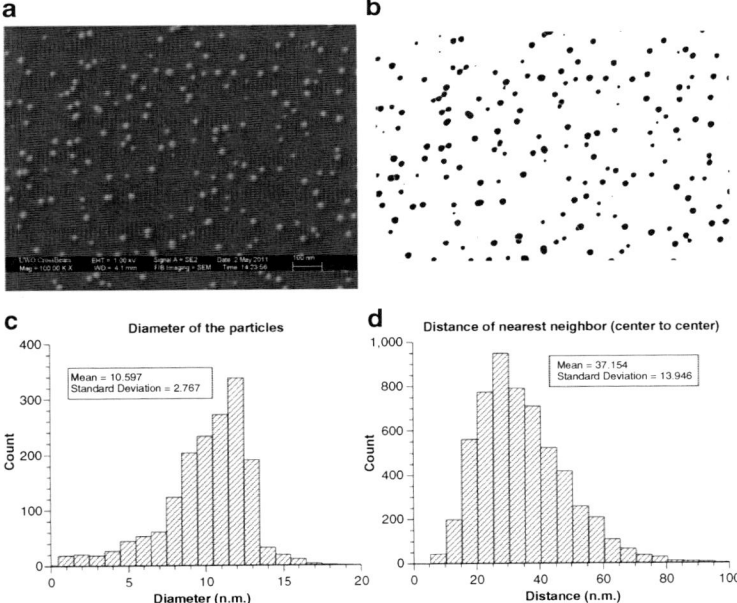

Fig. 9.7 (a) SEM image and (b) *black* and *white* mask image of a sample. (c) Size and (d) center-to-center nearest neighbor interparticle distance histograms of the nanoparticles in the sample

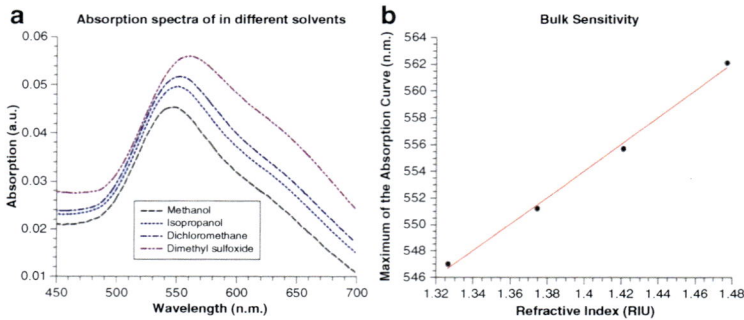

Fig. 9.8 (a) Absorption spectra of a sample immersed in various solvents with systematically increasing refractive index. (b) Wavelength of the maximum in the absorption curve versus the refractive index of the solvents

Immersing the sample in solutions with different refractive indices changes the LSPR frequency (Fig. 9.8a). An increasing refractive index causes a red shift in the maximum of the absorption curve (Fig. 9.8a). If the average interparticle distance is low enough for optical cross-talk, an increasing refractive index also increases the efficiency of the cross-talk. The cross-talk peak appears as a shoulder in the absorption spectra [35, 36].

9.3.3 Protein Sensing

Bulk sensing gives an idea about the sensing capabilities of the sensor platform. However, surface sensing would be more realistic for the practical applications. Biosensors are comprised of two components: a biomolecule that exhibits highly specific binding to an analyte (recognition), and a transducer that converts the biomolecular recognition to signal. The high affinity of gold to thiol (–SH) groups (which is strong but reversible) allows functional group attachments to biomolecules. The attachment of any organic material to nanogold produces a large change in its optical scattering properties. The combination of these two has lead to the development of GNP biosensors.

Biotin–streptavidin binding is one of the most common ways to evaluate protein sensing capabilities of a sensors. Streptavidin is a large protein (50,000 Dalton) which has four binding sites for biotin.

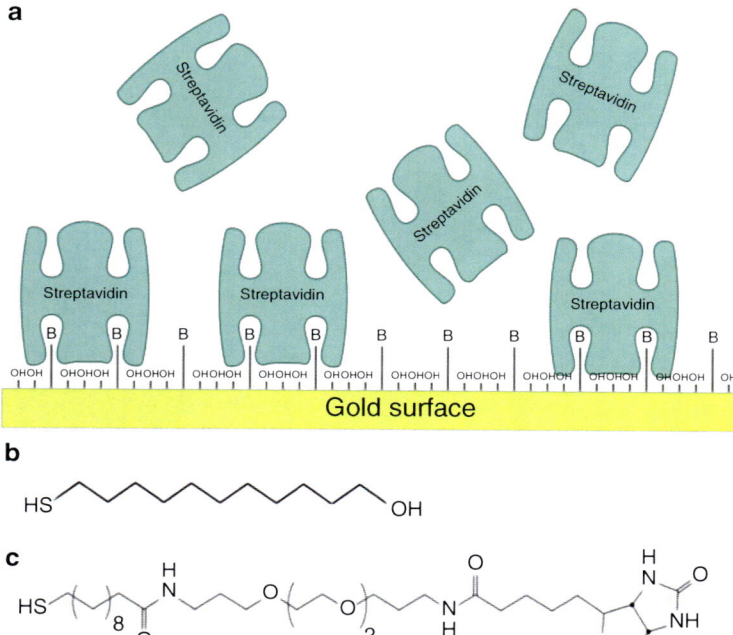

Fig. 9.9 (**a**) Illustration of streptavidin binding. (**b**) –OH terminated thiol and (**c**) biotinilated thiol

Biotin has the strongest biologically known affinity (K_a 10^{13} M^{-1}). Figure 9.9 shows the binding mechanism schematically. Streptavidin binds to a properly biotinilated surface selectively and with this above-mentioned ultra-high affinity. In all cases, some streptavidin molecules might bind only physically, which is termed unspecific binding and is undesirable.

First the surface of the GNP sample must be modified by a mixture of a hydroxyl terminated thiol (Fig. 9.9b) and a biotinilated thiol (Fig. 9.9c). Because of the geometrical structure of the streptavidin molecule, it is necessary to offer biotin on a spacer above the surface and to dilute the biotin on the surface to avoid steric hindrance during the binding process. The dilution is achieved by hydroxyl terminated thiol [38]. Samples were immersed in a mixture of 0.45 M hydroxyl terminated thiol ethanol solution and 0.05 M biotinilated thiol for self assembly for 45 min. Washing with copious amounts of ethanol

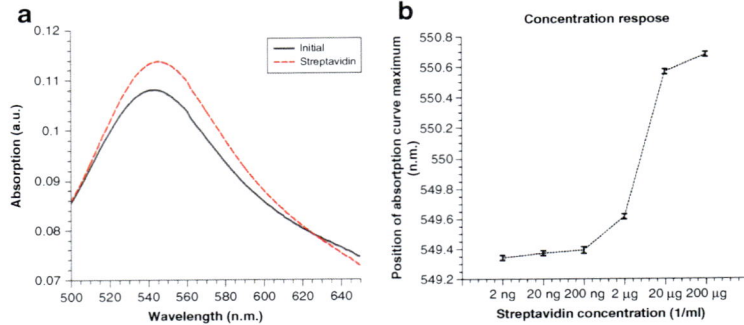

Fig. 9.10 (a) Absorption spectra of before and after streptavidin binding, (b) streptavidin concentration versus the wavelength position of the absorption maximum

removes the unbounded compounds from the surface. Biotin-modified samples were immersed in various concentrations of streptavidin solution in PBS buffer. After each immersion samples were rinsed with buffer solution. Streptavidin binding causes a red shift in the absorption spectrum (Fig. 9.10a).

Increasing the concentration causes increasing red shifts until a sensing saturation is reached. Figure 9.10b shows the streptavidin concentration response of the GNP sensor.

9.4 Modeling of Surface-Immobilized GNP Biosensors

The motivation for the theoretical study is to develop GNP biosensors of increasing sensitivity by varying the structure of the GNPs, the geometry of the arrangement, and ambient conditions. The optical response of GNPs is analyzed by calculating the LSPR spectrum in the presence of substrates, analytes, etc. The theoretical problem is the determination of the extinction spectra of an incident electromagnetic field upon interaction with a noble-metal nanoparticle and ambient structures.

For GNPs with diameters greater than 10 nm, we can ignore quantum effects and the problem reduces to the solution of Maxwells equations subject to the boundary conditions of a metal nanoparticle an object smaller than the wavelength of the incident field. The dielectric function of gold can be modeled by the Drude model or the exact values from data tables can be used. The methods we present include frequency-domain methods such as the Discrete-Dipole Approximation, Generalized Mie theory and GranFilm, and a popular time-domain method: the Finite-Difference Time-Domain (FDTD) approximation method.

Both gold and silver have small absorption losses in the optical regime. The dielectric function of gold has a negative real part and a small imaginary part in the visible range of the spectrum. For most metals, the absorption peak wavelength and width depend both on the real and imaginary part of the dielectric function. However, for noble metals, the LSPR peak wavelength depends only on the real part of the dielectric function and not on the imaginary part. The width of the resonance peak depends only on the imaginary part of the dielectric function and not on the real part [25].

Increasing the index of refraction of the dielectric environment both red-shifts the LSPR peak and increases its width. The change in LSPR peak wavelength with respect to ambient refractive index is proportional to the ratio of Re[$\varepsilon*$] and the slope of Re[$\varepsilon*$] at the plasmon wavelength [30]. For well-separated NPs, there is linear relation between the peak wavelength and the ambient refractive index.

When two or more nanoparticles are in proximity to each other, interactions between the induced multipoles within each particle become increasingly important. The LSPR peaks differ for a single nanoparticle and those having multiple neighbors, which exhibit collective dipole resonances. There are two types of electromagnetic interactions that take place between the nanostructures: near-field coupling and far-field dipolar interactions. When the elements are placed $d < \lambda$ apart, static dipolar interactions dominates with $1/d^3$ dependence on nanoparticle separation. For elements that are placed at a distance $d \simeq \lambda$ in the far-field, radiative dipolar coupling dominates with a $1/d$ dependence on nanoparticle separation. In this chapter, we only examine near-field coupling.

The polarization of the incident light that cause the induced multipoles also becomes important. Consider two spherical particles and the incident light propagating perpendicular to the interparticle axis. The peak shift when the electric field vector is parallel to the interparticle axis (i.e., longitudinal or s-polarized) is different from the peak shift when the electric field vector is perpendicular to the interparticle axis (i.e., transverse or p-polarized). An explanation for this effect is provided in [32, 35]. When the electric field vector is parallel to the interparticle axis (i.e., longitudinal or s-polarized), the induced local field between the two particles is in the same direction as the applied field as seen in Fig. 1.4a of [35]. The Coulomb forces subtract giving rise to red-shifted LSPR peaks. For cases where the electric vector is perpendicular to the interparticle axis (i.e., transverse or p-polarized), the induced local field between the two particles is in the opposite direction as the applied field as seen in Fig. 1.4b of [35]. The near-field electromagnetic fields interact with a nonbonding type of interaction, and the Coulomb forces add resulting in blue-shifted LSPR peaks. Calculations by Rooney et al. have shown that the effect of interparticle proximity can be achieved by coating with a dielectric material, an effect called optical clustering [36].

In the results presented, we give the extinction spectra of the various nanoparticle configurations. The extinction cross-section describes the effect of the interaction of the scattered field and particle with the incident field. Assume an incident beam of radiation is incident upon on isolated particle. The rate at which energy is lost due to the presence of the particle as the sum of the energy absorbed and scattered $W_{ext} = W_a + W_s$. The cross-sections will be defined as the fraction of the rate of energy deposition to the incident irradiance, I_i:

$$C_{ext} = \frac{W_{ext}}{I_i},$$

which will have dimensions of area. Similar expressions exist for C_a and C_s and accordingly $C_{ext} = C_a + C_s$. To avoid the note keeping of units we will work in terms of the efficiency factors defined to be unitless according to:

$$Q_{ext} = \frac{C_{ext}}{A},$$

where A is the cross-sectional area of the particle.

9.4.1 Macroscopic Theories

For a single spheroidal nanoparticle with dimension much smaller than wavelength of light, the absorption spectrum can be calculated to experimental accuracy using the well-known Mie theory [7, 29]. The Mie scattering method calculates the optical properties of single spherical particles of any radius using classical electrodynamics.

The incident light sets up the localized surface plasmon oscillation, and the induced potential is to a good approximation, a dipole. The spectrum of the reradiated light is calculated, and this has a peak whose wavelength depends on factors such as the nanostructure's size, shape, dielectric properties of the material, surrounding environment, and the incident field's polarization [47]. This analytical method provides an exact solution to Maxwell's equations for an electromagnetic wave scattered from a spherical particle, and is therefore used to benchmark numerical and approximation methods.

There are several approximate analytic methods for modeling point dipoles and their interactions. Some of them are: the electrostatic approximation method replaces Maxwell's equations with LaPlace equations and is valid for particles smaller than 10 nm [18]. The modified long-wavelength approximation is a perturbative correction of the electrostatic treatment and is relatively accurate for particles that are as large as 10% of the wavelength of the driving field. In the single-dipole approximation (SDA) method, each particle is treated as a single dipole scatterer [11]. The SDA method is accurate for small (dimension<<wavelength) and well-separated spherical particles. The coupled dipole approximation (CDA) is a method to include interparticle interactions. Each particle is treated as a single dipole, and two particles interact via dipole–dipole interactions [17]. All these methods work best for very small particles that are fairly well separated.

Other methods involve homogenization and these are sometimes called effective medium theories. Examples are Maxwell–Garnett [28] and the Marton–Schlesinger [27] methods. These well-known methods homogenize the solutions to the Maxwell equations in the presence of three-dimensional inhomogeneous media. Recently, Cheng, Rangan, and Sipe have developed an analytic theory for the homogenization of Maxwell's equations in media that are inhomogeneous only along two dimensions [9]. These methods also make the assumption that the nanoparticles are effectively point dipoles, although some corrections for finite size can be made.

To model the experimental situation of interest, OMCVD deposited nanoparticles on a substrate, a more accurate method is found to be required. We examine four methods: the Discrete Dipole Approximation (DDA), the Bedeaux–Vlieger method (GranFilm), the Generalized Multiparticle Mie (GMM), and FDTD in order to model this experimental geometry. In all methods, we use the dielectric function of gold from data tables. All three methods are benchmarked against Mie theory for a single spherical GNP of radius 20 nm.

9.4.2 GMM Theory

The extrapolation of Mie Theory to the multiparticle system is referred to as generalized multiparticle (sometimes referred to as multipole) Mie Theory, or GMM. While several variations of the theory currently exist, we use and evaluate the approach of Xu [46]. In this method, "partial scattering coefficients" account for the scattering of particle i due to the field from particle j.

The number of terms used in the harmonic expansions of the scattering coefficients is determined by Wiscombe's criterion [43] which poses an a priori method of estimating the number of terms to sufficiently model the system. The criterion relates the number of terms N to the size parameter of a single particle, $x = \pi a / \lambda$ where a is the radius of the particle and λ is the wavelength of incident radiation. It states that:

$$N = \begin{cases} x + 4x^{1/3} + 1 & 0.02 \leq x \leq 8 \\ x + 4.05x^{1/3} + 2 & 8 < x < 4,200 \\ x + 4x^{1/3} + 2 & 4,200 \leq x \leq 20,000 \end{cases} \quad (9.1)$$

For the calculations presented here, spectral convergence was achieved with $N = 7$.

9.4.3 Discrete Dipole Approximation

One numerical method that is suitable for the study of small clusters ($N = 2-10$) of nanoparticles (10–30 nm) is the well-known DDA. Developed by Draine and Flatau [12, 13] for modeling atmospheric phenomena, the DDA relies on the approximation of a continuous material by a discretized cubic grid of N point dipoles. One of the limitations of the method is the faithful representation of target surfaces. This problem could be circumvented by increasing dipole density in high-curvature surface regions, but this means giving up the use of the Fast Fourier transform algorithm that requires equally spaced grid points. We are interested in the extinction cross-section of the particles, or the sum of the absorption and scattering cross-sections.

The validity criterion for the DDA is the long-wavelength approximation: $|m|kd < 1$ where m is the complex refractive index, k is the wave number and d the grid spacing. We choose the grid spacing to be small enough so as to satisfy this criterion.

A second advantage of this method is that we can plot contour maps of the evanescent field around a metal nanoparticle. This gives us the ability to extract optoelectrical information at an extremely small surface. A limitation of this method is that computational resources (memory and time) place a limit on the size of the nanoparticles and/or the number of particles in a cluster.

9.4.4 Optical Properties of Nanoparticles on a Surface

The Bedeaux–Vlieger [5] method is a homogenization method that has provided quantitative calculations of the optical properties of nanoparticles on a surface. The GranFilm program developed by Lazzari and Simonsen [22] is designed to investigate the optical properties of granular thin films. The island polarizabilities are computed by solving the LaPlace equation in the quasi-static limit. This is accomplished through a multipole expansion with the presence of the surface taken into account through the method of images. The island polarizabilities are determined as the first order coefficients in the expansion. The presence of and interaction between islands are treated through a modification of the polarizability.

This method accounts for the multipolar coupling interaction between particle and substrate through the fulfillment of the boundary conditions, and the coupling effects between particles up to quadrupolar order. The advantage of this method is that single-particle-substrate interactions can be treated easily and accurately. The disadvantage is that for closely spaced particles the present implementation does not accurately account for higher multipolar interactions. Another limitation of these methods is that the electric field itself cannot be mapped, and more complex structures (such as nanoparticles made of concentric shells of materials) cannot be modeled. Note that later modifications of this method [23] have made it possible to visualize the multipolar potential yielding more physical insights into this problem.

A nice feature of GranFilm is that in addition to the square lattice arrangement (with a lattice constant input), the GranFilm program also has the capability to model random arrangements of particles through a mean field theory approximation. In the latter case, fraction of coverage is determined according to:

$$C = \frac{4\pi r^2}{(4r+d)^2}, \qquad (9.2)$$

where r denotes the particle radius and d the mean interparticle spacing.

9.4.5 The FDTD Method

The FDTD method is a time-domain numerical approach to model light-matter interactions based on electrodynamics calculations [39]. Maxwell's equations are solved by discretizing them with central-difference approximations that are accurate to the second-order for both space and time derivatives. The resulting finite-difference equations are solved numerically. By using Fourier transforms, the FDTD method can also be used to calculate quantities as a function of frequency such as normalized transmission and far-field projection. The FDTD software package used in this study is developed by Lumerical Solutions™. Electrodynamics calculations are solved based on Yee-cell mesh grids and a leap-frog update approach [45].

Since the E- and H-fields are computed at all points within the simulation volume at each time step, animated displays of the system's time-dependent electromagnetic response can be created. This technique is particularly valuable for applications that require a broad range of wavelengths for analysis such as interactions with short pulses. Some other advantages include the possibility of simulating infinitely long chains/arrays of nanoparticles (using periodic boundary conditions), producing contour plot visualizations of the electric and magnetic fields, integrating scripts compatible with Matlab, and running parallel computation for more sophisticated post processing techniques.

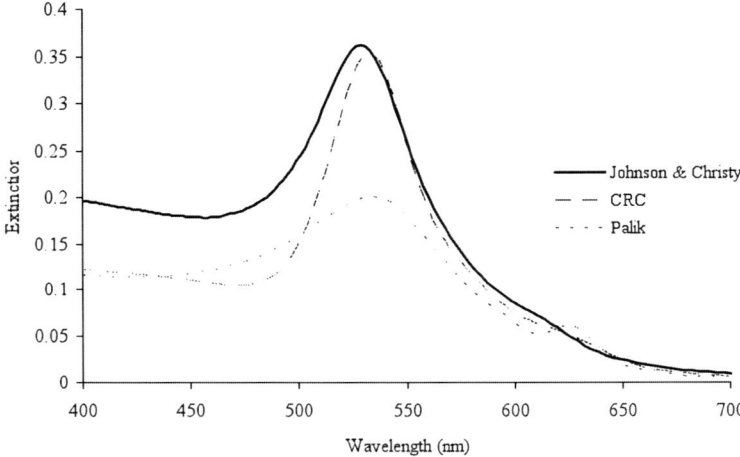

Fig. 9.11 Spectra of spherical gold nanoparticle of radius 20 nm calculated using optical constants from different data tables

A known issue of FDTD methods is that the decrease in the mesh size leads to convergence of the peak wavelength, but not of the spectrum itself. Thus, the peak wavelength of the LSPR wavelength was monitored for convergence testing. The FDTD method was calibrated by performing convergence tests for different Au particle sizes. It was found that the FDTD method yields accurate LSPR spectral peak values for particles with radius greater than 20 nm as compared with Mie theory.

9.4.6 Sensitivity to Gold Dielectric Constant Data

An unexpected finding of the FDTD calculations was the sensitivity of the spectra to the gold dielectric constant data input. Optical constants for gold were obtained from Palik [34], CRC optical data tables [33], and Johnson & Christy's optical tables [19]. Converged spectra using all three input data show LSPR peak positions at 529 nm, 533 nm, and 533 nm for Johnson & Christy, CRC, and Palik, respectively (see Fig. 9.11). The extinction represents the summation of absorption and scattering where the percentage of power transmitted is recorded by the field monitors.

We also note that the Lumerical's implementation of the FDTD program fits the input optical constant data to a polynomial, and all three programs are fit by quite different polynomials as seen in Fig. 9.12.

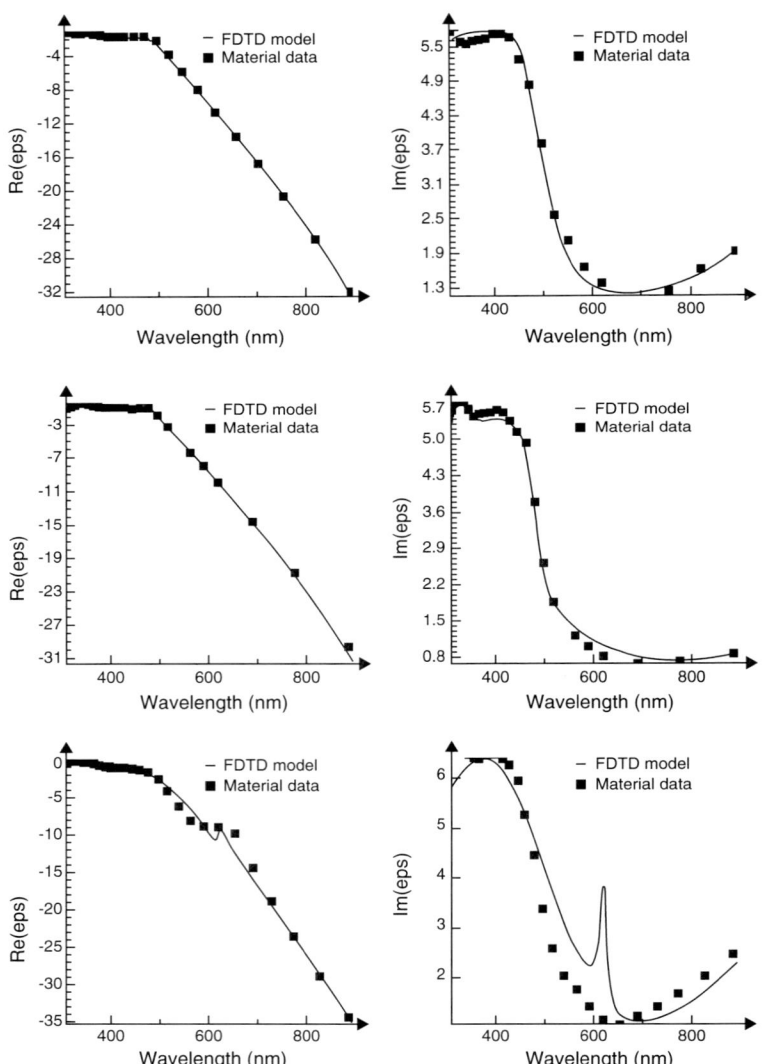

Fig. 9.12 The optical constants of gold from various data tables and their polynomial fits provided by the FDTD program. Johnson & Christy (*top*), CRC (*middle*), and Palik (*bottom*)

9 Experimental and Theoretical Issues of Nanoplasmonics in Medicine

9.5 Comparison Between Experimental Samples and Theoretical Models

9.5.1 OMCVD Experiments

The OMCVD experiment data were measured using UV-VIS spectroscopy at the University of Western Ontario by Dr. Silvia Mittler and Erden Ertorer. The data samples, referred to as L3, L6, L9, M9, M10, and M11, were constructed with the purpose of random sizes and spacings. After production, the particle sizes and spacings were measured and were received via several histograms as shown in Fig. 9.7. The data samples L3, L6, and L9 were bare nanoparticles on a glass substrate. The samples M9, M10, and M11 contained surface-immobilized nanoparticles that were coated with organic material of hydrodynamic radius of 1 nm.

To model the particle distributions, the system was modeled through the use of two GNPs separated by a finite center-to-center distance d. The size and spacings of the particles were varied and weighted according to the experimentally provided data. The coating of organic material on the nanoparticles in samples "M" were modeled by a concentric coating of thickness 1 nm and refractive index 1.35. The particles were subjected to two perpendicular polarizations directed parallel to and perpendicularly to the interparticle axis. We also calculated the spectra for unpolarized incident light.

For the above calculations, the radii of the nanoparticles were assumed to be identical and were varied, along with the interparticle spacing (center-to-center distance). To compare with experiments, the weighted averages of the spectra were calculated according to the following approximations.

First, we calculated the spectrum of two nanoparticles of the average radius, and separated by the average interparticle separation (these are labeled by "Avg.")

$$Q_{\text{ext,avg}} = q. \tag{9.3}$$

where q will be used to simplify the notation for the calculated extinction value. Next, we assumed that the spread in the particle radii could be neglected and therefore we took the particles to be of the average radius, but weighted the spectra according to the interparticle spacing histogram (labeled Ravgov)

$$Q_{\text{ext,Ravg}} = \frac{1}{\sum_{j} w_{j}} \sum_{j} w_{j} q_{j}, \qquad (9.4)$$

where summation is taken over all experimentally measured spacings [4 nm – 64 nm in 4 nm increments] and *wj* is the fraction of experimentally observed particles with spacing j. Since overlapping particles might be considered as larger particles, we calculated the same quantity as above, but omitting the overlapping particles in the weighting (Ravgno). The final level of approximation was to weight the spacings and radii according to the experimental histograms (labeled by "Allov"). For the same reason as above, we repeated the calculation without the overlapping particles (labeled by "Allno").

$$Q_{\text{ext,all}} = \frac{1}{\sum_{i,j} w_{i} w_{j}} \sum_{i,j} w_{i} w_{j} q_{i} q_{j} \qquad (9.5)$$

where the first summation is taken over all experimentally observed particle radii (2 nm – 20 nm in 2 nm increments), and the second is taken over all possible spacings as done previously.

These approximations were run for the DDA, GMM, and GranFilm methods with the exceptions that GMM and GranFilm cannot study overlapping particles as their fundamental theories break down [2, 21].

All results were then scaled to experimental measurements according to

$$s = \left| \frac{Q_{\text{exp},N} - Q_{\text{exp,max}}}{Q_{\text{theory},N} - Q_{\text{theory,max}}} \right| \qquad (9.6)$$

$$Q_{\text{ext,scaled}} = s Q_{\text{theory}} + \left(Q_{\text{exp,max}} - s Q_{\text{theory,max}} \right) \qquad (9.7)$$

where N is the number of frequencies used in the compution, i.e. $Q_{ext,N}$ is the final extinction value in the calculated spectrum.

9.5.2 Results of the Comparison

For all three methods, the third weighting approximation $Q_{\text{ext,all}}$ allowing for overlapping particles where possible produced the best agreement with experiment. Plots of experimentally measured and selected calculated spectra are shown in Figs. 9.13–9.30.

9 Experimental and Theoretical Issues of Nanoplasmonics in Medicine 367

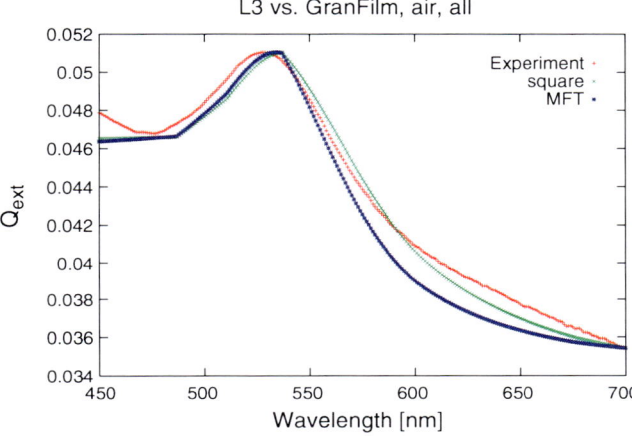

Fig. 9.13 Sample L3: OMCVD deposited GNPs. Comparison between experimental spectra and spectra calculated using GranFilm. In the calculations, the "Allno" averaging was employed for both square lattice GNP arrangements ("square") and random arrangements using mean field theory ("MFT")

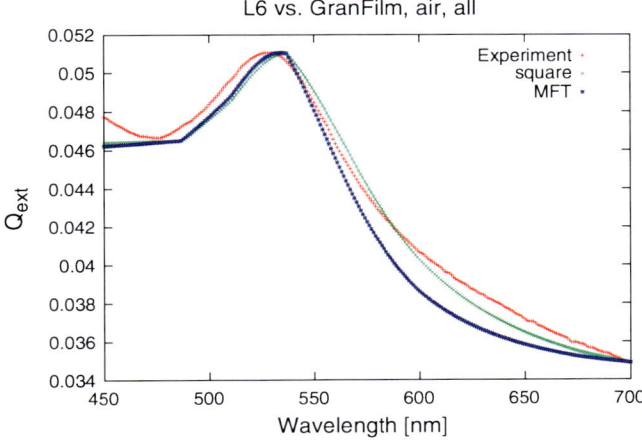

Fig. 9.14 Sample L6: OMCVD deposited GNPs. Comparison between experimental spectra and spectra calculated using GranFilm. In the calculations, the "Allno" averaging was employed for both square lattice GNP arrangements ("square") and random arrangements using mean field theory ("MFT")

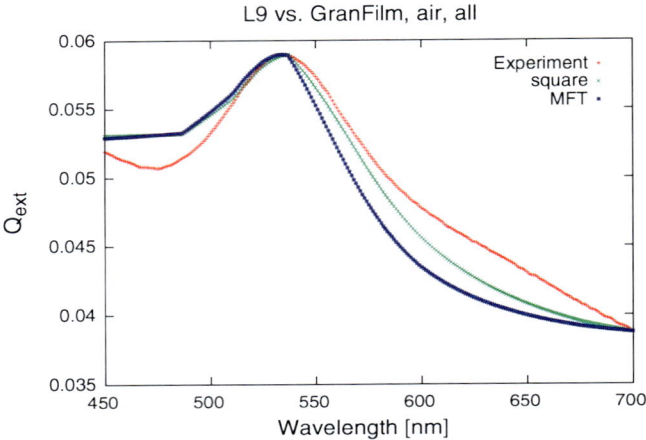

Fig. 9.15 Sample L9: OMCVD deposited GNPs. Comparison between experimental spectra and spectra calculated using GranFilm. In the calculations, the "Allno" averaging was employed for both square lattice GNP arrangements ("square") and random arrangements using mean field theory ("MFT")

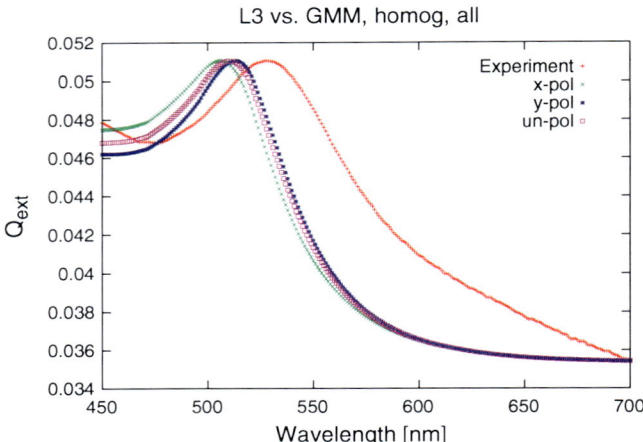

Fig. 9.16 Sample L3: OMCVD deposited GNPs. Comparison between experimental spectra and spectra calculated using the GMM method. In the calculations, the "Allno" averaging was employed for s-polarized ("x"), p-polarized ("y"), and unpolarized ("un-pol") incident light

9 Experimental and Theoretical Issues of Nanoplasmonics in Medicine 369

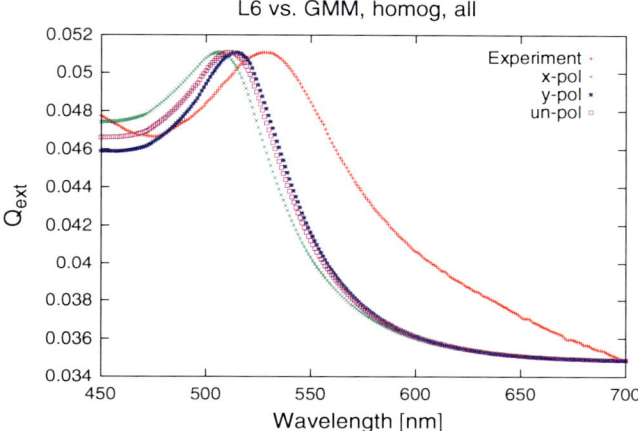

Fig. 9.17 Sample L6: OMCVD deposited GNPs. Comparison between experimental spectra and spectra calculated using the GMM method. In the calculations, the "Allno" averaging was employed for s-polarized ("x"), p-polarized ("y"), and unpolarized ("un-pol") incident light

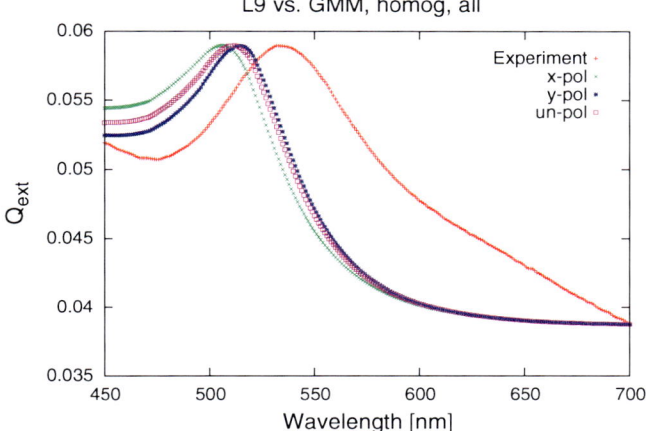

Fig. 9.18 Sample L9: OMCVD deposited GNPs. Comparison between experimental spectra and spectra calculated using the GMM method. In the calculations, the "Allno" averaging was employed for s-polarized ("x"), p-polarized ("y"), and unpolarized ("un-pol") incident light

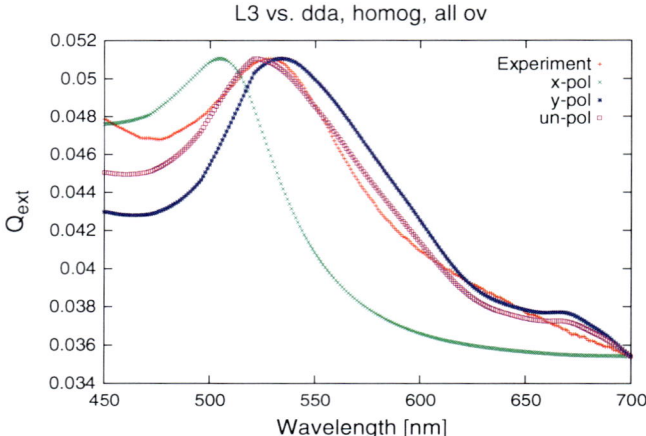

Fig. 9.19 Sample L3: OMCVD deposited GNPs. Comparison between experimental spectra and spectra calculated using the DDA method. In the calculations, the "Allov" averaging was employed for s-polarized ("x"), p-polarized ("y"), and unpolarized ("un-pol") incident light

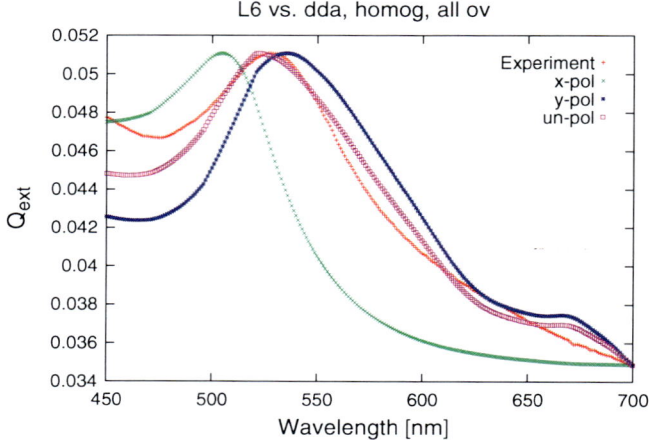

Fig. 9.20 Sample L6: OMCVD deposited GNPs. Comparison between experimental spectra and spectra calculated using the DDA method. In the calculations, the "Allov" averaging was employed for s-polarized ("x"), p-polarized ("y"), and unpolarized ("un-pol") incident light

9 Experimental and Theoretical Issues of Nanoplasmonics in Medicine

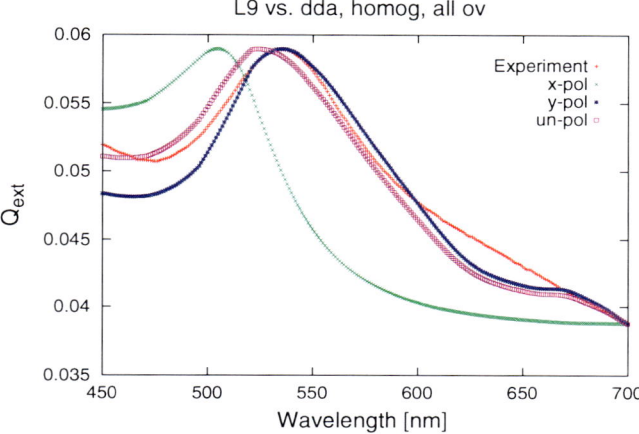

Fig. 9.21 Sample L9: OMCVD deposited GNPs. Comparison between experimental spectra and spectra calculated using the DDA method. In the calculations, the "Allov" averaging was employed for s-polarized ("x"), p-polarized ("y"), and unpolarized ("un-pol") incident light

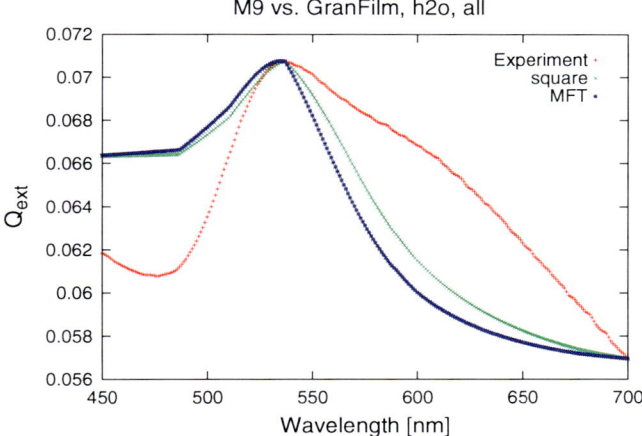

Fig. 9.22 Sample M9: OMCVD deposited GNPs [coated with a layer of ethanol]. Comparison between experimental spectra and spectra calculated using GranFilm. In the calculations, the "Allno" averaging was employed for both square lattice GNP arrangements ("square") and random arrangements using mean field theory ("MFT")

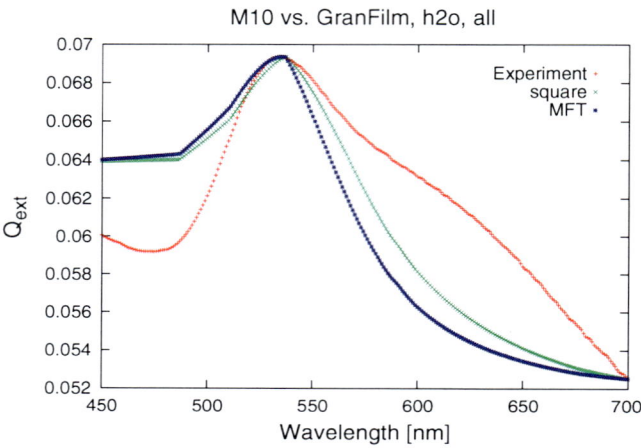

Fig. 9.23 Sample M10: OMCVD deposited GNPs [coated with a layer of ethanol]. Comparison between experimental spectra and spectra calculated using GranFilm. In the calculations, the "Allno" averaging was employed for both square lattice GNP arrangements ("square") and random arrangements using mean field theory ("MFT")

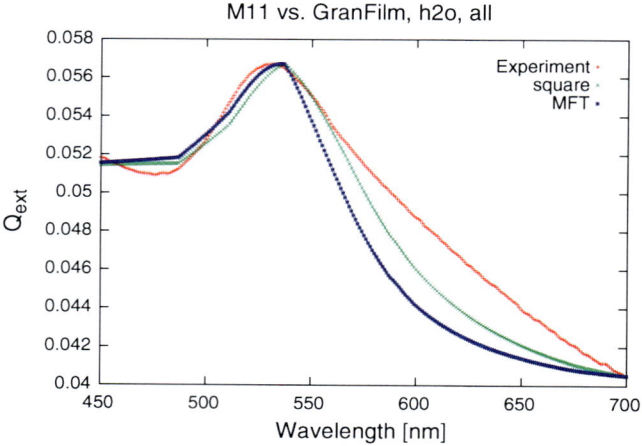

Fig. 9.24 Sample M11: OMCVD deposited GNPs [coated with a layer of ethanol]. Comparison between experimental spectra and spectra calculated using GranFilm. In the calculations, the "Allno" averaging was employed for both square lattice GNP arrangements ("square") and random arrangements using mean field theory ("MFT")

9 Experimental and Theoretical Issues of Nanoplasmonics in Medicine 373

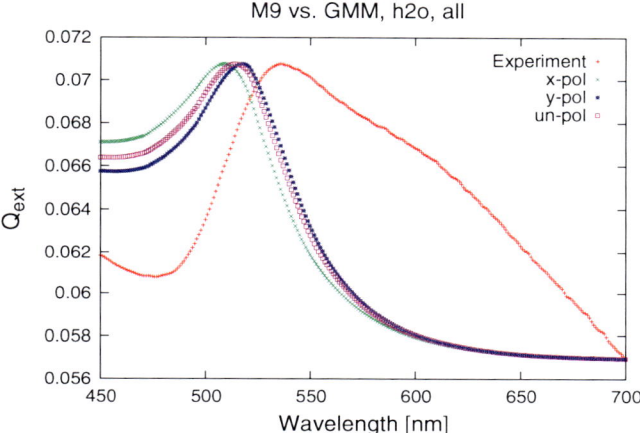

Fig. 9.25 Sample M9: OMCVD deposited GNPs [coated with a layer of ethanol]. Comparison between experimental spectra and spectra calculated using the GMM method. In the calculations, the "Allno" averaging was employed for s-polarized ("x"), p-polarized ("y"), and unpolarized ("un-pol") incident light

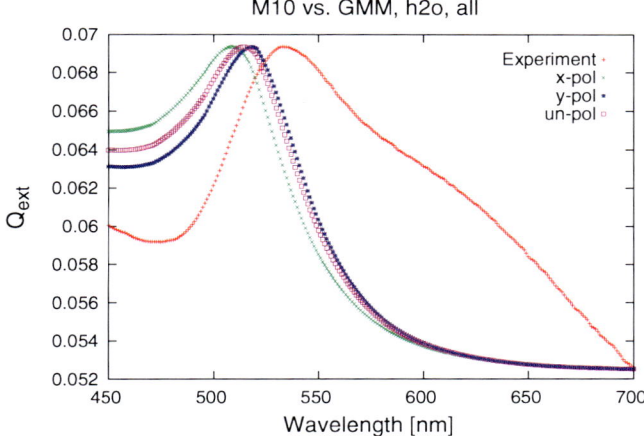

Fig. 9.26 Sample M10: OMCVD deposited GNPs [coated with a layer of ethanol]. Comparison between experimental spectra and spectra calculated using the GMM method. In the calculations, the "Allno" averaging was employed for s-polarized ("x"), p-polarized ("y"), and unpolarized ("un-pol") incident light

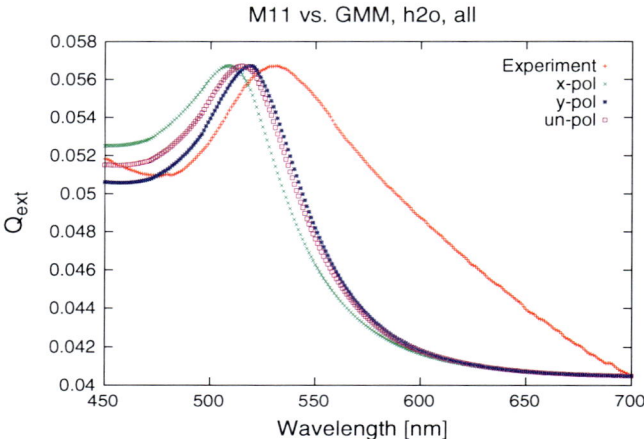

Fig. 9.27 Sample M11: OMCVD deposited GNPs [coated with a layer of ethanol]. Comparison between experimental spectra and spectra calculated using the GMM method. In the calculations, the "Allno" averaging was employed for s-polarized ("x"), p-polarized ("y"), and unpolarized ("un-pol") incident light

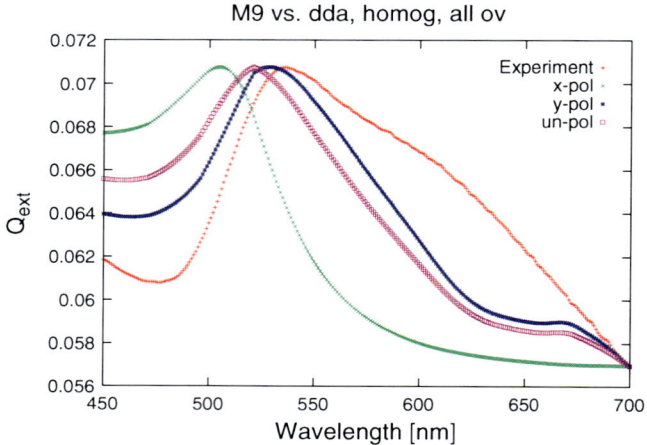

Fig. 9.28 Sample M9: OMCVD deposited GNPs [coated with a layer of ethanol]. Comparison between experimental spectra and spectra calculated using the DDA method. In the calculations, the "Allov" averaging was employed for s-polarized ("x"), p-polarized ("y"), and unpolarized ("un-pol") incident light

9 Experimental and Theoretical Issues of Nanoplasmonics in Medicine 375

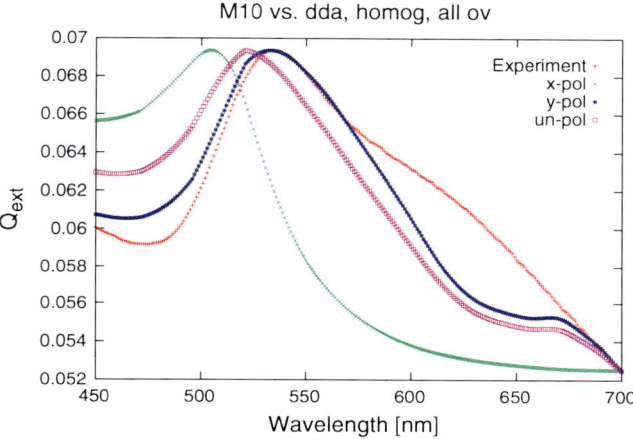

Fig. 9.29 Sample M9: OMCVD deposited GNPs [coated with a layer of ethanol]. Comparison between experimental spectra and spectra calculated using the DDA method. In the calculations, the "Allov" averaging was employed for s-polarized ("x"), p-polarized ("y"), and unpolarized ("un-pol") incident light

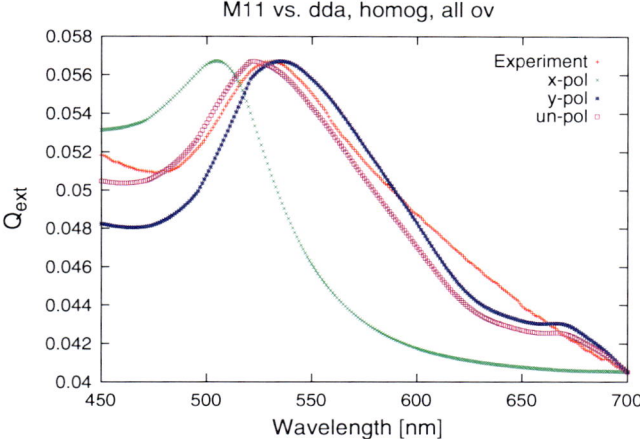

Fig. 9.30 Sample M9: OMCVD deposited GNPs [coated with a layer of ethanol]. Comparison between experimental spectra and spectra calculated using the DDA method. In the calculations, the "Allov" averaging was employed for s-polarized ("x"), p-polarized ("y"), and unpolarized ("un-pol") incident light

9.6 Summary

OMCVD is an easy and inexpensive method to fabricate GNPs on substrate surfaces. Flatness of the reactor bottom, heating method, and the quality of the surface functionalization are crucial for the stability and uniformity of the nanoparticles. Automatization of the processes increases the quality. SEM and UV-Vis absorption spectroscopy are helpful tool to characterize samples and control quality. SEM images were used to yield size and interparticle distance data. UV-Vis absorption spectroscopy can be used as a gauge of uniformity in particle size. Bulk and protein sensing experiments show that OMCVD grown GNPs are suitable for highly sensitive biosensor applications (Figs. 9.8–9.10).

From the modeling perspective, it was found that the DDA method gave the best comparisons with experiments. We find that it is necessary to have a weighted average of size and spacing distributions especially for the coated particles. This confirms the "optical clustering" hypothesis put forth in [35, 36]. The two-particle model works surprisingly well indicating that nearly all the physics is described by the single particle LSPR and the two-particle cross talk. The two particle cross-talk is mediated either by the proximity of the two bare GNP surfaces or by a coating on two not-so-proximate nanoparticles. The FDTD time-domain method showed a surprising sensitivity to input optical constants data. Even though the data from different sources themselves were similar, the fitting program gives completely different polynomial fits, and the best fit was for the data from Johnson and Christy.

A long-time desire of experimentalists is the ability to extract from UV-Vis absorption data of coated nanoparticles on surfaces, the thickness and refractive index of the analyte. This would be another alternative to planar surface plasmon spectroscopy. We are not there yet, but this work demonstrates that we are getting closer to this goal.

Acknowledgements Research support by the Natural Sciences and Engineering Research Council of Canada (Discovery Grants and NSERC Strategic Network on Bioplasmonic Systems) and Canada Foundation for Innovation is gratefully appreciated. Computations were done on the SharcNet (Compute Canada) supercomputing network. Silvia Mittler thanks the Canada Research Chairs program of the Canadian government. Ertorer was partly supported by the Ontario Graduate Scholarship and the Ontario Graduate Scholarship for Science and Technology.

References

1. Abramoff MD, Magalhes PJ, Ram SJ. Image processing with ImageJ. Biophoton Int. 2004;11(7):36–42.
2. Alivisatos AP. Science. 1996;271:933.
3. Aliganga AKA, Duwez AS, Mittler S. Binary mixtures of self-assembled monolayers of 1,8-octanedithiol and 1-octanethiol for a controlled growth of gold nanoparticles. Org Electron. 2006;7(5):337–50.
4. Aliganga AKA, Lieberwirth I, Glasser G, Duwez A-S, Sun Y, Mittler S. Fabrication of equally oriented pancake shaped gold nanoparticles by SAM templated OMCVD and their optical response. Org Electron. 2007;8:161–74.
5. Bedeaux D, Vlieger J. Optical properties of surfaces. 2nd ed. Singapore: World Scientific; 2004.
6. Bharathi S, Fishelson N, Lev O. Direct synthesis and characterization of gold and other noble metal nanodispersions in sol-gel-derived organically modified silicates. Langmuir. 1999;15(6):1929–37.
7. Bohren CF, Huffman DR. Absorption and scattering of light by small particles. New York: Wiley Interscience; 1983.
8. Chen C-D, Cheng S-F, Chau L-K, Wang CRC. Sensing capability of the localized surface plasmon resonance of gold nanorods. Biosensors Bioelectron. 2007;22:926–32.
9. Cheng T, Rangan C, Sipe JE. Metallic nanoparticles on waveguide structures: effects on waveguide mode properties, and the promise of sensing applications (manuscript); Journal of the Optical Society of America B (in press).
10. Daniel M, Astruc D. Gold nanoparticles: assembly, supramolecular chemistry, quantum-size-related properties, and applications toward biology, catalysis, and nanotechnology. Chem Rev. 2004;104:293–346.
11. Doyle WT. Optical properties of a suspension of metal spheres. Phys Rev B. 1989;39(14):9852–8.
12. Draine BT, Flatau PJ. J Opt Soc Am A. 1973;11:1491.
13. Draine BT, Goodman JJ. ApJ. 1993;485:685.
14. Fischer RA, Weckenmann U, Winter C, Kshammer J, Scheumann V, Mittler S. Area selective OMCVD of gold and palladium on self-assembled organic monolayers: control of nucleation sites. J Phys IV France. 2001;11(PR3):Pr3-1183–Pr3-1190.
15. Frederix F, Bonroy K, Laureyn W, Reekmans G, Campitelli A, Dehaen W, Maes G. Enhanced performance of an affinity biosensor interface based on mixed self-assembled monolayers of thiols on gold. Langmuir. 2003;19(10):4351–7.
16. Hampden-Smith MJ, Kodas TT. Chemical vapor deposition of metals: Part 1. An overview of CVD processes. Chem Vapor Deposition. 1995;1(1):8–23.
17. Haynes CL, McFarland AD, Zhao L, Van Duyne RP, Schatz GC. Nanoparticle optics: the importance of radiative dipole coupling in two-dimensional nanoparticle arrays. J Phys Chem B. 2003;107:7337–42.
18. Jensen T, Kelly L, Lazarides A, Schatz GC. Electrodynamics of noble metal nanoparticles and nanoparticle clusters. J Cluster Sci. 1999;10:295–317.

19. Johnson PB, Christy RW. Optical constants of the noble metals. Phys Rev B. 1972;6:4370–9.
20. Käshammer J, Wohlfart P, Wei J, Winter C, Fischer R, Mittler-Neher S. Selective gold deposition via CVD onto self-assembled organic monolayers. Opt Mater. 1998;9:406–10.
21. Kreibig U, Vollmer M. Optical properties of metal clusters. Berlin: Springer; 1995.
22. Lazzari R, Simonsen I. GRANFILM: a software for calculating thin-layer dielectric properties and Fresnel coefficients. Thin Solid Films. 2002;419(1–2):124–36.
23. Lazzari R, Simonsen I, Bedeaux D, Vlieger J, Jupille J. Eur Phys J B. 2001; 24:267.
24. Link S, El-Sayed MA. Size and temperature dependence of the plasmon absorption of colloidal gold nanoparticles. J Phys Chen B. 1999;103: 4212–7.
25. Maier SA. Guiding of electromagnetic energy in subwavelength periodic metal structures. Ph.D. Thesis, California Institute of Technology, Pasadena; 2003.
26. Manifar T, Rezaee A, Sheikhzadeh M, Mittler S. Formation of uniform self-assembly monolayers by choosing the right solvent: OTS on silicon wafer, a case study. Appl Surface Sci. 2008;254(15):4611–9.
27. Marton P, Schlesinger M. J Electrochem Soc. 1968;115:16.
28. Maxwell-Garnett JC. Philos Trans R Soc Lond. 1904;203:385; Ser A 1906;205:237.
29. Mie G. Beitrge zur Optik trber Medien speziell kolloidaler Goldlsungen. Ann Phys. 1908;25:377–445.
30. Miller MM, Lazarides AA. Sensitivity of metal nanoparticle plasmon resonance band position to the dielectric environment as observed in scattering. J Opt A: Pure Appl Opt. 2006;8:239–49.
31. Nicolas S, Dufour-Gergam E, Bosseboeuf A, Bourouina T, Gilles J-P, Grandchamp J-P. Fabrication of a gray-tone mask and pattern transfer in thick photoresists. J Micromech Microeng. 1998;8:95.
32. Noguez C. Surface plasmons on metal nanoparticles: the influence of shape and physical environment. J Phys Chem C. 2007;111:3806–19.
33. Weaver JH, Frederikse HPR. Optical Properties of Metals and Semiconductors, CRC Handbook of Chemistry and Physics, 74th Edition and subsequent printings (CRC Press, Boca Raton, Florida) pp. 12–109, 12–131.
34. Palik E. Handbook of optical constants of solids I–III. San Diego: Academic; 1998.
35. Rafsanjani SMH, Cheng T, Mittler S, Rangan C. Theoretical proposal for a biosensing approach based on a linear array of immobilized gold nanoparticles. J Appl Phys. 2010;107:094303.
36. Rooney P, Xu S, Rezaee A, Manifar T, Hassanzadeh A, Podoprygorina G, Bhmer V, Rangan C, Mittler S. Control of surface plasmon resonances in dielectrically-coated proximate gold nanoparticles immobilized on a substrate. Phys Rev B. 2008;77(23):235446.
37. Schott AG, 2007, Data Sheet N-BK7 [Online] Mainz, Germany: Schott. Available at http://www.schott.com/advanced_optics/english/abbe_data-sheets/schott_datasheet_n-bk7.pdf. [Accessed 03 January 2013].

38. Spinke J, Liley M, Schmitt F-J, Guder H-J, Angermaier L, Knoll W. Molecular recognition at self-assembled monolayers: optimization of surface functionalization. J Chem Phys. 1993;99(9):7012–19.
39. Taflove A, Hagness SC. Computational electrodynamics: the finite-difference time-domain method. 2nd ed. Boston: Artech House; 2005.
40. Thoma F, Langbein U, Mittler-Neher S. Waveguide scattering microscopy. Opt Commun. 1997;134:16–20.
41. Ulman A. An introduction to ultrathin organic films: from Langmuir-Blodgett to self-assembly, vol Xxiii. London: Academic; 1991. p. 442.
42. Weisser M, Thoma F, Menges B, Langbein U, Mittler-Neher S. Fluorescence in ion exchanged BK7 glass slab waveguides and its use for scattering free loss measurements. Opt Commun. 1998;153:27–31.
43. Wiscombe WJ. Improved Mie scattering algorithms. Appl Opt. 1980;19(9):1505–9.
44. Wohlfart P, Wei J, Kshammer J, Winter C, Scheumann V, Fischer R, Mittler-Neher S. Selective ultrathin gold deposition by organometallic chemical vapor deposition onto organic self-assembled monolayers (SAMs). Thin Solid Films. 1999;340:274–9.
45. Yee K. Numerical solution of initial boundary value problems involving Maxwell's equations in isotropic media. IEEE Trans Antenn Propag. 1966;14:302–7.
46. Yu-lin Xu. Electromagnetic scattering by an aggregate of spheres. Appl Opt. 1995;34(21):4573–88.
47. Zou S, Janel N, Schatz GC. Silver nanoparticles array structures that produce remarkably narrow plasmon lineshapes. J Chem Phys. 2004;120(23):10871–5.

Chapter 10
Modeling and Measuring Extravascular Hemoglobin: Aging Contusions

Oleg Kim, Collin Lines, Susan Duffy, Mark Alber,
and Gregory Crawford

10.1 Introduction

Pediatricians and other physicians who care for acutely injured patients are often asked for an opinion on the age of externally visible contusions for a variety of reasons. Commonly, medical expertise is sought to help delineate intentional from unintentional injury in cases

O. Kim
Department of Applied and Computational Mathematics and Statistics,
University of Notre Dame, Notre Dame, IN 46556, USA

C. Lines
Department of Physics, University of Notre Dame, Notre Dame, IN 46556, USA

S. Duffy
Department of Emergency Medicine and Pediatrics, Alpert Medical School
of Brown University, Providence, RI 02806, USA

M. Alber
Department of Applied and Computational Mathematics and Statistics,
University of Notre Dame, Notre Dame, IN 46556, USA

Department of Medicine, Indiana University School of Medicine, Indianapolis,
IN 46202, USA

G. Crawford (✉)
Department of Physics, University of Notre Dame, Notre Dame, IN 46556, USA

College of Science, University of Notre Dame, Notre Dame, IN 46556, USA
e-mail: crawford.52@nd.edu

involving child protection and interpersonal violence. In many circumstances, physicians are expected to render an opinion on whether or not the bruising pattern is consistent with abuse or if the appearance of bruising is consistent with the history provided by the patient or their caretaker. Sadly enough, physical abuse or inflicted trauma is one of the most common types of child maltreatment observed by physicians [1, 2] and the diagnosis is often dependent on physical findings and historical correlates. Accurately assessing the ages of bruises and differentiating bruises of different ages on the same patient are important components to validating historical accounts of injury. Law enforcement and child protection workers often rely on physician assessment of bruising and other injury to identify perpetrators and assess culpability.

Currently, there are no objective criteria for assessing bruises in living patients. In the past, forensic pathologists developed visible color-change models for bruises based on studies of cadavers. However, the findings from these studies have never been validated or replicated in living humans or animals. Despite their flaws, these visible color-change models have been incorporated into clinical practice and quoted as dogma in medical textbooks [3]. More recent research in dermatology and forensics has shown that visible color variation in skin is inadequate by itself for dating bruises because natural skin pigmentation, skin type and depth, and the involvement of other tissues impact the visible spectrum [4, 5]. Child welfare and criminal justice personnel make judgments of great consequence based on physician opinions regarding clinical evidence, but the medical literature and the experiences of physicians and researchers reveal that the dating of bruises is an inexact science. Investigators have demonstrated that physician estimates of bruises ages are highly inaccurate and appear to be no better than chance alone [6]. Even when a bruised area was examined directly, it was still difficult for physicians, independent of training and clinical experience, to reliably estimate the age of a bruise [6]. The fact that experienced clinicians and frontline social and health care workers who assess injured tissue on a regular basis exhibit inadequate accuracy in estimating the age of a bruise underscores the need for a device that can accurately and quantitatively measure the age of bruises. Although there have been rudimentary studies on spectrophotometric evaluation of bruises with conventional spectroscopic equipment [7] and new models developed to predict skin reflectance spectra [8], there is currently no instrument available for physicians with which to determine the age of bruises. Since there is so much at stake in assessing the nature and age of bruises, there is a compelling need for an accurate device to assess these injuries.

Generally, bruises develop as a result of rupture of blood vessels from compressive and shearing forces imposed on the body during blunt impact trauma. Depending on the affected tissue layer subcutaneous or intracutaneous bruises can develop [7]. Subcutaneous bruises result from blood extravasation of damaged vessels in deep tissue and gradually extend toward skin surface [9–11]. As hemoglobin degrades, the bruise may change color over time. The time course is variable. Some deep bruises are not visibly apparent for days. Intracutaneous bruises usually result from a strong impact and may be initially identified by a bright red color that often quickly appears. When skin is exposed to external pressure the damage of dermis and vessels occurs and blood leaking from damaged vessels tracks to regions not exposed. As hemoglobin diffuses and degrades bruise colors and appearance is altered.

The progression of hemoglobin metabolism has been well documented and is of utmost importance in optically monitoring bruise spectra progression. Extravascular hemoglobin caused by trauma to blood vessels is engulfed by macrophages and degraded into proteins and heme, which is further degraded into biliverdin (briefly visibly apparent in some bruises as a green color) and subsequently into bilirubin (leading to yellowish color) (Fig. 10.1). The strong spectral signatures of hemoglobin and bilirubin in particular allow a spectral estimation of the stage of breakdown and an indirect estimation of the age of contusions (Fig. 10.2).

To date several new methods are being explored that are believed to give more accurate assessment of the age of contusions. These methods are based on measurements of hemoglobin, one of the strongest chromophores of human skin. Hemoglobin as well as its breakdown products in dermis and subcutaneous tissue determines the spectrophotometric characteristics of the skin and its variations over time. Its concentration can serve as a metric when measuring diffuse reflective spectra of the skin to noninvasively screen contusion sites and assess the contusion age. Therefore, developing noninvasive device that measures total hemoglobin and the degradation products of hemoglobin would be of a great benefit to the biomedical community.

Currently, the methods used to assess hemoglobin include: (1) imaging and spectrophotometric imaging; (2) transmission and reflectance spectroscopy; (3) ultrasound and optoacoustic spectroscopy.

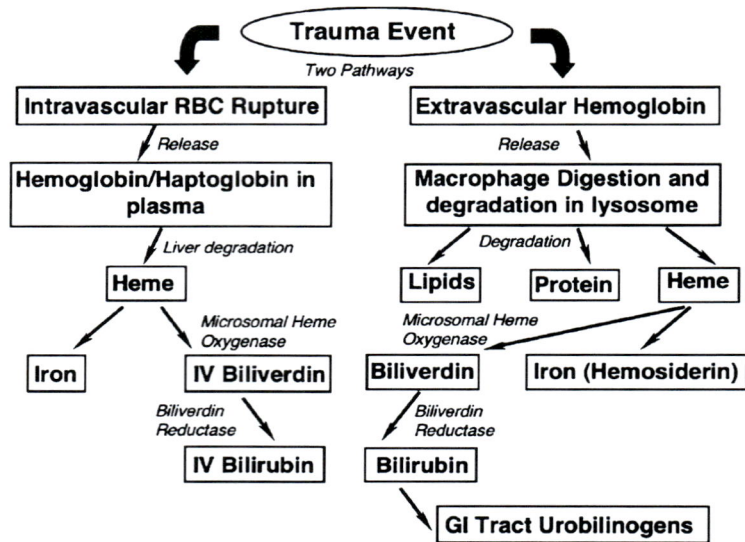

Fig. 10.1 The main steps of the biological breakdown processes following a contusion causing intravascular red blood cell (RBC) rupture and extravasated hemoglobin

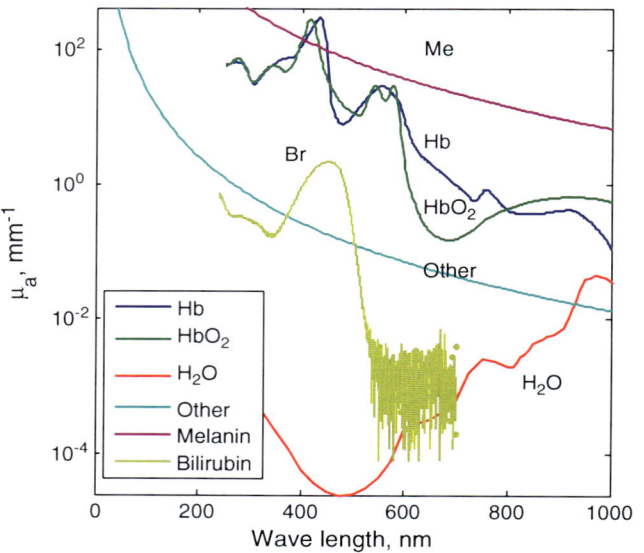

Fig. 10.2 Coefficient of absorbance of various skin chromophores [49]. *Me* melanin; *Hb* deoxyhemoglobin; HbO_2 hemoglobin; *Br* bilirubin; H_2O water

In what follows, different methods used to identify hemoglobin and its breakup products are reviewed including possible applications for assessing ages of contusions. Then modeling approaches are presented and integration with experimental methods is considered. Finally, challenges and potential applications for the future are summarized.

10.2 Imaging and Spectrophotometric Imaging

Several types of devices have been developed for estimating hemoglobin. One device uses the combination of near-infrared (NIR) vascular imagers that target blood vessels to estimate hemoglobin based on absorption characteristics of individual vessels. AstrimSysmex is one such noninvasive hemoglobin monitor that utilizes a near-infrared light source (660, 805, and 880 nm) positioned over a dorsal side of a finger and a CCD camera located underneath the finger. The finger is irradiated with light at different wavelengths and absorption measurements are taken, allowing an estimate of the amount of hemoglobin, when combined with direct measurements of the vascular diameter from the optical image. Although the device is reported [12] to have a diagnostic sensitivity of 78.3% and specificity of 69% when compared to clinically defined anemia, it is poorly suited to determine hemoglobin concentration from a single measurement because of the variability in efficacy with respect to finger position and therefore other access points must be used.

Another commercial instrument, Hemoscan, Cytometrics that uses orthogonal polarization spectral imaging (OPS), was implemented by Nadeau and Groner [13] to analyze vascular network of sublingual mucosa. These studies showed that OPS can be used to separate normal vs. anemic patients (12.0 g/dL cut-off) based on measurements having a correlation coefficient of 0.93 for predefined instrument calibration compared to automated hematology analyzer.

The standardized retinal imaging was used as an indicator for hemoglobin by Rice et al. [14]. The retina was illuminated through the pupil at isosbetic points for oxy- and deoxyhemoglobin, using several discrete visible/NIR light bands. The reflected light from vessels overlying the optic disc was analyzed. This technique provided a cross-validation coefficient of 0.89 based on calibrated reflected intensities when compared to an *in vitro* hemoglobin analyzer.

Although clinically tested and shown to be successful in predicting hemoglobin concentrations, all of the methods described use transmitted or reflected light with the finger tip or thin tissue regions of sublingual mucosa or retina as primary physiological locations. Thus, although imaging allows noninvasive measurement of hemoglobin, they have very limited use for monitoring bodily contusions. In addition, these complex devices may be cost prohibitive for medical applications as compared to traditional invasive methods [15].

10.3 Reflectance Spectroscopy and Hyperspectral Imaging

Reflectance spectroscopy and imaging can be advantageous for hemoglobin measurements in contusions because deep transmission through highly scattering medium is not necessary and blood vessel properties in the vicinity of the skin surface can be probed. Reflectance-based methods can be grouped as reflectance and spectral reflectance imaging methods and single point detection reflection spectroscopy. To analyze reflectance images certain color systems are used.

Colorimetry is the science that quantifies the persistence of color by human eyes. It can be used to measure color of the human skin and to assess the age of imaged bruises. Several color systems have been developed to quantitatively assess and interpret the color. These are based on the multidimensional qualities of light, which includes red/green/blue (RGB), cyan/magenta/yellow/black (CMYK), hue/saturation/brightness (HSB), and the Commission Internationale d'Eclairage (CIE). RGB system is used to display images in electronic systems, CMYK was developed for full color printing, HSB is a transformation of RGB method. CIE system has been successfully applied in dermatology and medicine to achieve consistent description of the skin color. CIE system expresses color as a point in a three dimensional sphere. Coordinates of this point are represented by lightness $L*$, and chromaticity parameters $a*$ and $b*$, showing the relative amounts of red/green and blue/yellow, respectively. CIE method has advantages over RGB, CMYK, and HSB because it allows close approximation of the interpretation of color by eyes unlike other visual assessment methods [16–18]. CIE method has been used alone, [16, 19] and in combination with reflectance spectrophotometric scans [20] to determine the age of human bruises.

Table 10.1 Correlation between physical and $L^*a^*b^*$ parameters [21]

	Oxy	Bd	Ec
L^*	0.72	−0.91	−0.81
a^*	–	0.74	0.92
b^*	0.85	−0.48	–

Oxy tissue oxygenation at 660/805 nm; *Bd* dermal blood volume fraction; *Ec* erythema index

Mimasaka et al. 2010 [16] measured color changes of bruises in 2–5-year-old children using a spectrophotometer (CM-2500d, Konica Minolta Holdings, Inc. Tokyo, Japan). The bruises on legs, knees, thighs, and elbows having size from 10×10 mm to 34×35 mm were studied. It was showed that changes in bruise color can be assessed using tristimulus method in terms of changes in L^*, a^*, and b^*.

CIE-based colorimetry was used in [20], to determine the age of bruising from 147 subjects (28 males and 119 females) with ages ranging from 18 to 72 years. The colorimetric measurement of the yellowness of a bruise was shown to account for 13% of the bruise age, which ranged from 0.5 to 231.5 h. Using General Linear Model, it was shown that yellowness could predict up to 32% of the age of a bruise, when combined with additional factors. Meanwhile, 68% of the variation was related to other factors, one of such factors being the underlying skin color or skin tone.

Colorimetry is a tool that allows objective description of color that can be used to assess skin color changes, particularly in bruises. The color coordinates can estimate the depth of bruises and evaluate if they are superficial or deep. Coordinate b^* can reflect the presence of yellow in the bruise, although it is not a reliable parameter for accurate determination of the bruise age. The increase of beta-carotene content due to specific dietary intake can increase the b coordinate, which can be misinterpreted as bilirubin. As shown by Randeberg et al. [21] porcine skin, the $L^*a^*b^*$ parameters (Table 10.1) are closely correlated to physical characteristics such as blood volume fraction, tissue oxygenation, and erythema index.

10.4 Single Point Reflection Spectroscopy

Single point reflectance spectroscopy is used as an alternative method to measure hemoglobin (Fig. 10.3). An advantage of this approach is that it assesses features close to the surface of the skin and avoids

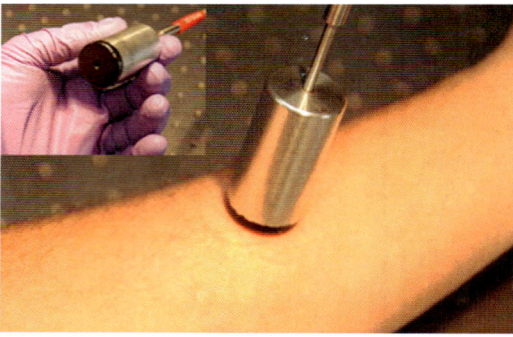

Fig. 10.3 Single point data collection probe and data collection technique

deep transmission through turbid tissue making this method very attractive for assessing normal and bruised skin conditions.

McMurdy et al. 2006 [22] demonstrated that spectroscopic measurements can be used to extract the concentration of hemoglobin using partial least squares multivariate regression model and discrete spectral region model. Hemoglobin concentration derived from both of these models was compared to *in vitro* measurements of Hgb. Root mean square errors of cross validation for these methods were 0.67 and 1.07 g/dL, respectively.

McMurdy et al. [23] studied temporal progression of reflection spectrum collected from accidentally inflicted contusions in adult and child using single point reflection spectroscopy. In their study, reflection spectra were collected using a portable fiber optic reflection spectrometer that tracked the increase in extravasated hemoglobin from trauma. The spectral sampling apparatus utilized a static grating spectrometer with 4 nm resolution (USB2000, Ocean Optics, Dunedin, FL) and a white LED (Ocean Optics, Dunedin, FL) to collect all reflectance spectra. A fiber optic sampling probe was introduced into the system to facilitate convenient data collection from each zone. The reflectance probe used six 200 µm LED irradiation fibers oriented around a single 200 µm collection fiber. This probe configuration allowed to measure the reflectance at a specified single point location and provided detailed distribution of chromophores' spectra over the trauma site area.

Skin changes following minor trauma were studied in [24] domestic pigs using integrating sphere setup (ISP-REF, SD2000, Ocean Optics,

Duiven, The Netherlands). Low speed blunt pendulum and high speed paint balls were used to controllably inflict injuries. It was found that the temporal development of the injury strongly depends on the weight and speed of the impacting object. Low speed (up to 4.1 m/s), blunt 2.5 kg objects did not cause immediate skin changes but resulted in deep muscular bleeding. High speed (100 m/s), light weight objects (3.15 g) resulted in rapidly developing, within 15–20 min, bruises. Deep muscular hemorrhages were not observed in these cases. In these preliminary studies reflectance spectroscopy was found to give useful results on immediate skin reaction to the trauma.

10.5 Hyperspectral Imaging

Hyperspectral imaging was originally developed for military applications and remote sensing, but recently found applications in other fields, including medicine. Hyperspectral imaging allows for the collection of both spatial and spectral information of an object. Every point on the object is imaged across several narrow and contiguous wavebands, thus forming a "hypercube" that comprises two spatial and one wavelength dimension. Hyperspectral imaging differs from multispectral imaging in the resolution achieved by the detector, a property with its own advantages and disadvantages [25]. Higher resolution allows for finer delineation between chromophores for more precise identification of composite materials. However, these detectors are costly and data processing can be time consuming for large samples. Hyperspectral imaging has applications ranging from remote sensing, agriculture, astronomy, and medicine, incorporating reflectance, fluorescence, and/or transmission imaging.

Radosevich and colleagues [26] developed a two-photon data acquisition and spectral unmixing technique capable of delineating the identity and concentrations of fluorophores *in vivo*. Using detectors spanning 350–650 nm and a Ti: sapphire laser tunable from 710 to 920 nm, they were able to extract physiologically meaningful structures with submicrometer resolution from live mice subjects.

De Beule and colleagues [27] developed a hyperspectral fluorescence probe for potential detection and diagnosis of skin cancer. Utilizing two ultrafast lasers at 355 and 440 nm, their probe excites autofluorescence in chromophores in skin tissue such as keratin, collagen, nicotinamide adenine dinucleotide (phosphate), and flavins.

Working on fresh ex vivo human skin lesions, they demonstrated potential to discriminate healthy skin tissue from cancerous tissue from the average fluorescence lifetime.

Randeberg et al. [11] used hyperspectral imaging to visualize skin vasculature and to monitor the development of skin bruises in a porcine model and humans. Images were collected using a push-broom hyperspectral camera VNIR 1600, Norsk Elektro Optikk AS, Lorenskog, Norway in the wave range of 400–1,000 nm and a spectral resolution of 3.7 nm. Images were classified using spectral angle mapping (SAM) based on minimum noise filtered (MNF) data, when pixels were classified according to statistical variance in spectral characteristics. Image analysis was done using principle component analysis (PCA) for data reduction and enhancement of spectral features.

Hyperspectral imaging allows collecting a large amount of data for a short period of time. Generally, Image analysis is based on statistical approaches to provide information from collected datasets, when pixels are classified according to spectral properties and statistical variance. Although these statistical tools are powerful in obtaining information from large datasets, it might be difficult to correctly retrieve absorbance and scattering parameters. The approach demonstrated promising results [11, 21, 28, 29] and should be explored further.

10.6 Ratiometric Approach in Reflectance Spectroscopy

Ratiometric method is one of the methods that estimates total hemoglobin concentration and hemoglobin saturation from optically measured diffusive reflectance spectra. Measurements of reflectance and/or fluorescence at two or more isosbestic points indicate wave lengths at which different chemicals have the same molar extinction. Using isosbestic and non-isosbestic points of oxy- and deoxyhemoglobin allows to quantify both hemoglobin saturation [30, 31] and hemoglobin concentration [32]. The ratio of 420 nm isosbestic points and 431 non-isosbestic points was used [30] to measure oxygen saturation in capillaries of the hamster cheek pouch. Pittman and Duling in 1975 [31] used two isosbestic wavelengths and a non-isosbestic point to determine the hemoglobin saturation. Recently, Phelps et al. in 2010 [33] developed a ratiometric method to assess total hemoglobin concentration from diffusive spectra in the ultraviolet–visible (UV–VIS) wavelength range. Three isosbestic wavelength ratio pairs of 545/390, 452/390, and 529/390 were shown to best correlate to total hemoglobin concentration independent of saturation and scattering. The correlation coefficients between ratiometric method and calibrated Monte Carlo model were shown to be 0.75, 0.76, and 0.88 for the three best ratios, 545/390, 452/390, and 529/390, respectively.

10.7 Optoacoustic Spectroscopy

Optoacoustic spectroscopy has been used as another noninvasive approach for measuring hemoglobin [34–36]. The principle of this technique is based on the absorption of electromagnetic radiation by tissue to produce sound waves. Characteristics of the sound waves produced depend on the absorption properties of tissue. Particularly, hemoglobin strongly absorbs radiation in the near infrared region which enhances thermal modulation of the signal. Esenaliev et al. [35] showed *in vivo* that optoacoustic waves effectively stimulated the superficial radial artery in the NIR because the vessels were close to the surface. With the saturation of 100%, errors associated with oxy- and deoxyhemoglobin absorptions were eliminated. The measured *in vivo* optoacoustic peak-to-peak intensity for various dilutions was shown to follow *in vitro* dilution experiments when hemoglobin concentration was measured directly. Results demonstrated that the slope of the optoacoustic waves linearly depends on the hemoglobin concentrations in the range from 6.2 to 12.4 g/dL.

Although the optoacoustic methods are interesting, the indications for bruising and comparison to optical measurements are not clear. In clinical settings these methods could be more difficult to implement than spectroscopic and imaging methods. Further evaluation and clinical testing will reveal its commercial potential.

10.8 Liquid-Crystal-Based Spectrometers

Liquid-crystal tunable filters (LCTFs) and spatial light modulators (SLMs) based on various material configurations represent efficient, low cost optical components. These components allow development of compact, bedside devices for a rapid noninvasive optical screening of cells, tissues, and other biomaterials. Extensive review of liquid-crystal materials and their biomedical applications was given by Woltman et al. [37]. LCTFs represent a class of optical components that can selectively reflect or transmit a wavelength band of light. LCTFs are used in one of the widely spread Lyot filter configurations in which stacks of variable optical retardance filters are made by combining fixed retardance bire fringent plates with tunable nematic liquid-crystal retarders. Tuning the retardance of each retarder film shifts the transmission profile. Integrating multiple stacks allows modulating superimposed sinusoidal transmission peak. Lyot filters can be fabricated for ultraviolet, visible, and infrared wavelengths. LCTFs-based nondispersive deformed helix-ferroelectric spectrometer has been recently used to image the capillaries beneath the mucosal membrane of the human conjunctiva and determine hemoglobin (Hgb) levels noninvasively [15]. The technique has led to the development of a small handheld system which may have further medical applications. Particularly such a device can be potentially useful in forensic medicine for hemoglobin determination while estimating contusion age.

10.9 Bruising Models

Mathematical models are important components of the determination of the ages of bruises. In order to model contusions the photon propagation in skin and transport of hemoglobin in dermis must be described. The optical photon transport model is essential in determining concentration of chromophores. Among them, hemoglobin and its breakup products are of the most interest since they can serve as markers of the bruise condition.

10.10 Light Propagation Models

Human skin is a complex and heterogeneous structure. Optical properties of the skin are such that scattering prevails absorption in the visible and NIR spectrum [38]. To model the propagation of electromagnetic waves in human tissue, one has to solve Maxwell's equations for a complex inhomogeneous medium combined with proper boundary conditions. However, the complexity of skin structure makes this task difficult, and instead of solving Maxwell's equations directly, simplified models are used. These models can be classified as radiative transport models, diffusion approximation, and Monte Carlo methods.

In radiative transport theory Boltzmann's transport equations are derived using conservation of energy transport rather than electric and magnetic field vectors in Maxwell's equations. However, solution of Boltzmann's equations can be difficult because of strong nonlinearity and various approximations are used [39–41].

Diffusion approximation (DA) was originally developed for analysis of port-wine birthmarks [42] and other skin conditions such as neonatal jaundice [43, 44]. DA was used to simulate reflectance spectra of bruises in human and porcine skins [11, 42, 45]. The last modification of the model [11] assumed three-layer skin model with two planar layers of a defined thickness resting on a semi-infinite layer. The first layer models epidermis, containing melanin and small amount of blood to correct for planar boundaries, which simplify the wavy shape boundary of the basal layer. The second and the third layer contain hemoglobin as the main chromophore. DA model was shown to reproduce experimentally measured reflectance spectra when oxygenation and concentration of blood and bilirubin were properly chosen.

Monte Carlo methods were applied to model photon propagation in normal skin [8, 46, 47] and in bruised tissue [48, 49].

The seven [8] and nine [46] skin layer models were successfully implemented to simulate reflectance of normal skins. In these models the reflectance spectra of human skin in visible and NIR spectral region were calculated using Monte Carlo technique. The model accounted for specular and internal reflection on skin layer surfaces and for variations in spatial distribution of blood, index of blood oxygen saturation, volume fraction of water, and chromophores content. The seven skin layers used in [8] are: stratum corneum, living epidermis, papillary dermis, upper blood net dermis, reticular dermis, deep blood net dermis, and subcutaneous fat. Maeda et al. [46] extended the skin model by including additional skin layers in epidermis: stratum granulosum, stratum spinosum, and stratum basal.

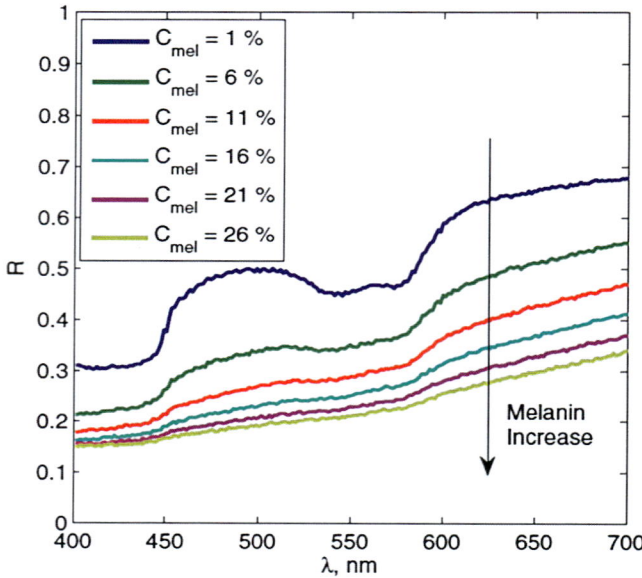

Fig. 10.4 Effect of melanin concentration on the skin reflectance spectrum

In addition, scattering coefficients were specified as a function of wave length for each layer.

Recently, Kim et al. [49] have demonstrated that Monte Carlo method can be used to model spectra of bruises in human skin. The reflectance spectrum of developed bruises in a healthy adult was simulated and the concentration of bilirubin, blood volume fraction, and blood oxygenation parameter S were determined for different times as the bruises progressed. In these simulations the thickness of stratum corneum was 10 μm, living epidermis: 70 μm, papillary dermis: 100 μm, upper blood net dermis: 100 μm, reticular dermis: 1.6 mm, deep blood net dermis: 120 μm and subcutaneous fat: 6 mm; S parameter was 60%.

To optimize model parameters, sensitivity analysis was performed by varying the thickness of skin layers, oxygenation parameter S, and melanin concentration. The epidermal melanin concentration was varied from 0 to 31% to simulate different skin types (Fig. 10.4): from amelanotic to darkly pigmented. For higher melanin content, the hemoglobin signal became weaker and was significantly suppressed when melanin concentration became greater than 20%. These results

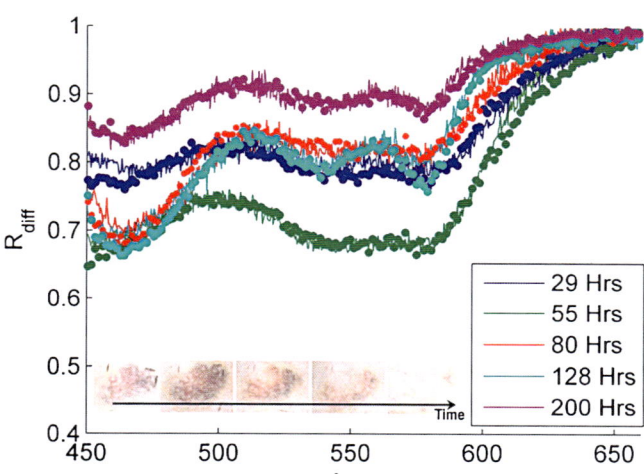

Fig. 10.5 Example of the experimental and simulated reflectance spectra of a bruised skin at different time moments. Experimental data are shown by solid line, and modeling results are shown by *circles*

were consistent with *in vivo* reflectivity measurements Svaasand et al. [42] for different types of normal human skin using an integrating sphere.

Before simulating bruised skin, the MC model was optimized using normal skin spectra. The reflectance spectra of bruised skin then were simulated by varying blood oxygenation and by increasing the blood and bilirubin content. To simulate intracutaneous bruises, the content of extravascular blood was varied from 1 to 60%. Bilirubin absorption was modeled by adding 0.05–0.4 g/L of bilirubin to the dermal layers, assuming equal bilirubin and extravascular blood volume fractions. The oxygenation parameters were varied from 40 to 80%. Simulations were performed in wavelength ranges from 440 to 660 nm at intervals of 1.4 nm and agreed well with the experimental data (Fig. 10.5).

In the model cases, both bilirubin and blood volume fraction achieved their peak values of 0.4 g/L after 80 h and of 0.6 after 55 h of bruising, respectively (Fig. 10.6). These results are consistent with findings by Yajima et al. [19] and Randeberg et al. [50] who measured bilirubin index in bruise as a function of time in adults. The blood oxygenation parameter was found to be 33% lower than normal value

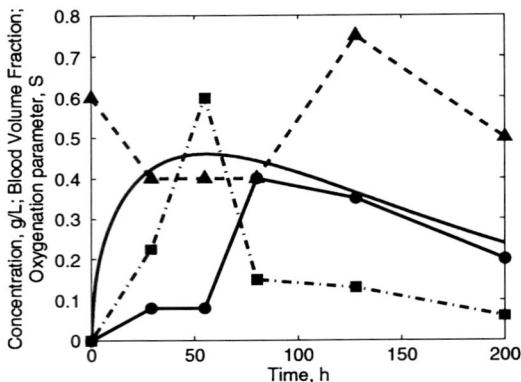

Fig. 10.6 The change of bilirubin concentration (*circles*), extravascular blood volume fraction (*squares*), and oxygenation parameter S (*triangles*) as a function of time. The solid line shows bilirubin concentration as a function of time predicted by an analytical blood diffusion model by Randeberg et al. [51]

after 80 h of bruise formation. These results demonstrated that Monte Carlo-based model can provide not only details of tissue optics but can also aid in predicting the age of bruises when coupled with blood transport models.

10.11 Blood Transport Models

To improve the reflectance spectroscopy method for contusion problems, the physical model for a contusion tissue has to be developed. Excessive skin pressure or impact can traumatize the tissue and results in blood vessels damage and rapid blood sipping in subcutaneous skin layers. Currently a few studies have been done to describe blood transport inside the skin tissue after injury. In these models Darcy's law and diffusion equations for Hgb transport are solved analytically, which results in expression for the chromophore concentration as a function of space and time [43, 51].

Randenberg et al. [43] used analytical model to predict bruise aging. The model was verified using data from skin hematomas in cardiothoracic patient for which reflectance spectra were measured in the wave range from 400 to 850 nm from normal and bruised skin using integrating sphere setup. In these studies the skin hematoma was caused by external trauma, cardiothoracic examinations, or surgery. The model was shown to predict the data with an accuracy of approximately 1 day.

Stam et al. [52] developed numerical model to simulate bruise healing process. The model was an improvement of Randeberg's model [51] and involved solution of Darcy's law-based convection and Fick's first law-based diffusion equations combined with Michaelis–Menten kinetics of the enzymes involved in conversion of hemoglobin to bilirubin. The model used three layers skin model consisting of the top layer of the dermis, the bottom layer of the dermis, and subcutaneous tissue layer. The initial condition of the bruise was modeled as an instantaneous circular or non-homogeneous pool of hemoglobin in subcutaneous layer. The concentration of hemoglobin was changed due to convection in vertical direction from subcutaneous layer into dermis, vertical and horizontal diffusion between and along layers, and conversion of hemoglobin to bilirubin. Results of the simulation were shown to simulate a non-homogeneous bruise for a given set of parameters. Although simplified and needed improvement, the model represents promising approach to determine bruising age. For this, further parameter optimization can be done by combining simulations with spatially resolved reflectance spectroscopy measurements and with photon propagation model for the skin.

10.12 Summary and Challenges

Although the spectroscopic measurements allow extraction of quantitative information about hemoglobin concentration in the human tissue [22], derivation of bruise age remains a challenging problem. The spatial heterogeneity of bruising spectral breakdown makes the estimation of the age of a bruise by collecting signal from a single spot difficult. Because skin vessels are inconsistently damaged during trauma, the distribution of hemoglobin and its breakdown components also distribute non-uniformly and diffuse inside the skin tissue both laterally and in depth. The location to use for predicting the age of the bruise has not yet been determined. Therefore, it is important to account for these effects when interpreting the measured spectra. To determine spatial variations of chromophores over the contusion site region, the use of spectral imaging might be more efficient rather than the collection of spectra from single point measurements. Although giving higher space resolution, these devices can be less cost-effective and mobile and might require additional skills from examining doctors.

As well, one should be careful when using bilirubin spectrum to determine the age of the bruise. In large bruises, the hemoglobin absorption can be high enough to shield bilirubin spectral signature and impede its detection even when using spectroscopy [42]. Bilirubin photobleaching has to be accounted for in sun exposed bruises which will behave differently from nonexposed bruises [50].

Another factor that makes the bruise age estimation difficult is the type of human skin or the amount of melanin. Most of the works on human bruises were done for Caucasian skin type. Further efforts are required to extent the methods to other, more pigmented skin types in which bilirubin and hemoglobin detection can be difficult in VIS and NIR spectrum. The implementation of inverse photon transport model is required to obtain true absorption coefficients of bruised skin.

Determining the age of a bruise might require estimation of the depth of the contusion. This can be achieved by using ultrasound or optical coherence tomography to identify tissue regions of vessels rupture. The obtained information can be incorporated into the spectral prediction model to define location of blood extravasation.

The combination of mathematical models of photon and blood transport in skin tissue and optical tools is believed to provide fundamentals in tissue optics and evaluation of contusions. Development of a cost effective medical tool that would allow for reflectance measurements of contused skin in VIS and NIR spectral ranges would be of great benefit for forensic and medical expertise. One of such potential tools can use LCTFs, which are low cost optical components that allow development of compact, bedside spectroscopic devices for a rapid noninvasive optical screening of cells and tissues. Thus, there is a need in further studies of bruise physiology and in development of optical tools to fully understand the process of tissue damaging and healing and to improve noninvasive bruise age determination methods.

Acknowledgements Authors were partially supported by the Gerber Foundation and NSF DMS grant 0800612.

References

1. Maguire S. Bruising as an indicator of child abuse: when should I be concerned? Paediatr Child Health. 2008;18(12):545–9.
2. Murty OP, Jia MC, Asyraf EM, Kim YP, Chee YT. Physical injuries in fatal and non-fatal child abuse cases: a review of 16 years with hands on experience of 2 years in Malaysia. Int J Med Toxicol Legal Med. 2006;9:33–43.
3. Langlois NE, Gresham GA. The aging of bruises: a review and study of the color changes with time. Forensic Sci Int. 1991;50:227–38.
4. Munang LA, Leonard PA, Mok JYQ. Lack of agreement on colour description between clinicians examining childhood bruising. J Clin Forensic Med. 2002;9:171–4.
5. Grossman SI, et al. Can we assess the age of bruises? An attempt to develop an objective technique. Med Sci Law. 2011;51:170–6.
6. Bariciak ED, Plint AC, Gaboury I, Bennett S. Dating of bruises in children: an assessment of physician accuracy. Pediatrics. 2003;112:804–7.
7. Bohnert M, Baumgartner R, Pollak S. Spectrophotometric evaluation of the color of the intra- and subcutaneous bruises. Int J Legal Med. 2000;113:343–8.
8. Meglinski IV, Matcher SJ. Computer simulation of the skin reflectance spectra. Comput Methods Programs Biomed. 2003;70:179–86.
9. Berg S. Grundriss der rechtsmedizin. Munchen: Muller and Steinicke; 1976.
10. Knight B. Forensic pathology. London: Arnold; 1996.
11. Randeberg LL, Larsen EP, Svaasand LO. Characterization of vascular structures and skin bruises using hyperspectral imaging, image analysis and diffusion theory. J Biophotonics. 2010;3(1–2):53–65.
12. Kanashima H, Yamane T, Takubo T, Kamitani T, Hino M. Evaluation of non-invasive hemoglobin monitoring for hematological disorders. J Clin Lab Anal. 2005;19:1–5.
13. Nadeau RG, Groner W. The role of a new noninvasive imaging technology in the diagnosis of anemia. J Nutr. 2001;131:1610S–4.
14. Rice MJ, Sweat RH, Rioux JM, Williams WT, Routt W. Non-invasive measurement of blood components using retinal imaging. 2002; United States Patent No. US 6,477,394 B2.
15. McMurdy J, Jay GD, Suner S, Crawford G. Noninvasive optical, electrical, and acoustic methods of total hemoglobin determination. Clin Chem. 2008;54(2):264–72.
16. Mimasaka S, Ohtani M, Kuroda N, Tsunenari S. Spectrophotometric evaluation of the age of bruises in children: measuring changes in bruise color as an indicator of child physical abuse. Tohoku J Exp Med. 2010;220:171–5.
17. Westerhof W. CIE colorimetry. In: Jemec GBE, Serup J, editors. Handbook of non-invasive methods and the skin. Boca Raton: CRC; 1995. p. 385–97.
18. Weatherall IL, Coombs BD. Skin color measurements in terms of CIELAB color space values. J Invest Dermatol. 1992;99:468–73.
19. Yajima Y, Funayama M. Spectrophotometric and tristimulus analysis of the colors of subcutaneous bleeding in living persons. Forensic Sci Int. 2006;156:131–7.

20. Langlois VK, Hughes NE. Use of reflectance spectrophotometry and colorimetry in a general linear model for the determination of the age of bruises. Forensic Sci Med Pathol. 2010;6(4):275–81.
21. Randeberg LL, Winnem AM, Larsen ELP, Haaverstad R, Haugen OA, Svaasand LO. In vivo hyperspectral imaging of traumatic skin injuries in a porcine model. Progress in biomedical optics and imaging—proceedings of SPIE. 2007;6424:642408. doi:10.1117/12.699380.
22. McMurdy JW, Jay GD, Suner S, Trespalacios FM, Crawford GP. Diffuse reflectance spectra of the palpebral conjunctiva and its utility as a noninvasive indicator of total hemoglobin. J Biomed Opt. 2006;11:014019-1–8.
23. McMurdy JW, Duffy S, Crawford GP. Monitoring bruise age using visible diffuse reflectance spectroscopy. In Biomedical Optics (BiOS). International Society for Optics and Photonics. 2007;6464:643426. doi:10.1117/12.701592.
24. Randeberg LL, Winnem AM, Langlois NE, Larsen ELP, Haaverstad R, Skallerud B, et al. Skin changes following minor trauma. Lasers Surg Med. 2007;39(5):403–13.
25. Gowen AA, O'Donnell CP, Cullen PJ, Downey G, Frias JM. Hyperspectral imaging—an emerging process analytical tool for food quality and safety control. Trends Food Sci Technol. 2007;18:590–8.
26. Radosevich AJ, Bouchard MB, Burgess SA, Chen BR, Hillman EMC. Hyperspectral in vivo two-photon microscopy of intrinsic contrast. Opt Lett. 2008;33:2164–6.
27. De Beule PA, Dunsby C, Galletly NP, Stamp GW, Chu AC, Anand U, et al. A hyperspectral fluorescence lifetime probe for skin cancer diagnosis. Rev Sci Instrum. 2007;78:123101-1–7.
28. Payne G, Langlois N, Lennard C, Roux C. Applying visible hyperspectral (chemical) imaging to estimate the age of bruising. Med Sci Law. 2007;47:225–32.
29. Randeberg LL, Baarstad L, Løke T, Kaspersen P, Svaasand LO. Hyperspectral imaging of bruised skin. Proceedings of SPIE. 2006;6078:60780O. doi:10.1117/12.646557.
30. Ellsworth ML, Pittman RN, Ellis CG. Measurement of hemoglobin oxygen saturation in capillaries. Am J Physiol. 1987;252:H1031–40.
31. Pittman RN, Duling BR. Measurement of percent oxyhemoglobin in the microvasculature. J Appl Physiol. 1975;38:321–7.
32. Liu Q, Vo-Dinh T. Spectral filtering modulation method for estimation of hemoglobin concentration and oxygenation based on a single fluorescence emission spectrum in tissue phantoms. Med Phys. 2009;36(10):4819–29.
33. Phelps JE, Vishwanath K, Chang VTC, Ramanujam N. Rapid ratiometric determination of hemoglobin concentration using UV-VIS diffuse reflectance at isosbestic wavelengths. Opt Express. 2010;18:18779–92.
34. Deyo DJ, Esenaliev RO, Hartrumpf O, Motamedi M, Prough DS. Continuous noninvasive optoacoustic monitoring of hemoglobin concentration. Anesthesiol Analgesia. 2001;92:139.
35. Esenaliev RO, Petrov YY, Hartumpf O, Deyo DJ, Prough DS. Continuous, noninvasive monitoring of total hemoglobin concentration by an optoacoustic technique. Appl Opt. 2004;43(17):3401–7.

36. Petrova IY, Esenaliev RO, Petrov YY, Brecht HPE, Svensen CH, Olsson J, et al. Optoacoustic monitoring of blood hemoglobin concentration: a pilot clinical study. Opt Lett. 2005;30(13):1677–9.
37. Woltman S, Jay DG, Crawford GP. Liquid-crystal materials find a new order in biomedical applications. Nature. 2007;6:929–38.
38. Tuchin VV. Light scattering study of tissues. Phys-Uspekhi. 1997;40:495–515.
39. Ishimaru A. Wave propagation and scattering in random media. New York: Academic; 1978.
40. Farrell TJ, Patterson MS, Wilson B. A diffusion theory model of spatially resolved, steady-state diffuse reflectance for the noninvasive determination of tissue optical properties in vivo. Med Phys. 1992;19(4):879–88.
41. Kienle A, Patterson MS. Improved solutions of the steady-state and the time-resolved diffusion equations for reflectance from a semi-infinite turbid medium. J Opt Soc Am. 1997;14(1):246–54.
42. Svaasand LO, Norvang LT, Fiskerstrand EJ, Stopps EKS, Berns MW, Nelson JS. Tissue parameters determining the visual appearance of normal skin and port wine stains. Lasers Med Sci. 1995;10:55–65.
43. Randeberg LL, Roll EB, Nilsen LT, Christensen T, Svaasand LO. In vivo spectroscopy of jaundiced newborn skin reveals more than a bilirubin index. Acta Paediatr. 2005;94(1):65–71.
44. Randeberg LL, Bonesrønning JH, Dalaker M, Nelson JS, Svaasand LO. Methemoglobin formation during laser induced photothermolysis of vascular lesions. Lasers Surg Med. 2004;34(5):414–9.
45. Svaasand T, Spott L. Collimated light sources in the diffusion approximation. Appl Opt. 2000;39:6453–65.
46. Maeda T, Arakawa N, Takahashi M, Aizu Y. Monte Carlo simulation of spectral reflectance using a multilayered skin tissue model. Opt Rev. 2010;17:223–9.
47. Wang L, Jacques SL, Zheng L. MCML—Monte Carlo modeling of light transport in multi-layered tissues. Comput Methods Programs Biomed. 1995;47:131–46.
48. Lines C, Kim O, Alber M, Crawford G. Modeling and measuring extravascular hemoglobin: aging contusions. Proceeding of SPIE. 2011;8087:80872T. doi:10.1117/12.896610.
49. Kim O, McMurdy J, Lines C, Duffy S, Crawford G, Alber MS. Reflectance spectrometry of normal and bruised human skins: experiments and modeling. 2012;33:159–75. doi:10.1088/0967-3334/33/2/159.
50. Randeberg LL, Skallerud B, Langlois NEI, Haugen OA, Svaasand LO. The optics of bruising. In: Welch AJ, van Gemert MJC, editors. Optical-thermal response of laser-irradiated tissue. 2nd ed. Berlin: Springer; 2011. p. 825–8.
51. Randeberg LL, Haugen OA, Haaverstad R, Svaasand LO. A novel approach to age determination of traumatic injuries by reflectance spectroscopy. Lasers Surg Med. 2006;38(4):277–89.
52. Stam B, van Gemert MJC, van Leeuwen TG, Aalders MCG. 3D finite compartment modeling of formation and healing of bruises may identify methods for age determination of bruises. Med Biol Eng Comput. 2010;48:911–21.

Chapter 11
Model of Tumor Growth and Response to Radiation

L.J. Liu, S.L. Brown, and M. Schlesinger

11.1 Introduction

Cell growth environment may be taken as a micro-electrochemical system. Interstitial fluid is the solution including electrolytes. With the surrounding aqueous medium, surfaces such as biological membranes usually develop a potential difference, which has been related to some fundamental physiological processes such as adenosine triphosphate (ATP) synthesis and growth yield. External electric fields that are applied to eukaryotic cells can modulate some cellular processes, such as development, regeneration, and motility [1]. In fact, cell growth and development are related to electrochemical processes. Understanding the mechanism of cell growth and development, as well as cell response to therapy, is helpful in applying these processes to their best advantage. The purpose of this chapter is to study the laws of cell (including tumors) growth and cell response to therapy.

Tumors have properties of organisms possessing specific characteristics, such as growth and metastasis. Many mathematical models for tumor growth are derived from a reaction–diffusion differential equation. For many solid tumors, there is a necrotic core where most cells are dead and there are no functional exchange vessels. According to some views, tumor cords are one of the fundamental microarchitectures of solid tumors, consisting of a microvessel nourishing nearby tumor cells. By viewing a large number of human bronchial cancer samples, Thomlinson and Gray [2] found that the necrotic core enlarges when the tumor chord grows larger, since the distance for oxygen diffusion is limited. Therefore, the thickness of the sheath of viable tumor cells remains constant [3]. Based on experimental results, Thomlinson and Gray presented a mathematical model of the diffusion and consumption of oxygen [2]. Burton [4] developed this model by considering both the distribution of oxygen in a spherical tumor and the "relative radius of the central zone to the total radius" [5]. To explain the existence of a steady-state tumor size, Greenspan [6] extended these models by considering "a surface tension among the living cancer cells in order to maintain a compact, solid mass" and that "necrotic cellular debris continually disintegrates into simpler chemical compounds that are freely permeable through cell membranes" [5]. However, these models are limited to avascular solid tumors whose nutrition and oxygen are diffused from the stroma. The modeling of avascular tumors is just the first step toward building models for fully vascularized tumors since there are three distinct stages (avascular, vascular, and metastatic) to cancer development [7]. Later, many other models were developed by modifying or extending Greenspan's model. Some new models based on diffusion–reaction differential equation and mass balance equation also exist [6–15]. In addition, many other empiric and phenomenological models are presented based on random onset and statistics [14–21], as well as curve fittings and experimental data analysis [22–25]. However, these models can only qualitatively reflect the process of tumor growth with three distinct phases: "an initial exponential growth phase, followed by some degree of retardation, culminating in a final phase where retardation by both mitotic inhibition and cell death ultimately gave rise to dormancy" [5]. This may explain why Gompertz's function is still a viable option for describing tumor growth in some cases today.

Though some models have many parameters, they might still be unable to sufficiently describe the growth process. One may ask as to how many parameters should be considered and what the exact pattern for each specific parameter should be. In fact, no matter how many parameters are used to determine the growth of an organism, they should embody the production and death rates, which reflect

11 Model of Tumor Growth and Response to Radiation

metabolism. Usually, for a biosystem, the number density of cells does not vary a whole lot, but the size (volume) does change as it grows.

Ever since Laird [26] first applied Gompertz's function to fit the growth of tumors in 1964, theoretical and experimental studies have much advanced the development of Gompertzian functions to tumor growth [22, 25, 27]. The Gompertzian model for tumor growth by mass may be written as:

$$m(t) = M\exp[\ln(m_0/M)\exp(-\upsilon t)] \quad (11.1)$$

where $m_0 = m(0)$, M represents the maximum mass that can be reached with the available nutrients and is expressed as $M = \lim_{t\to\infty} m(t)$, and υ is a constant related to the proliferative ability of the cells. However, Steel [28] and Wheldon [29] noted that Gompertz's function might not be adequate for modeling the growth of small tumors or cell systems. Wheldon [29] proposed a critical size when applying Gompertz's function.

West et al. [30] presented a general model to describe organism growth (mass-time relation) by combining energy conservation with Kleiber's law [31], which reads [30]

$$\frac{dm}{dt} = am^{3/4}\left[1-\left(\frac{m}{M}\right)^{1/4}\right] \quad (11.2)$$

The solution of this is [30]

$$1-\left(\frac{m}{M}\right)^{1/4} = \left[1-\left(\frac{m_0}{M}\right)^{1/4}\right]\exp\left(-\frac{at}{4M^{1/4}}\right) \quad (11.3)$$

where m_0 is the initial mass, M is the asymptotic mass that corresponds to the maximum size, and a is a parameter which is related to the growth characteristics of the organism or tumor [30, 32]. When $m = (3/4)^4 M \approx M/3$, $d^2m/dt^2 = 0$, the growth rate is maximum. It seems to be a universal law of organism growth. Mammals, birds, fish, molluscs [30], and their tumors all follow this growth law. West et al.'s model is based on a more fundamental biological law and shows that it is a "universal law" for ontogenetic development. The maximum growth rates for Gompertz's curve and West's model occurs at $(1/e)M \approx 0.37M$ and $(81/256) \approx 0.32M$, respectively. The larger the mass, the less difference there is between these two models. We compare the growth rates of Gompertzian model and West et al.'s model in Table 11.1 ($\upsilon = aM^{-1/4}/4$).

Table 11.1 The growth rates of Gompertzian model and West et al.'s model

m(t)/M		1/100	1/32	1/16	1/8	1/4	1/3	1/2	3/5	4/5	0.9	1
Growth rate (υM)	Gompertzian	0.046	0.108	0.173	0.260	0.347	0.366	0.347	0.306	0.179	0.095	0
dm/dt	West et al.'s	0.086	0.172	0.250	0.341	0.414	0.421	0.378	0.327	0.184	0.096	0

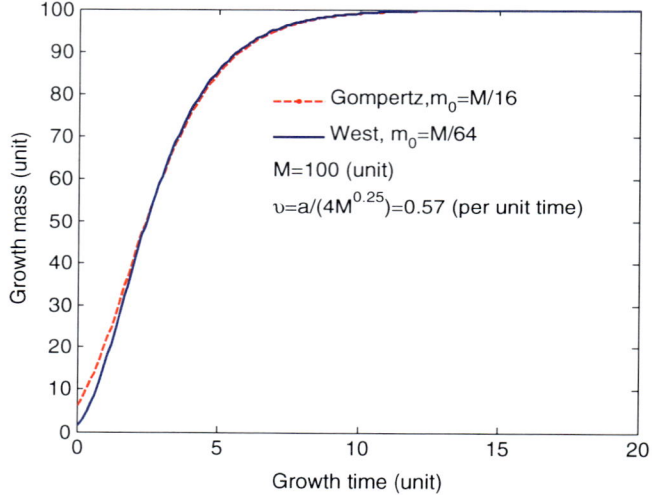

Fig. 11.1 Tumor growth curves for Gompertz and West models

From Table 11.1, we see that the growth rate of Gompertzian model from $M/16$ to $M/4$ fit that of West's model from $M/32$ to $M/8$ quite closely. This means that these two models can fit well in a certain time period by choosing some specific parameters m_0, M, and υ. Figure 11.1 shows that Gompertzian model and West's model can fit well after $m(t) > M/3$ when we choose specific m_0.

Both Gompertzian and West's models are widely used. Which one reflects the actual growth of an organism or cell system? Based on the mass balance equation, we derive our general equation for organism growth (including tumor growth) and show that it is the same as West et al.'s. In this model, we see that $a = \beta^\circ M^{1/4} = 4\upsilon M^{1/4}$, where $\beta^\circ = 4\upsilon$ is a constant which controls the growth of a given organism.

Guiot et al. [32] applied West's model for organism growth to tumor study and found that tumor growth also follows this "universal law."

Even for some small tumors such as 9L and U118, the growth curves fit West's model quite well. Tumor possesses some characteristics of an organism with specific properties in the growth process. Various tumors and their specific growth process make tumors rather complicated. The growth of a tumor mimics the growth of an organism and mathematical models have potential to predict their curative and subcurative response to therapies.

West's model idealized the entire organism as an identical cell system. Obviously, this is too inexact to describe multicellular organisms. Different tissues/organs have different kinds/types of cells. Malignant tumors are characterized by tumor cell heterogeneity [33, 34]. Also, it is recognized that in one single tumor there are many different regions that cannot be described by a simplified identical cell system. It is well established at present that tumor cells can be categorized by three conditions: (a) viable oxygenated cells, (b) viable hypoxic cells, and (c) dying or necrotic cells which stay in the tumor and form a necrotic region. These three types of cells have distinctly growth rates. When a tumor metastasizes, the model cannot be applied. Also, a good model should reflect the response of an organism or cell system to treatment. The survival response to therapy can be used to verify that it is a model for organism growth. Although no theory, at present, can uniformly explain tumor growth, tumor metastasis, and the results of tumor therapies, we provide the formalism for a more general equation for these cases. A general model which possesses the attributes of West's model and can describe practical application is the goal of this chapter.

11.2 The Model

We assume initially that all cells in a specific organ/tissue are identical and treat them as a system. We use the subscript k to identify this system. In this small system, cells have a constant number density C_k and the mass of a cell is m_{ck}. When they grow, the number density does not change or changes only a little. This causes both the volume and mass to increase. Organism growth is the process of biological metabolism in terms of cell division. Some fluids such as blood and interstitial fluid in normal tissue may circulate or flow in and out in a balanced way, maintaining the metabolism by supplying nutrients and draining waste. Suppose at time t, the volume of this system is V_k, the number of cells that died in a unit volume in unit time is E_k, and the average number of cells produced in a unit volume in unit time is $\overline{S}_k$. The mass is $m_k = V_k C_k m_{ck}$. In the natural growth process, cell location in one specific tissue/organ is relatively steady. Cells do not flow from one tissue to another. Thus, the cell flux is 0. The only exchange with the outside is when nutrition is supplied or when wastes are drained, maintaining the metabolism. This process embodies the production and death rates of cells, which determines the growth of the system. Since the cell flux is 0, the growth rate of this system can be written as:

$$\frac{dm_k}{dt} = m_{ck} V_k (\overline{S}_k - E_k) \tag{11.4}$$

A tumor can also be divided into different systems since it may be composed of various tumor cells [33, 34], each made of identical cells. Similarly, we can use (11.4) to describe the growth of each small system in a tumor before metastasis since there is no tumor cell flux during this growth stage. For an identical system, we can use the number of cells (which is the most common method), mass, or volume to express its growth. Once we know the percentage of specific cells in a tumor at a certain time, we can estimate the amount of those cells later on. Tumor growth is angiogenesis-dependent. Tumor blood vessels are abnormal and leaky. They are longer, larger, and denser than normal microvessels [35]. On the one hand, this results in elevated tumor interstitial fluid pressure; on the other hand, it may be a source of nutrition for tumor growth, which reflects the production and death rates. Equation (11.4) can be applied before metastasis occurs. According to the activity of cells, a tumor may be divided into three regions: a necrotic core where most tumor cells are dead, a quiescent cell region where most cells are non-proliferating, and a well-vascularized region where tumor cells are active [35]. Tumor cells have different growth environments and nutrition supplies in different regions. Consequently, parameters such as $\overline{S}_k$, E_k are not the same. Some environments may even cause additional cell death rate. In the quiescent cell region, $\overline{S}_k \approx E_k$ (the production and death rates are almost balanced). In the necrotic core, it is probable that $\overline{S}_k < E_k$ (cells are dying or dead). In the well-vascularized region, $\overline{S}_k > E_k$ (where tumor cells contribute the most towards tumor growth), which means we should concentrate on them. Nevertheless, in tumor therapy, the quiescent cells cannot be ignored since quiescent cells may be radio-resistant and become active again under certain conditions.

In fact, (11.4) is a mass balance equation. The system's net growth rate is proportional to the difference between the production and death rate, as well as the volume of cell system. We will show the advantage of introducing the production and death rates in application to tumor therapies.

Now, let us determine the death rate E_k and production rate $\bar{S}_k$ in (11.4). For convenience, we ignore the subscript k from now on unless we need to use it for specific labeling. We adopt West's viewpoint, which is simply logical, and describe it as: "the cell death rate is proportional to the number of cells present" [30]. In a specific system, the average lifetime of every single cell is fixed, and so is the number of cells in a unit volume C. Thus, it is reasonable to assume that the death rate E of cells in a unit volume in unit time is a constant in this case. The most important matter is to determine the production rate $\bar{S}$. Most biological phenomena scale as a quarter power of the mass [30, 31, 36–41]. Though this is an empirical law that is similar to but more general than Kleiber's law, it is ubiquitous in biological allometry. Based on the ¼ power law, some empirical equations fit the ontogenetic growth trajectories for organisms better than any biological model [30, 42–45]. West et al. indicated that "rates of cellular metabolism and heartbeat (scale) as $M^{-1/4}$ and whole-organism metabolism rate as $M^{3/4}$." [36] Considering the general ¼ power law in biological allometry, the production rate of cells during metabolic processes should be the reciprocal of the ¼ power of the mass, namely,

$$\bar{S} \propto 1/m^{1/4} \quad \text{or} \quad \bar{S} = \lambda m^{-1/4}$$

where λ is a proportionality factor. Also, considering the condition that $m = M$ at $dm/dt = 0$, we have $\lambda = EM^{1/4}$. Thus,

$$\bar{S} = E\left(\frac{M}{m}\right)^{1/4} \tag{11.5}$$

where M is the maximum mass reached asymptotically among species within a system [30], which corresponds to the mass when $dm/dt = 0$. The production rate is much greater than the death rate at an early growth stage. It slows down gradually while the mass increases. When the mass reaches the asymptotic mass, the production rate equals the death rate. Substituting (11.5) into (11.4) results in

$$\frac{dm}{dt} = \frac{m}{C}\left[E\left(\frac{M}{m}\right)^{1/4} - E\right] \tag{11.6}$$

Next we show that (11.6) is exactly the same as West's equation [30]. In that reference, West et al. discussed the death rate of cells and discovered that death rate in the entire system is $aM^{-1/4}N_c$, where N_c is the total cell number. It should be equal to EV in our model. In this case, $EV = aM^{-1/4}N$. Since $C = N_c/V$,

$$\frac{E}{C} = aM^{-1/4} \text{ and } \frac{\lambda}{C} = a$$

Thus, (11.6) is the same as (11.2). Therefore, (11.3) is also a solution of (11.6). For a tumor, m_0 and M are the initial and final masses [32]. If the number density of cells, the cell mass, the average production, and death rates are all fixed in the absence of therapy for a specific cell system, then its growth process is fixed.

West et al. view all cells in an organism as the same. However, as we have remarked above, cells in one tissue are different from those of another. Therefore, the number density, the production and death rate, and the mass may all be different. Each specific tissue has its own asymptotic maximum and initial mass. Generally, for tissue i, we write the governing equation for growth as:

$$\frac{dm_i}{dt} = \frac{E_i}{C_i}m_i\left[\left(\frac{M_i}{m_i}\right)^{1/4} - 1\right] \tag{11.7}$$

Then, the growth equation for the entire organism is expressed as:

$$\frac{dm_t}{dt} = \frac{d}{dt}\left(\sum_{i=1}^{N} m_i\right) = \sum_{i=1}^{N} \frac{E_i}{C_i}m_i\left[\left(\frac{M_i}{m_i}\right)^{1/4} - 1\right] \tag{11.8}$$

Here N is the number of different tissue types in a given organism. Let us assume that all types of tissue grow isotropically and harmoniously, and that there is a standardized mass $m°(t)$. The relationship between it and the mass of each tissue is $m_i(t) = k_i m°(t)$. Thus, $M_i = k_i M°$ and $m_t = m°(t)\sum_{i=1}^{N} k_i$, which can lead to $M_t = M°\sum_{i=1}^{N} k_i$. We call this the similarities of growth. Here k_i is the proportionality coefficient of growth of tissue i and $M°$ is the asymptotic mass of $m°(t)$.

Since the death rate of cells is proportional to the number of cells, cells in a tissue with a higher number density also have a higher death rate. In each specific tissue, the number density C_i is fixed. It is reasonable to suppose that $E_t/C_i = \beta°$ (Different tissues share the same $\beta°$). For a specific organism, $\beta°$ of its various tissues should remain constant; otherwise, the entire system could not grow harmoniously or isotropically. Based on these two assumptions, we have:

$$\frac{dm_t}{dt} = \beta° m° \left[\left(\frac{M°}{m°}\right)^{1/4} - 1\right] \sum_{i=1}^{N} k_i = \beta° m_t \left[\left(\frac{M_t}{m_t}\right)^{1/4} - 1\right] \quad (11.9)$$

By comparing (11.2) with (11.6), $\beta°$ can be determined by the relation $\beta° = E/C = aM^{-1/4}$. For the entire organism to follow the growth law, it is necessary for different tissues to share a similarity in growth. $M°$, $m°$, and $\beta°$ should be the parameters of the original cells, which are often stem cells, of an organism. According to the explanation of West et al. [30], the asymptotic maximum mass M of an organism is constant at extreme nutrition condition. However, the asymptotic maximum mass of each tissue might be affected by the nutrition supply. This makes the growth of an entire organism complicated.

Based on the rules for similarity of growth and (11.3), it is easy to prove that the fractions of different cells in a system are invariable if assorted cells have the same $\beta° = E/C$ and their asymptotic masses share the same fractions. For example, if a system is composed of two kinds of cells with fractions $x_1 = 9x_2$, then $M_1 = 9M_2$ and $a_1 \approx 1.732 a_2$ (since $\beta°_1 = \beta°_2$).

Nutrition supply might be spatially dependent. Tumor cells proliferate and grow actively when they are close to nutrition (including oxygen) [2]. Thomlinson and Gray [2, 3] recognized that the necrotic center enlarges when a tumor cord grows larger, so that the thickness of the sheath of viable tumor cells remains essentially constant [2, 3]. In fact, the asymptotic mass is nutrition-dependent. We can take it as a function of nutrition. Cells that are close to a nutrition source have plenty of nourishment. Correspondingly, they have a larger asymptotic mass M and growth parameter a than the cells that are farther away. Based on (11.7), it is easy to conclude that they grow faster. Even so, it does not mean that these cells will crowd around the nutrition source and be scattered if they are away from the source. In fact, all cells still have the same size and density (Here is the case for one type of cell system.). They are distributed evenly and, following (11.9). This is a very important point. It makes a given system have a constant (or relatively steady) density and death rate, whereas the production rate, and correspondingly the net growth rate, changes spatially based on the distribution of nutrition. This is the real meaning of harmonious growth. In fact, (11.9) is average effect. According to the rules for similarity of growth, if the nutrition supply is spatially dependent, the proportionality coefficient k will also be spatially dependent. The sum of k_i %$_k$ i) becomes an integral ($\int dk$). The total asymptotic mass is $M_t = M° \int dk$ and total instant mass is $m_t = m_t° \int dk$. If we know the distribution function of k, we can find the integral. The M_t and m_t still follow (11.9). Similarly, for multicellular population systems, $M_t = M° \sum \int dk_j$ and $m_t = m_t° \sum \int dk_j$. If these cells follow the rules for similarity of growth, (11.9) is satisfied. Things might become complicated if the identical cells are separated by other cells spatially (e.g., tumor metastasis). They may grow at different times and growth rates.

11.3 Application

11.3.1 Application to Tumor Growth

The theory that tumors originate from tumor stem cells is gaining acceptance [46–49]. This, in turn, implies that tumor growth is an ontogenetic process in an independent system. Therefore, tumor growth also follows the "universal" law for organism growth [32]. Multicellular organisms are composed of various organs/tissues. Consuming nutrients and excreting waste are necessary in order to maintain a regular metabolism. They are included in the production and death rates. West's model applies to organism growth in an unrestricted dietary condition, which makes the asymptotic mass M maximum [30, 32]. Any lack of nutrition causes the asymptotic mass to decrease. Correspondingly, the production rate $\overline{S}$ decreases. Once the asymptotic mass becomes smaller than the present mass $m(t)$, the production rate becomes smaller than the death rate ($\overline{S} < E$); then the growth rate is less than zero ($dm/dt < 0$).

In the case of a primary tumor, no tumor cell flows or diffuses into it from the outside. Also, no tumor cells flow out before metastasis. Therefore, (11.6) can be applied. If we let $E_k / C_k = \beta_k°$, the equation can be rewritten as:

$$\frac{dm_k}{dt} = \beta_k° m_k \left[\left(\frac{M_k}{m_k} \right)^{1/4} - 1 \right] \qquad (11.6')$$

11 Model of Tumor Growth and Response to Radiation

In a tumor mass, there may be both normal and tumor cells. For normal cells, no matter how many different types there are, they all have the same $\beta°$ and follow (11.9). We use subscripts "N" and "T" to identify the normal cells and tumor cells, respectively. Assume there are G kinds of tumor cells. They all follow (11.6) or (11.3), but may or may not follow (11.9). Therefore, the total mass of the entire tumor m_t is: $m_t = m_N + m_T = m_N + \sum_{j=1}^{G} m_j$. Here, we only discuss two special cases:

1. All normal and tumor cells of the tumor follow the rules for similarities of growth. Therefore, tumor growth follows (11.9) and can be expressed as:

$$\frac{dm_t}{dt} = \beta °m_t \left[\left(\frac{M_t}{m_t} \right)^{1/4} - 1 \right] \tag{11.9'}$$

where $\beta° = \dfrac{\overline{E}}{\overline{C}} = \dfrac{E_N + E_T}{C_N + C_T}$. This may be the case for benign tumor or hyperplasia. The entire system still grows harmoniously and is under control. The solution for the equation above is:

$$1 - \left(\frac{m_t}{M_t} \right)^{1/4} = \left[1 - \left(\frac{m_{0t}}{M_t} \right)^{1/4} \right] \exp\left(-\frac{\beta°_t}{4} \right) \tag{11.10}$$

2. All tumor cells obey the rules for similarities of growth, but it is different for normal cells.

$$\frac{dm_T}{dt} = b_T °m_T \left[\left(\frac{M_T}{m_T} \right)^{1/4} - 1 \right] \tag{11.11}$$

The solution for tumor cells is:

$$1 - \left(\frac{m_T}{M_T} \right)^{1/4} = \left[1 - \left(\frac{m_{T0}}{M_T} \right)^{1/4} \right] \exp\left(-\frac{\beta°_T t}{4} \right) \tag{11.12}$$

We ignore the mass of the normal cells in the tumor and use (11.12) to approximate the growth of the entire tumor when tumor cells dominate the tumor growth ($m_T \gg m_N$). $m_{0T} \approx m_{0t}$, $M_T \approx M_t$. Most tumors probably belong to this case. We may use (11.12) to express the growth of tumors modeled in reference [32]. The cases are: tumor cells *in vitro* (9L, SNB19, and U118), as well as tumor cells *in vivo* for rodents (Fibro, Walker, KHJJ, C3H, EMT6, NCTC2472, Osteo, C33ISS) and for patients (breast tumor and prostate tumor).

In other cases, since different tumor cells do not have the same $\beta°$, they cannot be unified by the rules of similarity of growth. We cannot express the total tumor mass in a uniform equation. Instead, we must find the solutions separately and then add them together. In this case, the growth process is complex, though each set of identical cells still follows the growth law for organisms in its own way. It is even more complicated when tumor cells metastasize to different locations and have different growth rates in various systems.

Despite the fact that no tumor cell can flow or diffuse into a primary tumor from the outside, some may flow out or diffuse into either blood vessels or lymphs once the tumor reaches a certain size. Some of them will be carried out of the primary tumor or even transported to other normal organs where the conditions are better for their growth (i.e. passive and active transport). This process should be related to the transport path and the environment. However, once a "seed" is planted into good "soil," the growth process of this secondary tumor should follow the same pattern/equation as that of primary tumor growth. However, parameters such as number density, death, and production rates may not be the same. Demidem et al. [27] studied the growth of secondary tumors and found that their growth curves also fit the "Gompertz" function. We use (11.12) to perform curve fitting for the data in reference [27]. The result is shown in Fig. 11.2. The symbol "*" represents data that we obtained from Fig. 6 of reference [27] by measuring the coordinates (We choose this particular one as an example of a complex system). The curve represents (11.12). They fit rather well when some parameters are chosen. Here we let the first experimental point be the initial point.

The complexity is that the suitable "soil" and paths between the source (primary tumor) and the "soil" may cause more than one "seed" to grow in that location. Also, the growth of these metastases will likely not follow the rules for similarities of growth.

11.3.2 Application to Cell Response to Radiation

When ionizing radiations such as X-rays or α-particles interact with cells in a bio-system, some kinetic energy is being deposited in the system causing possible damage. Linear energy transfer (LET) is

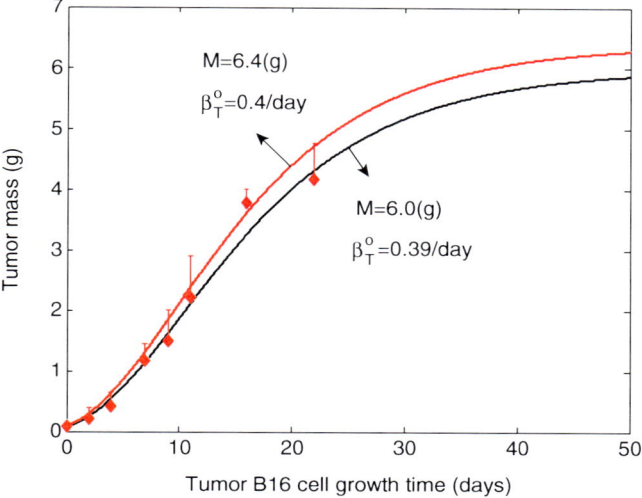

Fig. 11.2 Growth curves of a human prostate tumor without metastasis

the measure of the energy deposited per unit length (kev/μm) of the radiation/particles' track [50]. Depending on the value of LET, ionizing radiation may be divided into two types: low and high LET emissions. For cell response to a low LET radiation such as X-ray, the linear quadratic (LQ) model is widely used [50–53]. Many survival curves fit the LQ or two component LQ model [51]. However, many researchers noted that the LQ model does not fit the survival curves well, especially in the high dose region [50, 52, 54–57]. Astrahan [52] showed that the survival response for Chinese hamster cells in culture [52, 58] does not fit the LQ model in the high dose region ($D>7$ Gy), while Human HeLa cells in culture [50, 52] and DU145 cells *in vitro* [52, 55, 59, 60] do not fit even in the low dose regions ($D>2$ Gy). Garcia et al. [55] studied the behavior of α, β, and α/β in different dose regions by dividing the survival curves into three regions: (1) low dose range (LR), (2) linear quadratic or middle dose region (LQR), and (3) high dose region (HR). These regions in terms of Gy are different for various cells. The conclusion reached is that α, β, and α/β have different values in different regions. It shows that the LQ model fits the survival curves in the middle dose region, but not in the low and high dose regions. They indicated that the mechanism for cell survival in the low dose region might be different from the LQ model. Astrahan modified the LQ model and

Table 11.2 Biological effect of human cells after α-particles radiation

Energy (Mev)	2.5	3.4	4.0	5.1	8.3	25
LET (kev/μm)	166±20	140±20	110±10	88±6	61±5	26±2
OER	1.0±0.1	1.1±0.05	1.3±0.1	1.7±0.15	2.05±0.25	2.4±0.3
$\log_{10}(SF) \sim D$ fit	Linear	Linear	Linear	Linear	Linear	LQ

presented an LQL model [52]. This model can fit some experimental data to an extent. However, it cannot fit the Human HeLa and DU145 cell survival curves in some dose regions, especially for the DU145 cells. Also, the LQL model cannot be applied to the survival curves for EMT-6/Ro tumors under aerobic conditions [61].

Unlike X-ray, alpha particle emission causes a linear survival curve in most cases [50, 62, 63]. Barendsen et al. [62] studied human cell response and corresponding oxygen-enhancement ratios (OER, which is the ratio of doses administered under hypoxic to aerated conditions needed to achieve the same biological effect) to α-particles with different energy. Instead of using air, Barendsen et al. used nitrogen for hypoxic condition. The results are summarized in Table 11.2.

LET increases to 166 kev/μm while the kinetic energy of each particle decreases to 2.5 Mev. Correspondingly, OER decreases to 1.0. The survival curves fit almost linearly when the kinetic energy is less than 8.3 Mev and LET is greater than 61 kev/μm. An α-particle has a typical energy of 5 Mev (usually between 3 and 7 Mev) and is composed of two protons and two neutrons. Relative to other particles such as neutrons, protons, and electrons, α-particle is of a large size and mass, which make it have low penetration depth. Therefore, the energy of α-particles is absorbed by tissues (cells) within a short range. This makes it extremely dangerous once the source is ingested or inhaled since all the particles stay. The damage of biological effects is about 20 times as that caused by an equivalent amount of γ or β radiation (e.g., number of cells killed in 1 Gy = 1 J/kg dose). While interacting with cells, α-particle transfers most of its energy in a small region. Therefore, the localized DNA damage is difficult to repair or is even irreparable. Studies [50, 62] show that the OER is 1.0 when LET is greater than 165 kev/μm. Generally, the OER decreases with the increase of LET [64]. When LET is between 61 and 110 kev/μm, the OER is from 2.0 to 1.3, though the survival response is still linear. When LET reduces to 26 kev/μm, the OER increases to 2.4 and the survival curve shows features of the LQ model.

Assuming that a double strand break in the DNA helix is the critical damage, Chadwick and Leenhouts statistically derived the LQ equation and approximated it as [53]:

$$\text{SF} = \exp[-k_1 \Delta D - k_2 (1-\Delta)^2 D^2] \qquad (11.13)$$

where SF represents cell surviving fraction, k_1, k_2 are two constants, D is the absorbed dose; Δ is a proportion of dose D that is inactivated via single event killing (meaning both strands of the DNA double helix are broken in one radiation event), and $1-\Delta$ is the proportion that is inactivated via double event killing (meaning each strand of DNA double helix is broken independently during different radiation events). Here Δ is LET related. This model implied that Δ increases when LET increases. The linear term $\exp(-k_1 \Delta D)$ dominates survival at low doses. With increase in LET or Δ, the quadratic term $\exp[-k_2(1-\Delta)^2 D^2]$ plays an increasing role [53]. The LQ model may be applied for cell response to high LET radiations such as α-particles if the quadratic term is ignored.

However, there are some experimental data that do not fit the LQ model (including the linear limitation) and its multicomponent forms. For cell response to α radiation, many survival curves fit a linear line and some might follow the LQ model, while other survival curves do not belong to either of these two cases. For example, Hieber et al. [65] showed that the response of C3H 10T1/2 cells to α radiation deviates upward from the linear line after 1.5 Gy. Beaton et al.'s experiment for A-549s's cell response to α radiation [66] showed a similar result. According to the description for setting up the experiments, the results could not be merely explained as "unattached mitotic cells not reached by the α-particles." From the experimental result given by Durante et al. [67] for H184B5 F5-1M/10 cell response to α-particles, we can also summarize that the curve is different from either the linear or the LQ model. Some additional experiments [63, 68–72] also showed that the survival curves deviate upwards from a linear line. Some experimental results, such as the survival of asynchronous V79 cells vs. DNA-incorporated activity [73] after 30-min exposure to ^{211}AtdU (5-[211At]astato-2′-deoxyuridine), may be explained by introducing the two-component exponential model [73, 74]. However, the experimental data in references [65] and [66] do not fit the linear model, its two-component form, or any other previous model. For cell response to X-ray radiation, Powers and Tolmach [75] reported the survival of subcutaneous 6C3HED mouse lymphosarcomas and suggested that the survival kinetics were determined mainly by two cell populations. Even so, no existing models fit the survival curve, including the two-population LQ model. Explaining all these survival curves consistently may lead us to propose a general model for cell response to radiation. In fact, all previous models are derived from statistical theory. In the LQ model, the surviving fraction is expressed as $SF = e^{-\alpha D - \beta D^2}$, where α and β are two parameters that are determined by cell response. It is assumed that the factor $e^{-\alpha D}$ comes from single event killing, and the factor $e^{-\beta D^2}$ is derived from double event killing [50, 53]. This implies that high LET radiations such as α-particles cause mainly single event killing. In fact, α radiation causes more serious damage to cell nuclei (mainly DNA) than other kind of radiations. High LET radiation increases the complexity of lesions due to the formation of multiply damaged sites [76]. The damage caused by DNA double strand breaks are considered important and can be seen from evidence relating to cell lethality [76, 77]. In addition, when dose or dose rate is 0, a model for cell response to therapy should be compatible with the natural growth or death process before treatment is applied. However, previous models did not relate to this aspect.

As we have discussed above, (11.6) is a general equation describing growth of a cell system before treatment in terms of production and death rates. When no radiation is applied, cells or organisms grow and die/shrink with a lower death rate when compared to the case where radiation is applied. Treatments cause the existence of an additional death rate. If the production rate is greater than the total death rate, a cell system or organism grows; otherwise, it shrinks or even disappears. When we apply (11.6) to radiotherapy, the physical meaning is as follows: therapy causes an additional cell death rate K, which is defined as the number of cells "killed" in a unit volume in unit time. Thus, when treatment is applied, (11.6) is amended to read:

$$\frac{dm}{dt} = \frac{m}{C}\left[E\left(\frac{M}{m}\right)^{1/4} - E - K\right] \qquad (11.6')$$

This is a general expression for cell response to treatment. For radiation such as X-ray and α-particles, the dose in a unit time (dose rate) can be selected/controlled and set as a constant since the energy per

11 Model of Tumor Growth and Response to Radiation

particle and the distance of the target are fixed (dose rate $\dot{D} = dD/dt$ is set as a constant [57]). In a specific treatment period, the average dose rate is practically set as a constant. Therefore, the dose absorbed by an organism (or a cell system) is proportional to the amount of time it is exposed to radiation ($D = \dot{D}t$). We employ a general relation $t = \lambda D$, where λ is the reciprocal of dose rate. Thus, (11.6') reads:

$$\frac{dm}{dD} = \frac{\lambda m}{C}\left[E\left(\frac{M}{m}\right)^{1/4} - E - K\right] \quad (11.14)$$

K is a key parameter for determining cell survival response. A correct pattern of K should reflect the experimental results. Considering the properties of α-particles, we divide it into three different cases:

(a) When LET is greater than a specific value L_0 (LET $> L_0$) in a given cell system, the energy deposited is large enough to destroy all localized cells present equally (regardless whether the cells are hypoxic or oxygenated). The damage done to most cells is irreparable, or more accurately, the number of repairable damaged cells is much less than that of the irreparable ones. Therefore, the number of repairable cells may be ignored. The additional death rates are proportional to the quantity of cells present, which should be constant since the cell density is constant. In this case, the OER is equal to 1.0.

(b) When LET is smaller than L_0 but greater than a given value L_1 ($L_1 <$ LET $< L_0$), the energy deposited is still large enough to cause serious damage immediately while interacting with cells and most damage is still irreparable. However, the energy deposited is not large enough to destroy all localized cells with the same effect. It is oxygen-dependent. The damage under well-oxygenated condition is larger than that under hypoxic condition. The ratio of damage varies from onefold to nearly threefolds. The exact value for a given cell system is LET-dependent [50, 62, 64]. The number of cells killed is still proportional to the number of cells present, but the proportionality varies for cells under normoxic and hypoxic conditions. Therefore, the additional death rate is still constant (though the value is oxygen-dependent) since the number density of cells is constant. In this case, the OER is greater than 1.0.

(c) When LET is smaller than L_1 (LET$<L_1$), the energy deposited is not large enough to destroy all localized cells immediately. The mechanism of cell killing is the same as that for cells under low LET radiation. There is a time accumulating effect. We classify it as a linear quadratic case.

In cases (a) and (b), the LET is large enough such that the localized nuclei of cells are damaged and the DNA double strands are broken immediately. There is no time accumulating effect. In this case, the additional death rate caused by α-particles is proportional to the cells present. For a specific kind of cell, the number density C is fixed. Thus, the additional death rate is a constant ($K=v$). The solution of (11.14) is:

$$E-(E+\nu)\left(\frac{m}{M}\right)^{1/4} = \left[E-(E+\nu)\left(\frac{M_0}{M}\right)^{1/4}\right]\exp\left(-\frac{E+\nu}{4C}\lambda D\right) \quad (11.15)$$

The surviving fraction is

$$\text{SF} = \frac{m}{M_0} = \frac{M}{M_0(E+\nu)^4}\left\{E-\left[E-(E+\nu)(\frac{M_0}{M})^{1/4}\right]\exp\left(-\frac{E+\nu}{4C}\lambda D\right)\right\}^4 \quad (11.15')$$

where M_0 is the mass of cells when radiation starts. In actuality, M_0/M can be determined theoretically by comparing two identical cell systems. Radiation is applied to one but not the other. Usually, the additional death rate caused by α radiation is much greater than the production rate in a cell system, namely $(E+\nu)(m/M)^{1/4}>>E$ is satisfied. In this case, it gives an approximately linear cell survival curve and is expressed as

$$\text{SF} = \frac{m}{M_0} = \exp\left(-\frac{E+\nu}{C}\lambda D\right) \quad (11.16)$$

Figure 11.3 depicts log(SF) vs. D with various values of $E\lambda/(4C)$, $v\lambda/(4C)$, and M_0 based on (11.15′).

For case (c), the transferred energy is not large enough to damage some of the targeted nuclei immediately. The DNA damage of nuclei has a time accumulating effect. Since the energy deposited is not large enough, one attack does not break DNA double strands and cause serious damage to some cells. With time accumulating, the number of DNA double strand breaks increases. Therefore, the damage

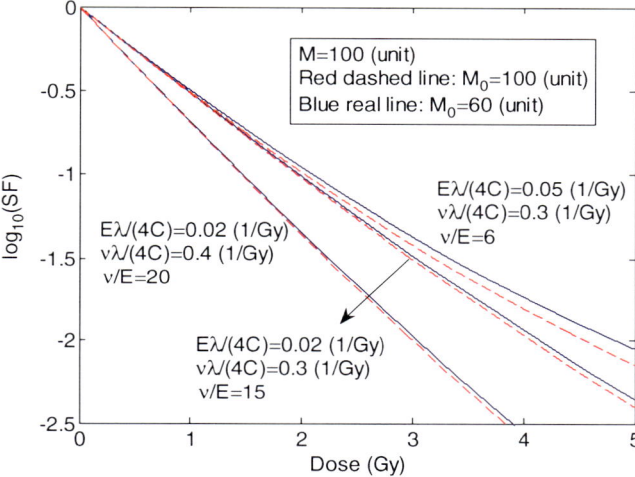

Fig. 11.3 General cell response to α-particle radiation with different parameters

of some cells is proportional to the time. Therefore, the additional death rate also has a time accumulating effect. Mathematically, the LQ model is equivalent to the equation $dN/N = -(\alpha + 2\beta D)dD$. The constant term α creates surviving fraction proportional factor $e^{-\alpha D}$, which is assumed to originate from single event killing. The term $2\beta D$ causes surviving fraction proportional factor $e^{-\beta D^2}$ and is assumed to be derived from double event killing [50, 53]. In fact, the term $2\beta D$ contains the time accumulating effect since $D = \dot{D}t$. In fact, the view of multiply damaged sites [50, 76] agrees our model. Also, accumulation of DNA single strand break can cause DNA double helix break. Studies show that α-particles cause much more serious damage in the nuclei than low LET radiations, such as X-ray [50]. DNA double strand breaks are the primary critical lesion for cell killing [76, 77]. Considering these effects, we assume that some of the nuclei are damaged immediately and others are damaged gradually under low LET radiation. With time or radiation dose accumulating, more and more DNA double strands are broken and nuclei are damaged correspondingly, which causes on average an additional death rate κD. Therefore the total additional death rate in this case is expressed as $K = \kappa D + \nu$, where κ and ν are two constants for a specific radiation and target. Then (11.14) becomes:

$$\frac{dm}{dD} = \frac{\lambda m}{C}\left[E\left(\frac{M}{m}\right)^{1/4} - E - \kappa D - \nu\right] \qquad (11.17)$$

The complete solution can be expressed by using the confluent hypergeometric function or given numerically. However, when condition $\kappa D + \nu + E \gg E(M/m)^{1/4}$ is satisfied, (11.17) is reduced to:

$$\frac{dm}{dD} = -\frac{\lambda}{C}m(\kappa D + \nu + E) \qquad (11.18)$$

The solution in such case is:

$$\text{SF} = \exp\{-\lambda[\tfrac{1}{2}\kappa D^2 + (\nu + E)D]/C\} \qquad (11.19)$$

Let $\alpha = (\nu + E)\lambda/C$, $\beta = \tfrac{1}{2}\kappa\lambda/C$, $\alpha/\beta = 2(\nu+E)/\kappa$, we have:

$$\text{SF} = \exp(-\alpha D - \beta D^2) \qquad (11.19')$$

This is the formula for the LQ model. For a specific target, the α, β, and C are fixed. Therefore, smaller λ (corresponding to larger dose rate $\dot{D}$) causes larger κ and ν.

Regarding cell response to α-particle radiation, the present model is compatible with all existing models. However, for C3H 10T1/2 cell response to α radiation given in reference [65], the experimental data cannot be fitted by any previous models, even considering the form for two component cell system. The energy of α-particles is 2.7 Mev, their dose mean is LET 147 kev/μm, and frequency mean of LET is 144 kev/μm. When we apply (11.15′) to the experimental data, it fits well. According to the description of the experiment, the abnormal experimental result cannot be explained due to the fact that unattached mitotic cells cannot be reached by the α-particles. Here we choose $M_0/M = 25\%$ since the plating efficiency of control cultures in the experiment was 20–30%. The result is shown in Fig. 11.4.

Similarly, for A-549s's cell response to α-particle radiation in reference [66], the survival curve deviates away from the linear line. The last experimental point does not fit a linear response at all. However, (11.15′) fits the experimental points well, which is shown in Fig. 11.5.

For H184B5 F5-1M/10's cell response to α-particle radiation in reference [67], we can see that a linear line does not fit some experimental points well. Equation (11.15′) fits the experimental points better, as show in Fig. 11.6.

11 Model of Tumor Growth and Response to Radiation

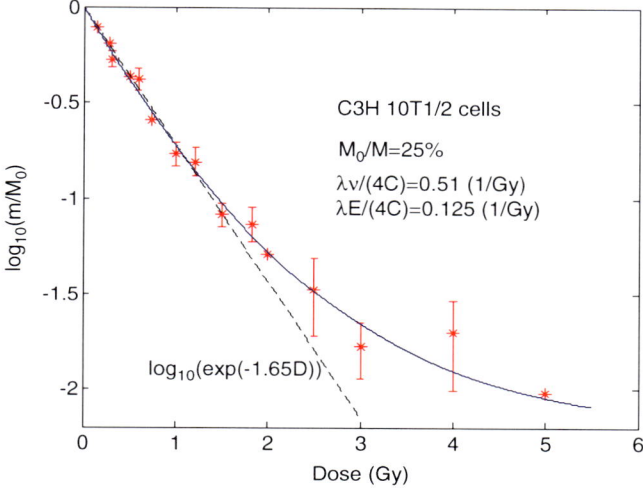

Fig. 11.4 C3H 10T1/2 cell response to α-particle radiation

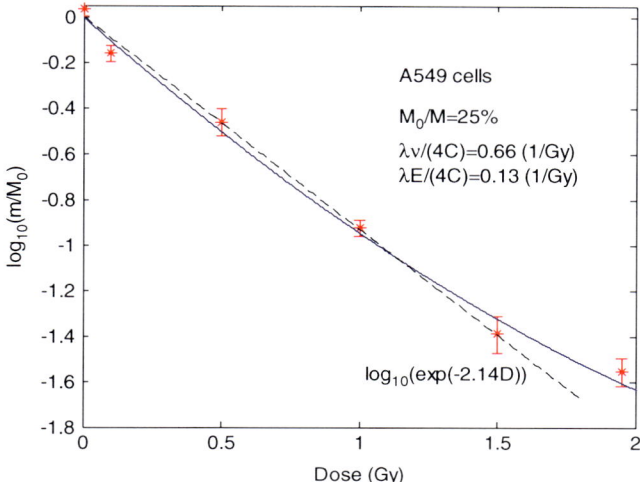

Fig. 11.5 A549s cell response to α-particle radiation

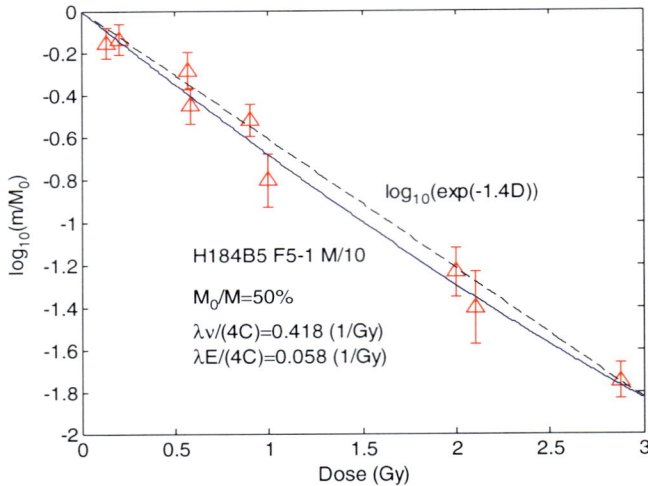

Fig. 11.6 H184B5 F5-1M/10 cell response to α-particle radiation

Table 11.3 The goodness of linear fit and the fit of the present model

Cells	The present model (free fit)		The linear fit	
	RMS	χ^2	RMS	χ^2
A549s	0.0464	5.69	0.110	22.28
H184B5	0.0730	5.416	0.0888	5.717

For comparison, here we calculate the root-mean-square (RMS) differences based on the formula $\text{RMS} = \sqrt{1/m \sum_{i=1}^{m} (\log \text{SF}_i - \log \text{SF}_{\text{fit}})^2}$ [51]. We also calculate $\chi^2 = \sum_{i=1}^{m} (\log \text{SF}_i - \log \text{SF}_{\text{fit}})^2 / \sigma_i^2$ if we know the variance of each experimental point, where σ_i represents the variance of $\log \text{SF}_i$. We do so since the increments of logSF are equally spaced and the differences of $\log \text{SF}_i - \log \text{SF}_{\text{fit}}$ are "more easily correlated to the goodness of fit that is visually apparent in the usual semi-log plot of SF vs. D [51]." The results are listed in Table 11.3. It shows that the present model fits the curves better.

Besides using α-particles, Barendsen et al. [62] also used deuterons and neutrons to irradiate T-1 g cells in culture under oxygenated condition (with air) and hypoxic condition (with nitrogen). For deuterons, when the LET is 20 kev/μm, the survival curves are almost straight lines. However, when the LET is 5.6 kev/μm, they are not straight anymore. Here we apply the LQ model (11.19′) to fit the

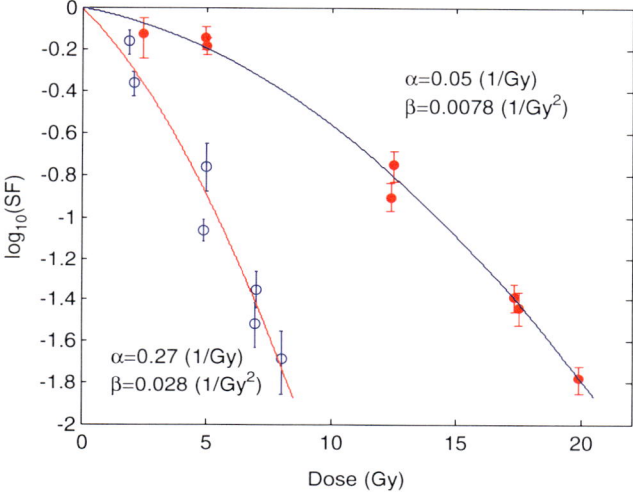

Fig. 11.7 T-1g cell response to deuterons (14.9 Mev) with air (*open circles*) and nitrogen (*closed circles*). LET: 5.6 ± 0.3 kev/μm; OER: 2.6 ± 0.3

experimental data. The result is shown in Fig. 11.7. Similarly, for the survival curves under 25 Mev α-particles with LET 26 ± 2 kev/μm, they cannot be fitted linearly, especially for the survival curve under hypoxic condition. We use the LQ model to fit the survival curves. The result is shown in Fig. 11.8. Obviously, it shows that the survival curves' response to heavy particles such as α-particles and deuterons also has the characteristics of X-ray when LET is low.

Since an organism has various production rates at different growth stages, the response to a therapy should be different. The present model reflects this effect. However, the LQ, as well as other models, cannot. Here we take the two most radio-resistant cases, EMT-6 mouse tumor and MO16 human glioblastoma, which were thought to be incompatible with the LQ models [52], as an example and use (11.17) to fit the experimental data. The results are shown in Fig. 11.9. The survival curves are different for various M_0. In these two cases, the change of ν (correspondingly the α) is not sensitive enough to change the survival curves since κ (correspondingly the β) is relatively large, which places $\kappa D \gg \nu$ in the experimental dose region. This might be a common property for radio-resistant cases. In this case, the "radiation shoulder" will be wide. The two most radio-resistant cases in Fig. 11.9 might have this condition.

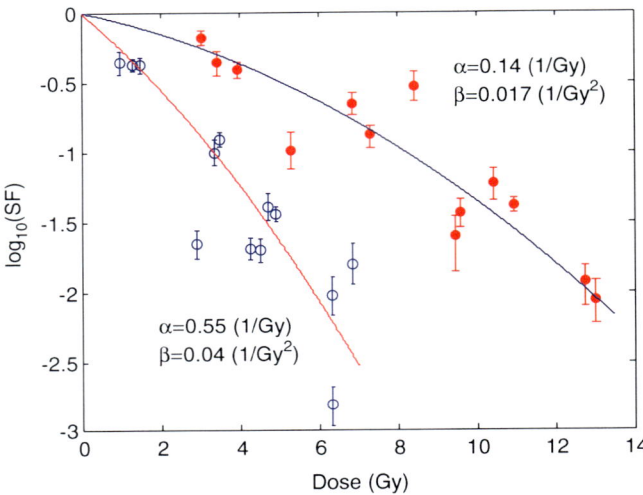

Fig. 11.8 T-1g cell response to α-particles (25 Mev) with air (*open circles*) and nitrogen (*closed circles*). LET: 26 ± 2 kev/μm; OER: 2.4 ± 0.3

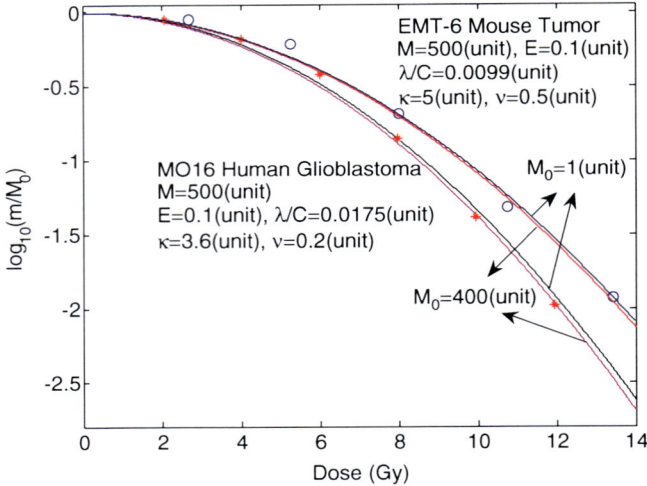

Fig. 11.9 The curvature of surviving fraction vs. radiation dose for radio-resistant cases

11.4 Discussion

As we mentioned above, the additional death rate K should be the difference between the number of damaged and repaired (self-regeneration) cells in a unit volume in a unit time. There are three cases: (1) LET is large. In this case, most damaged cells are irreparable and there is no difference between hypoxic and oxygenated. (2) LET is medium. Though the damage is mostly irreparable, the number of damaged cells is oxygen- and LET-dependent. Since the LET is high enough to cause a serious damage to the cells present immediately in these two cases, there is no time accumulating effect. Generally, the amount of cells damaged in a unit volume in unit time is the sum of the number of irreparable and repairable cells in a unit volume in unit time. It is expressed as $N_d = N_i + N_r$. The number of repaired cells should be proportional to the number of repairable ones. Therefore, the number of cells repaired in a unit volume in unit time is $n_r = cN_r$, where c is the proportionality constant. The additional death rate should be the difference between the number of cells in a unit volume in unit time that are damaged and that of repaired cells. This is expressed as $K = N_d - n_r = N_i + (1-c)N_r$. The larger the LET, the larger the number of irreparably damaged cells, N_i. If LET does not change, N_i is larger under aerobic condition than under hypoxic condition. When LET $> L_1$, $N_i \gg N_r$. Most damaged cells are irreparable and the number of cells that can be repaired may be ignored. When LET $> L_0$, the energy deposited is large enough to kill all localized cells with the same effect. There is no difference in cell damage between hypoxic and aerated condition. (3) LET is low. In this case, some cells are damaged by radiation gradually. The number of damaged cells has a time accumulating effect: $n_d = v_d + \Delta_d t$ (v_d includes irreparable and repairable damaged cells). Considering $t = \lambda D$, we rewrite: $n_d = N_i + N_r + \kappa_d D$, which represents number of damaged cells in unit volume in unit time. The number of repaired cells should be proportional to the number of repairable damaged ones: $n_r = c_1 N_r + c_2 \kappa_d D$, c_1 and c_2 are two proportionality constants. Since these two terms are derived from distinct killing mechanism, c_1 and c_2 are not the same ($c_1 \neq c_2$). Figures 11.5 and 11.6 also show that the values of α and β are not proportional when cells are under hypoxic and oxygenated conditions. Therefore, the additional death rate is $K = n_d - n_r = \kappa D + v$. According to the discussion above, κ and v (correspondingly α and β) change when the oxygenated condition varies. How to determine the change pattern needs further study.

If the cell system is composed of two different kinds of cells (one resistant, the other regular), the relationship between the surviving fraction and radiation dose is complicated. It is easy to show that the fractions of different cells in a system before therapy ($K=0$) are constant if various cells have the same E/C and their asymptotic masses share the same fractions. These two conditions (E_i/C_i is constant and $M_i = x_i M$) ensure that the cells grow harmoniously before treatment. The different responses to therapy enable us to detect the existence of various cell subpopulations. Assuming that the resistant cells have x fraction (regular cells $1-x$ fraction). For a system composed of identical cells, the mass is proportional to the volume or the number of cells ($m = VCm_c = Nm_c$). The surviving fraction of the entire system is $SF = m/M_0 = (m_{res} + m_{reg})/M_0$. If the LET is large enough to kill all localized cells equally, the additional death rates are proportional to the amount of cells present and have a relationship of $v_{res}/C_{res} = v_{reg}/C_{reg} = \theta$ for different kinds of cells. For cells in the same tissue or from the same source which grow harmoniously before therapy, condition $E_{res}/C_{res} = E_{reg}/C_{reg} = \beta°$ should be satisfied. Also, $M_{res} = xM$ and $M_{0res} = xM_0$. Thus we have,

$$SF = \frac{m_{res} + m_{reg}}{M_0} = \frac{M}{M_0(\theta + \beta°)^4} \{\beta° - \left[\beta° - (\theta + \beta°)\left(\frac{M_0}{M}\right)^{1/4}\right]$$
$$\exp[-(\theta + \beta°)\lambda D / 4]\}^4 \qquad (11.20)$$

In this case, the consequence is the same as (11.15′) for single cell component system. When the total death rates are much greater than the production rates for both cell populations, we have:

$$SF = \exp[-(\theta + \beta°)\lambda D] \qquad (11.20')$$

Equation (11.20′) shows a linear relation of $\log(SF) \sim D$. For cells that do not grow harmoniously before therapy, we have to use (11.15) for calculating the masses separately and then obtain the total surviving fraction of cells $SF = (m_{res} + m_{reg})/M_0$. Since M, M_0, and $\beta°$ are not related for different cell populations, the situation is complicated. If LET is not large enough, $v_{res}/C_{res} \neq v_{reg}/C_{reg}$ since the number of reparable cells for one or even both subpopulations cannot be ignored. In this case, (11.20) may not be satisfied. If LET is low, the total surviving fraction may be determined by using the two-component LQ model [51]. In this case, how should the values of α and β for different types of cells be constrained for the best fit? Some models for the response of mixed cell populations [51, 78–80] fit the experimental data rather well, though they are empiric models. Skarsgard et al. [51] found that a two-population LQ model gave the best fit. However, there are no constraints between the αs and βs of different cell populations. This may cause the parameters to be redundant and the choices to be more than one set—α, β as well as cell fraction x—for the same survival response. The two-population LQ model is expressed as [51]:

$$SF = (1-x)\exp[-(\alpha_s + \beta_s D)D] + x\exp[-(\alpha_r + \beta_r D)D] \quad (11.21)$$

where subscript s represents response parameters for sensitive cells and r represents those for resistant cells. Since these response parameters are not restrained, the best fit is not unique. Different kinds of cells (which grew harmoniously before therapy) may have different λ/C, but they have the same or proportional value of ν to κ (corresponding to α/β) under the same radiation field with the same environmental conditions. The additional death rates of different populations should be equal or proportional when they are under the same radiation field, expressed as $\alpha_s / \beta_s = \alpha_r / \beta_r$. For convenience's sake, we use α and β to represent the response parameters of sensitive cells. Thus, the response parameters of resistant cells are $\gamma\alpha$ and $\gamma\beta$ (γ is a proportionality constant). Therefore, (11.21) becomes:

$$SF = (1-x)\exp[-(\alpha + \beta D)D] + x\exp[-\gamma(\alpha + \beta D)D] \quad (11.22)$$

In ref. [51], authors expected approximately comparable numbers of sensitive and resistant cells in their experiment. Setting $x=0.5$, they found that $\alpha_s=0.2414$, $\beta_s=0.1006$, $\alpha_r=0.0886$, and $\beta_r=0.0345$. Thus, $\alpha_s/\beta_s=2.40$ and $\alpha_r/\beta_r=2.57$, which are close. The difference might be because the fractions of the two components are not exactly the same. In the free fitting case, they found that $x=0.63$, $\alpha_s=0.2064$, $\beta_s=0.1218$, $\alpha_r=0.1402$, and $\beta_r=0.0318$, which gives $\alpha_s/\beta_s=2.69$ and $\alpha_r/\beta_r=4.41$. The authors also tried a forced fit by letting $x=0.37$ and found that $\alpha_s=0.2492$, $\beta_s=0.0874$, $\alpha_r=0.0231$, and $\beta_r=0.0378$, which gives $\alpha_s/\beta_s=2.85$ and $\alpha_r/\beta_r=0.611$. The values of α_s/β_s and α_r/β_r in these two cases are distinctly different.

Here we apply (11.22) to fit some experimental data from literature and see the results. Astrahan positioned Elkind and Sutton's experimental data [58] into an enlarged logarithmic coordinate system to show that

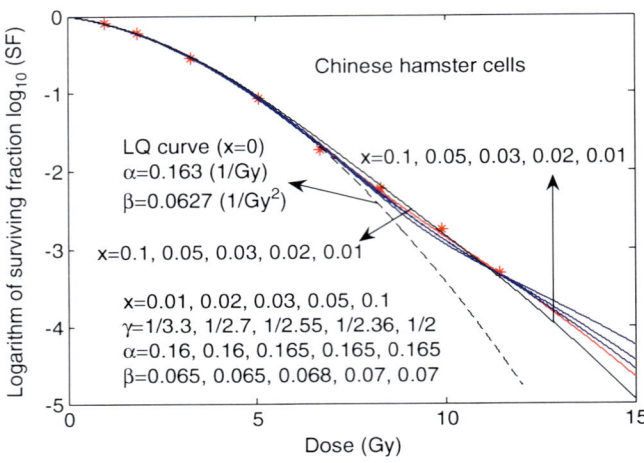

Fig. 11.10 Survival curves of Chinese hamster cells with different components of "radio-resistant" cells

the Chinese hamster cells in culture closely fit the LQ model up to a dose of 6 Gy but then continues to deviate from the LQ curve [52]. Here we adopted the data from reference [52] and fit the curve based on (11.22). The result is shown in Fig. 11.10 with the best fit curves. We keep the error bar for an experimental point in the figures if there is one from literature. Fig. 11.10 shows that the curves for $x \leq 0.03$ do not fit the experimental data well. We estimate that the fraction x of "resistant" cell population is between 0.04 and 0.1. We also apply (11.22) to Human HeLa cells in culture (the data is from Fig. 3B in ref. [52]) and DU145 cell line *in vitro* (the data is from Fig. 4A in Ref. [52] though it was from ref. [55]). The curves fit better than the LQ model and the LQL model, especially for DU145 (relative to the dots if there are error bars). The results and corresponding parameters are shown in Fig. 11.11.

Figures 11.10 and 11.11 clearly show that some "radio-resistant" cells exist in culture and *in vitro* for both normal cells and cancer cells. Studies show that cancer stem cells are more resistant to radiation than other cancer cells. A similar argument can be made for any subpopulation of cells that is radio-resistant. Currently there is considerable debate over the importance of tumor stem cells to radiation response. If tumor stem cells are indeed resistant to the damaging effects of ionizing radiation, the discussion above for radio-resistant cells may

11 Model of Tumor Growth and Response to Radiation

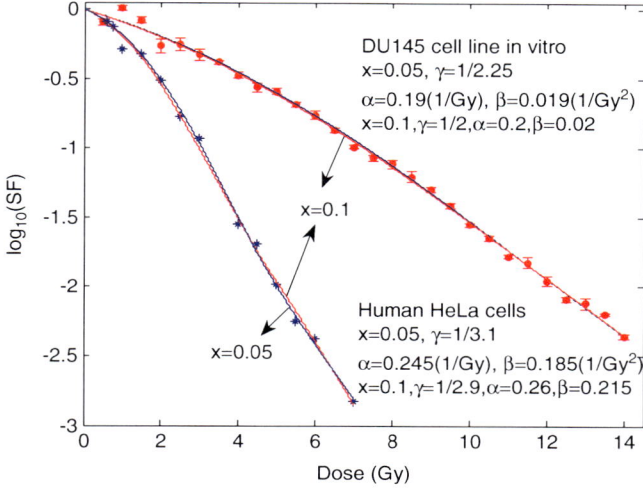

Fig. 11.11 Survival curve of human HeLa cells in culture DU145 cell line *in vitro*

Table 11.4 The goodness of fit for different models

	The present model ($x=0.1$)			The LQL model	
	RMS	χ^2		RMS	χ^2
Human HeLa	0.04725			0.05651	
DU145	0.03580	90.26	Upper dashed blue line	0.05590	168.76
			Lower dashed red line	0.04536	113.37

be similar to the one for stem cells. As we mentioned above, different cell populations in one system can grow harmoniously before therapy and may not be identified. After a therapy, different cell populations respond to the therapy differently and can be discovered.

For comparison, we calculate the RMS and χ^2 of $\log_{10}(SF)$ for Human HeLa and DU145 cells' response to x-ray by using the present and the LQL models. The χ^2 can be calculated only when we know the variance of each experimental point. The results are listed in Table 11.4, and we can see that the present model fits better than the LQL model. We can also see that the surviving fraction of the second point in the graph below for DU145 is 1.0215, which is most

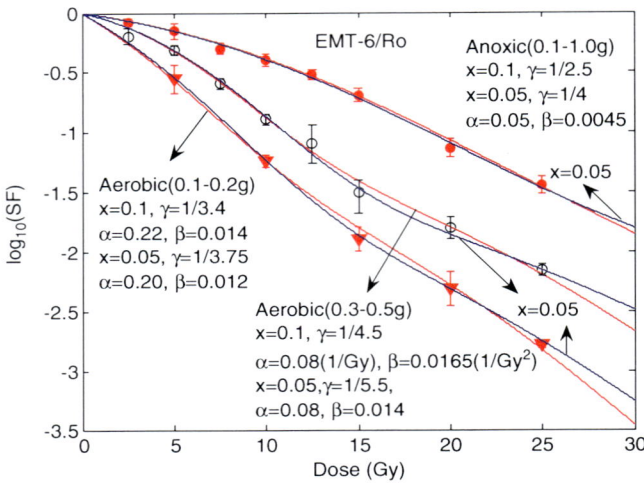

Fig. 11.12 Survival curves for EMT-6/Ro tumors under aerobic or anoxic conditions

likely an impossible value. If we take this point out, the fit for the present model will be even better than it does now.

We also employed (11.22) to fit the survival curve for C3H10T1/2 [52], CP3 (a human prostate carcinoma cell line) and U-373MG (a human glioblastoma cell line) [55]. Generally, the survival curves fit these experimental data quite well.

When we apply (11.22) to fit EMT-6/Ro tumors under aerobic and anoxic conditions, we find that some more resistant cells exist even under anoxic condition. Since a small population of "resistant" cells always exist under both well-oxygenated and anoxic conditions, it implies that these specific cells are inherent. Obviously, the survival curves under aerobic condition do not satisfy the LQL model. However, (11.22) fits it well. The data are digitized from Fig. 3 in reference [61]. The results are shown in Fig. 11.2.

In (11.22), because the value of γ has an exponential effect on the surviving fraction, it may have a strong influence on the surviving fraction of the entire system. If we rewrite (11.22) in the form of the LQ equation,

$$e^{-(\alpha' + \beta' D)D} = (1-x)e^{-(\alpha + \beta D)D} + xe^{-\gamma(\alpha + \beta D)D}$$

we have:

$$\alpha'D + \beta'D^2 = \alpha D + \beta D^2 - \ln[1 - x + xe^{(1-\gamma)(\alpha+\beta D)D}]$$

Obviously, α' and β' are the functions of α, β, γ, x, and D. For a specific species and radiation field, α, β, γ, and x are fixed. Therefore, α' and β' are functions of D. Garcia et al. [55] analyzed some experimental data in different dose regions and noticed that the dependence of the fitted α and β on the dose range has an impact on the α/β ratio itself determined from the survival data. This may explain the results of Garcia et al. regarding the behaviors of α, β, and α/β [55]. From the above equation, we can see that the contribution of $\exp[(1-\gamma)(\alpha D + \beta D^2)]$ is significant if γ is small, especially in the high dose region. In this case, the change in x may not affect the variation in SF as much as a change in α or β, especially β.

Equations (11.19) and (11.22) are the results of $K \gg \overline{S}$ and $K \gg E$, respectively. These two conditions may be satisfied in most cases. However, there may be some cases where these two conditions may not be satisfied in the low dose region or at the early growth stage in which M_0 is small. Therefore, we have to solve (11.17) completely for $K = \kappa D + v$. Powers and Tolmach [75] reported the survival response of 6C3HED mouse lymphosarcoma cells to radiation *in vivo*. They plotted several sets of experimental data in one graph and fitted the curve by eye. It is impossible to fit the data when we try other models such as the LQ, the LQL, and the multicomponent LQ model. However, when we apply (11.17) to this case, it fits well. The result is shown in Fig. 11.13. The points are digitalized from reference [75]. Here we did not use various symbols to identify the data for different experiments. The response of the same cells to a specific radiation is similar, though the data are from different experiments.

Analogous to radiation, chemotherapy also causes DNA damage. For cell survival experiment *in vitro*, there are usually two cases: (1) cells are in the growing media with various drug concentrations for the same length of time; and (2) cells are in the growing media with the same drug concentration for different length of time. In any case, the additional death rate K should relate to the absorbed dose, which is defined as the product of drug concentration and the amount of time the cells are immersed, $D = n \cdot t$ [81, 82]. Similar to low LET radiation, chemotherapy also has a time accumulating effect since the chemotherapeutic agent is absorbed through diffusion. The absorbed dose in cells should be proportional to time and drug concentration. We assume that the additional death rate has a linear relation to the

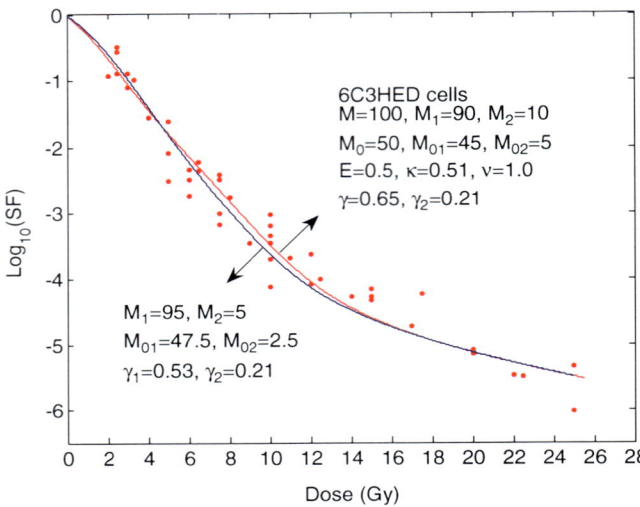

Fig. 11.13 Survival of 6C3HED mouse lymphosarcoma cells irradiated *in vivo*. The *dots* represent experimental data that are digitalized from Fig. 1 in ref. [75].

absorbed dose, $K = \kappa D + \nu = \kappa n t + \nu$. Here, ν should be related to drug concentration $\nu = \nu(n)$, and $\nu(0)$ should be 0. We might be able to estimate $\nu(n)$ through experimentation. For a fixed concentration, ν is a constant. Therefore, (11.6′) becomes

$$\frac{dm}{dt} = \frac{m}{C}\left[E\left(\frac{M}{m}\right)^{1/4} - E - \kappa n t - \nu(n)\right] \quad (11.23)$$

Usually, (11.23) only has a numerical solution and can be expressed as a confluent hypergeometric function. However, when condition $E + \kappa n t + \nu(n) Ð E(M/m)^{0.25}$ is satisfied, (11.23) is reduced to

$$\frac{dm}{dt} = -\frac{m}{C}[E + \kappa n t + \nu(n)] \quad (11.24)$$

The solution of (11.24) is $m = M_0 \exp[-([E + \nu(n)]t/C) - (\kappa n t^2/2C)]$.

When the time for cells treatment in the drug medium is fixed, the survived cell mass or number is a function of drug concentration. Therefore, $m = M_0 \exp[-\alpha n - \beta(n) - \delta]$, where Δ is a constant that is

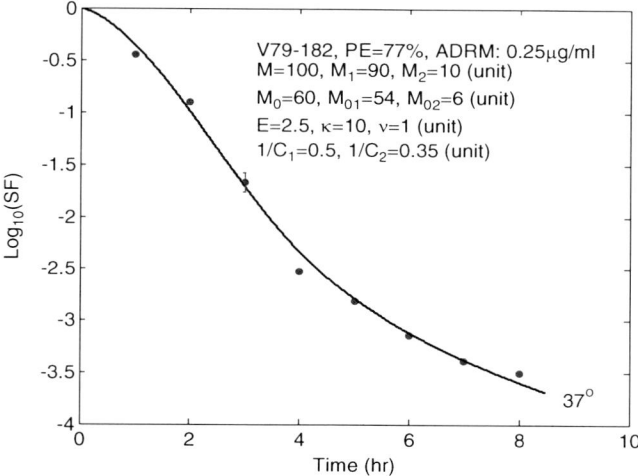

Fig. 11.14 Survival response of V79 cells to ADRM (0.25 μg/mL)

derived from term E. Similarly, we assume that there is x fraction of "resistant" cells. The surviving fraction of the entire cell system is

$$\text{SF} = (1-x)\exp[-\alpha n - \beta(n) - \delta] + x\exp\{-\gamma[\alpha n + \beta(n) + \delta]\} \quad (11.25)$$

However, for the survival response of V79-182 Chinese hamster cells to Adriamycin [50, 83], we have to use (11.23) to provide a complete numerical solution. Belli and Piro [83] gave a surviving fraction vs. time graph for V79 cell under a constant concentration of Adriamycin agent. The experimental data is from Chart 2 in ref. [83]. The result is shown in Fig. 11.14. Equation (11.23) fits the experimental data fairly well.

When metastasis happens in a primary tumor, some cells flow out of the system. Therefore, a mass flux exists and (11.4) must be modified as (assuming that all tumor cells spread out of the system, no cells can flow in):

$$\frac{m_{ck}C_k \mathrm{d}V_k}{\mathrm{d}t} = -C_k m_{ck} \int \nabla_k \cdot \vec{u}_k \mathrm{d}V_k + V_k m_{ck}(\overline{S}_k - E_k) \quad (11.26)$$

$$-C_k m_{ck} \int \nabla_k \cdot \vec{u}_k \mathrm{d}V_k = -C_k m_{ck} \oiint \vec{u}_k \cdot \mathrm{d}\vec{A}_k + \text{cons. (Divergence theorem)}$$

where $\vec{u}_k$ is the velocity of cells spreading out of the tumor in system k, A_k is the closed surface of V_k, and the direction of $\vec{A}_k$ is the outward direction of the normal to A_k. Equation (11.5) should still be satisfied. Since $\vec{u}_k$ is usually spatial and time-dependent, there is no analytic solution.

Here we take a simple case, $\nabla \cdot \vec{u} =$ cons. μ and there is only one kind of tumor cell, as an example. Suppose the tumor begins to metastasize when it grows to a specific size/mass M_{T0}. Equation (11.26) can be rewritten as:

$$\frac{dm_T}{dt} = -\mu m_T + \beta_T^\circ m_T \left[\left(\frac{M_T}{m_T}\right)^{1/4} - 1 \right] \quad (11.27)$$

This equation does not have an analytic solution. There is a equilibrium mass M_{eq} in this case. When $m_T = M_{eq}$, $dm_T/dt = 0$, the primary tumor will not grow. When $m_T > M_{eq}$, $dm_T/dt < 0$, the primary tumor decreases. The "unknown primary origin" cancers [84, 85] may relate to this case. When $m_T < M_{eq}$, $dm_T/dt > 0$, the primary tumor increases until it reaches a specific size. M_{eq} satisfies the following equation:

$$M_{eq} = \left(\frac{\beta_T^\circ}{\beta_T^\circ + \mu}\right)^4 M_T \quad (11.28)$$

11.5 Conclusion

From the discussion above, we know that there is a "universal" governing equation for an organism's growth and development. It includes the biological (growth terms) and physical (mass flux terms) activities and can be applied to tumor growth and cell response to therapy. It might even be extended to tumor metastasis.

The present model is based on the condition of constant number density. It is reasonable since the energy for maintaining the metabolism of a single cell is fixed. This means that the size of a single cell (m_{ck} and r_{ck}) is constant. Correspondingly, the number density of cells C_k is a constant. In West's model, M is the asymptotic mass for organism growth in unrestricted dietary conditions. Relation $E/C = aM^{-1/4}$ shows that the growth parameter a decreases when M decreases, which slows down the growth rate dm/dt. The system can still grow harmoniously.

We sort identical cells into a specific system (tissue/organ) and use the rules of similarities of growth to unite the systems harmoniously. It is noteworthy that different kinds of normal cells follow the rules and grow harmoniously, but different kinds of tumor cells may not (cells in a benign tumor may but not those in a malignant one). In fact, M must depend on nutrition concentration c_n and oxygen concentration c_o, which are very important for organism growth. We need further studying in order to determine the function $M = M(c_n, c_o)$. Nutrition supply might make organism growth complicated.

Therapies cause an additional death rate (relative to the natural death rate). The present model connects the consequence of radiation with the growth process of a biosystem. This model explains the response of cell population under both high and low LET radiations consistently. Traditionally, high and low LET radiations are classified by the types of ions or particles of radiation. Usually, X-ray, γ-ray, and β-particles are categorized as low LET radiations, whereas α-particles, protons, neutrons, etc. (heavy particles) are taken as high LET radiations. Based on this model, we now suggest that the categories of high and low radiations should be classified according to the exact values of LET. When LET is high enough, radiation causes a linear survival response; while LET is low, the cell response follows the linear quadratic pattern. The present model explains that the OER decreases while LET increases consistently. It also shows that the LQ model is an approximation of the present model under a specific condition: the total death rate, which includes additional death rate caused by radiation and natural death rate, is much greater than the production rate.

References

1. Busalmen JP, de Sánchez SR. Electrochemical polarization-induced changes in the growth of individual cells and biofilms of Pseudomonas fluorescens (ATCC 17552). Appl Environ Microbiol. 2005;71:6235–40.
2. Thomlinson RH, Gray LH. The histological structure of some human lung cancers and the possible implications for radiotherapy. Br J Cancer. 1955;9:539–49.
3. Hall EJ, Giaccia AJ. Radiobiology for the radiologist. 6th ed. Philadelphia: Lippincott Williams & Wilkins; 2006. p. 90–103.
4. Burton AC. Rate of growth of solid tumours as a problem of diffusion. Growth. 1966;30:157–76.

5. Araujo RP, McElwain DLS. A history of the study of solid tumor growth: the contribution of mathematical modelling. Bull Math Biol. 2004;66:1039–91.
6. Greenspan HP. Models for the growth of a solid tumor by diffusion. Stud Appl Math. 1972;52:317–40.
7. Tiina Roose S, Chapman J, Maini PK. Mathematical models of avascular tumor growth. SIAM Rev. 2007;49:179–208.
8. Ferreira Jr SC, Martins ML, Vilela MJ. Reaction–diffusion model for the growth of avascular tumor. Phys Rev E Stat Nonlin Soft Matter Phys. 2002;65:021907.
9. Bertuzzi A, Fasano A, Gandolfi A, Marangi D. Cell kinetics in tumor cords studied by a model with variable cell cycle length. Math Biosci. 2002;177:103–25.
10. Ward J, King J. Mathematical modelling of avascular-tumour growth. IMA J Math Appl Med Biol. 1997;14(1):39–69.
11. Fernández Slezak D, Suárez C, Soba A, Risk M, Marshall G. Numerical simulation of avascular tumor growth. J Phys Conf Ser. 2007;90:012049.
12. Chaplain MAJ, Graziano L, Preziosi L. Mathematical modelling of the loss of tissue compression responsiveness and its role in solid tumor development. Math Med Biol. 2006;23:197–229.
13. Basan M, Risler T, Joanny J-F, Sastre-Garau X, Prost J. Homeostatic competition drives tumor growth and metastasis nucleation. HFSP J. 2009;1:98.
14. Deroulers C, Aubert M, Badoual M, Grammaticos B. Modeling tumor cell migration: From microscopic to macroscopic models. Phys Rev E Stat Nonlin Soft Matter Phys. 2009;7(9):031917.
15. Lo Stochastic CF. Gompertz model of tumor cell growth. J Theor Biol. 2007;248:317–21.
16. Kim PS, Lee PP, Levy D. A PDE Model for Imatinib-Treated Chronic Myelogenous Leukemia. Bull Math Biol. 2008;70:1994–2016.
17. Gatenby RA, Vincent TL. Application of quantitative models from population biology and evolutionary game theory to tumor therapeutic strategies. Mol Cancer Ther. 2003;2:919–27.
18. Albert PS, Shih JH, Tumor M. Growth with random onset. Biometrics. 2003;59:897–906.
19. Delsanto PP, Condat CA, Pugno N, Gliozzi AS, Griffa M. A multilevel approach to cancer growth modeling. J Theor Biol. 2008;250:16–24.
20. Byrne H, Drasdo D. Individual-based and continuum models of growing cell populations: a comparison. J Math Biol. 2009;58:657–87.
21. Barbolosi D, Benabdallah A, Hubert F, Verga F. Mathematical and numerical analysis for a model of growing metastatic tumors. Math Biosci. 2009;218:1–14.
22. Hahnfeldt P, Panigrahy D, Folkman J, Hlatky L. Tumor development under angiogenic signaling: a dynamical theory of tumor growth, treatment response, and postvascular dormancy. Cancer Res. 1999;59:4770–5.
23. Jain RK, Tong RT, Munn LL. Effect of vascular normalization by antiangiogenic therapy on interstitial hypertension, peritumor edema, and lymphatic metastasis: insights from a mathematical model. Cancer Res. 2007;6(7):2729.
24. Sutherland RM. Cell and environment interactions in tumor microregions: the multicell spheroid model. Science. 1988;240:177–84.
25. Yorke ED, Fuks Z, Norton L, Whitmore W, Ling CC. Modeling the development of metastases from primary and locally, recurrent tumors: comparison with a clinical database for prostatic cancer. Cancer Res. 1993;53:2987–93.

26. Laird AK. Dynamics of tumor growth. Br J Cancer. 1964;18:490–502.
27. Demidem A, Morvan D, Papon J, De Latour M, Madelmont JC. Cystemustine induces redifferentiation of primary tumor and confers protection against secondary tumor growth in a melanoma murine model. Cancer Res. 2001;6(1):2294.
28. Steel GG. Growth kinetics of tumors. Oxford: Clarendon; 1977.
29. Wheldon TE. Mathematical models in cancer research. Bristol: Adam Hilger; 1988.
30. West GB, Brown JH, Enquist BJ. A general model for ontogenetic growth. Nature. 2001;413:628–31.
31. Kleiber M. Body size and metabolic rate. Physiol Rev. 1947;27:511–41.
32. Guiot C, Degiorgis PG, Delsanto PP, Gabriele P, Deisboeck TS. Does tumor growth follow a "universal law"? J Theor Biol. 2003;225:147–51.
33. Colombo F, Baldan F, Mazzucchelli S, Martin-Padura I, Marighetti P, et al. Evidence of distinct tumour-propagating cell populations with different properties in primary human hepatocellular carcinoma. PLoS One. 2011;6:e21369. doi:10.1371/journal.pone.
34. González-García I, Solé RV, Costa J. Metapopulation dynamics and spatial heterogeneity in cancer. Proc Natl Acad Sci USA. 2002;99:13085–9.
35. Liu LJ, Brown SL, Ewing JR, Schlesinger M. Phenomenological model of interstitial fluid pressure in a solid tumor. Phys Rev E Stat Nonlin Soft Matter Phys. 2011;84:021919.
36. West GB, Brown JH, Enquist BJ. The fourth dimension of life: fractal geometry and allometric scaling of organisms. Science. 1999;284:1677–9.
37. West GB, Brown JH, Enquist BJ. A general model for the origin of allometric scaling laws in biology. Science. 1997;276:122–6.
38. Savage VM, Allen AP, Brown JH, Gillooly JF, Herman AB, Woodruff WH, et al. Scaling of number, size, and metabolic rate of cells with body size in mammals. Proc Natl Acad Sci USA. 2007;104:4718–23.
39. Schmidt-Nielsen K. Scaling: why is animal size so important. Cambridge: Cambridge University Press; 1984.
40. Calder III WA. Size, function and life history. Cambridge: Harvard University Press; 1984.
41. Feldman HA, McMahon TA. The 3/4 mass exponent for energy metabolism is not a statistical artefact. Respir Physiol. 1983;52:149.
42. Charnov EL. Life history invariants: some explorations of symmetry in evolutionary ecology. Oxford: Oxford University Press; 1993.
43. Stearns SC. The evolution of life histories. Oxford: Oxford University Press; 1992.
44. Reiss MJ. The allometry of growth and reproduction. Cambridge: Cambridge University Press; 1989.
45. von Bertalanffy L. Quantitative laws in metabolism and growth. Q Rev Biol. 1957;32:217–31.
46. Al-Hajj M, Wicha MS, Benito-Hernandez A, Morrison SJ, Clarke MF. Prospective identification of tumorigenic breast cancer cells. Proc Natl Acad Sci USA. 2003;100:3983.
47. Reya T, Morrison SJ, Clarke MF, Weissman IL. Stem cells, cancer, and cancer stem cells. Nature. 2001;414:105–11.
48. Lapidot T, Sirard C, Vormoor J, Murdoch B, Hoang T, Caceres-Cortes J, et al. A cell initiating human acute myeloid leukaemia after transplantation into SCID mice. Nature. 1994;367:645–8.

49. Singh SK, Clarke ID, Terasaki M, Bonn VE, Hawkins C, Squire J, et al. Identification of a cancer stem cell in human brain tumors. Cancer Res. 2003;63:5821–8.
50. Hall EJ, Giaccia AJ. Radiobiology for the Radiologist. 6th ed. Philadelphia: Lippincott Williams & Wilkins; 2006.
51. Skargard LD, Skwarchuk MW, Wouters BG. The survival of asynchronous V79 cells at low radiation doses: modeling the response of mixed cell populations. Radiat Res. 1994;138:S72–5.
52. Astrahan M. Some implications of linear-quadratic-linear radiation dose–response with regard to hypofractionation. Med Phys. 2008;3(5):4161.
53. Chadwick KH, Leenhouts HP. A molecular theory of cell survival. Phys Med Biol. 1973;18:78–87.
54. Bradly G. Cells at intermediate oxygen levels can be more important than the "hypoxic fraction" in determining tumor response to fractionated radiotherapy. Radiat Res. 1997;147:541–50.
55. Garcia LM, Wilkins DE, Raaphorst GP. α/β ratio: a dose range dependence study. Int J Radiat Oncol Biol Phys. 2007;67:587–93.
56. Guerrero M, Carlone M. Mechanistic formulation of a linear-quadratic-linear (LQL) model: split-dose experiments and exponentially decaying sources. Med Phys. 2010;3(7):4173.
57. Curtis SB. Lethal and potentially lethal lesions induced by radiation—a unified repair model. Radiat Res. 1986;106:252–70.
58. Elkind MM, Sutton H. X-ray damage and recovery in mammalian cells in culture. Nature. 1959;184:1293–5.
59. Stone K, Wunderli H, Mickey G, Paulson D. Isolation of a human prostate carcinoma cell line (DU-145). Int J Cancer. 1978;21:274–81.
60. Leith J, et al. Radiobiological studies of PC-3 and DU-145 human prostate cancer cells, x-ray sensitivity in vitro and hypoxic fractions of xenografted tumors in vivo. Int J Radiat Oncol Biol Phys. 1993;25:283–7.
61. Siemann DW. Tumour size: a factor influencing the isoeffect analysis of tumour response to combined modalities. Br J Cancer. 1980;41(Suppl IV):294.
62. Barendsen GW, Koot CJ, van Kersen GR, Bewley DK, Field SB, Parnell CJ. The effect of oxygen on impairment of the proliferative capacity of human cells in culture by ionizing radiations of different LET. Int J Radiat Biol. 1966;10:317–27.
63. McNally NJ, de Ronde J, Folkard M. Interaction between X-ray and α-particle damage in V79 cells. Int J Radiat Biol. 1988;53:917–20.
64. Wenzl T. and Jan J Wilkens, Modelling of the oxygen enhancement ratio for ion beam radiation therapy. Phys Med Biol. 2011;56:3251–68.
65. Hieber L, Ponsel G, Roos H, Fenn S, Fromke E, Kellerer AM. Absence of a dose-rate effect in the transformation of C3H 10T1/2 cells by α-particles. Int J Radiat Biol. 1987;52:859–69.
66. Beaton LA, Trevor B, Stocki TJ, Vinita C, Wilkins RC. Development and characterization of an in vitro alpha radiation exposure system. Phys Med Biol. 2011;56:3645–58.
67. Durante M, Grossi GF, Gialanella G, Pugliese M, Nappo M, Yang TC. Effects of α-particles on survival and chromosomal aberrations in human mammary epithelial cells. Radiat Environ Biophys. 1995;34:195–204.

68. Martin SG, Miller RC, Geard CR, Hall EJ. The biological effectiveness of radon-progeny alpha particles. IV. Morphological transformation of Syrian hamster embryo cells at low doses. Radiat Res. 1995;142:70–7.
69. Thomas P, Tracy B, Ping T, Baweja A, Wickstrom M, Sidhu N, et al. Relative biological effectiveness (RBE) of alpha radiation in cultured porcine aortic endothelial cells. Int J Radiat Biol. 2007;83:171–9.
70. Lloyd EL, Gemmell MA, Henning CB, Gemmell DS, Zabransky BJ. Transformation of mammalian cells by alpha particles. Int J Radiat Biol. 1979;36:467.
71. Zhou H, Randers-Pehrson G, Waldren CA, Vannais D, Hall EJ, Hei TK. Induction of a bystander mutagenic effect of alpha particles in mammalian cells. Proc Natl Acad Sci USA. 2000;97:2099–104.
72. Hei TK, Wu LJ, Liu SX, Vannais D, Waldren CA, Randers-Pehrson G. Mutagenic effects of a single and an exact number of α particles in mammalian cells. Proc Natl Acad Sci USA. 1997;94:3765.
73. Walicka MA, Vaidyanathan G, Zalutsky MR, James Adelstein S, Kassis AI. Survival and DNA damage in Chinese hamster V79 cells exposed to alpha particles emitted by DNA-incorporated astatine-211. Radiat Res. 1998; 150:263–8.
74. Hall EJ, Gross W, Dvorak RF, Kellerer AM, Rossi HH. Survival curves and age response functions for Chinese hamster cells exposed to x-ray or high LET alpha-particles. Radiat Res. 1972;52:88–98.
75. Powers WE, Tolmach LJ. A multicomponent x-ray survival curve for mouse lymphosarcoma cells irradiated in vivo. Nature. 1963;197:710–1.
76. Newman HC, Prise KM, Folkard M, Michael BD. DNA double-strand break distributions in X-ray and α-particle irradiated V79 cells: evidence for non-random breakage. Int J Radiat Biol. 1997;71:347–63.
77. Blöcher D. DNA double-strand break repair determines the RBE of α-particles. Int J Radiat Biol. 1988;54:761–71.
78. Wouters BG, Skarsgard LD. The response of a human tumor cell line to low radiation doses: evidence of enhanced sensitivity. Radiat Res. 1994;138:S76–80.
79. Skarsgard LD, Hill AA, Acheson DK. Evidence for two forms of substructure in the cell survival curve. Acta Oncol. 1999;38:895–902.
80. Howell RW, Neti PV. Modeling multicellular response to nonuniform distribution of radioactivity: Differences in cellular response to self-dose and cross-dose. Radiat Res. 2005;163:216–21.
81. Rich JN, Stem C. Cells in radiation resistance. Cancer Res. 2007;67:8980–4.
82. Liu G, Yuan X, Zeng Z, et al. Analysis of gene expression and chemoresistance of CD133[+] cancer stem cells in glioblastoma. Mol Cancer. 2006;5:67. doi:10.1186/1476-4598-5-67.
83. Belli JA, Piro AJ. The interaction between radiation and adriamycin damage in mammalian cells. Cancer Res. 1977;37:1624–30.
84. Pentheroudakis G, Briasoulis E, Pavlidis N. Cancer of unknown primary site: missing primary or missing biology? Oncologist. 2007;12:418–25.
85. Briasoulis E, Pavilidis N. Cancer of unknown primary origin. Oncologist. 1997;2:142–52.

Index

A

ABSA. *See* Aminobenzenesulfonic acid (ABSA)
Activation energy, 166, 206, 211, 217–221, 226, 228
Activation threshold, 192
Adhesion, 25–28, 66–75, 88, 161, 165, 166, 189–238, 346
Advanced calorimetry, 190
Ag/AgCl, 104, 130, 134–136, 142, 151, 157, 162–165, 168, 170, 172, 235, 265, 272
Aliphatic peroxide, 253
Allergenic, 11, 12
Allergy, 38
Allometry, 409
Alloys, 5–8, 10, 11, 15, 20–23, 26, 28, 38, 45–47, 65, 67, 139, 174, 308
Alpha particles (α particles), 414, 416–422, 424–426, 437
Aluminum, 2, 5, 15, 20, 49
Aminobenzenesulfonic acid (ABSA), 283
Aminopropyl-triethoxysilane (APTS), 75
Amperometric, 42, 43, 100, 104, 108, 109, 135, 137–140, 142, 145, 146, 153, 156, 158–174, 261, 299, 303–305, 307, 323
Analytes, 84–86, 92–97, 100, 105, 108, 110, 111, 250, 286, 289–291, 297, 299, 300, 304, 307, 311–314, 316, 319, 320, 326–329, 343, 344, 354, 356, 376
Anemia, 385
Angina, 59
Angina symptoms, 59
Angiographic examination, 59
Anode, 2, 21, 26. 122–124, 135, 171–174
Antibiotics, 27, 70, 221
Antimony, 261
Antiproliferative, 62, 64
Antiproliferative agents sirolimus (Rapamycin), 62
APTS. *See* Aminopropyl-triethoxysilane (APTS)
Arrhenius equation, 206, 226
Arrhythmia, 36
Ascorbic acid (AA), 92, 101, 157, 290–299, 311, 318–320, 324
Atomic force microscopy (AFM), 72, 109, 190, 194

B

Basal ganglia, 37
Basecoating, 69
Battery, 39, 154, 161, 171

Benzendiazonium tetrafluoroborate, 68
Biliverdin, 383
Bioabsorbable polymers, 17–19
Biocompatibility, 2, 4–16, 20–27, 37–39, 46, 47, 49, 50, 55, 62, 64, 65, 67, 69, 70, 76, 100, 229
Biological membranes, 189, 190, 198, 226, 403
Bio-mechanical support devices, 36, 44–49
Bio-medical, 8, 24, 37, 74, 76, 87, 104, 287, 383, 392
Biomimetic, 16, 47, 48, 197
Biosensors, 4, 36, 40–44, 83, 87, 88, 99, 104, 105, 108, 143, 193, 249, 250, 298–300, 303, 343, 345, 354, 356–358, 376
Bio-stimulating devices, 36
Bipyridine, 129, 133, 134
Bisphophonate, 48
Blood
  glucose, 41, 42, 108, 109, 126, 128, 129, 136, 143–160, 167, 169, 175, 251
  vessels, 3, 7, 10, 21, 23, 25, 47, 311, 383, 385, 386, 396, 408, 414
Brain, 36, 37, 40, 309–311, 329
Bromoethyl, 68
Bruising, 382, 387, 391, 392, 395, 397

C
CA. *See* Chronoamperometry (CA)
CABG. *See* Coronary artery bypass grafting (CABG)
Calcium hydroxyapatite, 16, 27
Capacitive spike, 198
Capillary, 109, 147–150, 152–153, 167, 168, 170, 308
Carbon black, 84
Carbon nanotubes (CNT), 98, 100, 139, 300, 303, 312, 315, 318, 319, 326, 329
Carbon paste electrode (CPE), 101, 127, 146, 292, 321

Carcinogenic, 12, 96
Cardiac electrode, 39
Cardiolipins, 192, 193, 226, 231–233
Cardiovascular, 10, 24, 41, 44, 55, 56, 227, 310
Catalysis, 132, 304
Catalyzing enzymes, 123
Cathode, 2, 21, 26, 37, 135, 141, 171–174
Chalcogen, 252–258, 273, 274
CHEMFET, 140
Chemical vapor deposition (CVD), 66, 346, 348
Chemiluminescence (CL), 97
Chemotherapy, 62, 433
Child protection workers, 382
Cholesterol, 157, 192, 193, 219, 223–227, 304
Chromatography, 84, 93, 97, 254, 308
Chromophores, 383, 384, 388, 389, 392, 393, 397
Chronoamperometry (CA), 146, 190, 193, 197–204, 207
CL. *See* Chemiluminescence (CL)
CNT. *See* Carbon nanotubes (CNT)
Coating, 10, 15, 21–28, 46–49, 55, 62–76, 152, 172, 264, 312, 323, 325, 344, 358, 365, 376
Cognitive, 36, 308, 310
Collagen, 7, 48, 62, 290, 389
Colony forming unit (CFU), 259
Conducting polymers (CP), 39, 70, 139, 284–292, 295, 301–303, 306, 313, 314, 316, 319, 326, 328, 329
Contusions, 381–398
Copolymerized, 69
Co-polymers, 18, 19, 24, 63, 70–73, 153
Coronary artery bypass grafting (CABG), 60, 62
Corrosion, 2, 4–6, 8–12, 17, 21, 22, 24, 38, 45–47, 172
Coulometry, 42, 43, 146
CP. *See* Conducting polymers (CP)
CPE. *See* Carbon paste electrode (CPE)

Crystalline, 17, 18, 27, 96, 167, 203, 221, 223, 226, 228, 295
Curative response, 407
Cyclic voltammetry (CV), 69, 70, 251
Cyclodextrin, 98, 99, 329
Cytochalasin, 259
Cytotoxic, 12, 27, 28, 260, 275

D
DBS. *See* Deep brain stimulation (DBS)
Deep brain stimulation (DBS), 36–40
Delamination, 64, 68, 70, 71, 75
Density (ρ(g cm^{-3})), 191
Dermatology, 382, 386
Deuterons, 424, 425
Diabetes, 41, 43, 50, 62, 63, 83, 84, 109, 121–175, 307
Diagnostic, 43, 44, 60, 111, 142, 144, 161, 263, 285, 307, 310, 327, 385
Diazonium, 68–70
Differential pulse polarography (DPP), 251, 257, 270, 273
Differential pulse voltammetry (DPV), 293, 297, 299, 306, 313–317, 321, 324, 326, 328
Diffusion barrier, 71
Dihydrogen trioxide, 253
Dipping, 23, 66, 70
Diseases, 36, 37, 40, 41, 43, 44, 50, 55, 56, 60, 84, 108, 109, 143, 145, 159, 192, 230, 233, 273, 290, 298, 310, 344
Disposable sensors, 108
Disulfide, 250, 254, 256, 257, 267–271
Docking, 209, 212, 213
Dopamine (DA), 37, 40, 92, 99, 127, 157, 297, 299, 308–311, 313, 318–325
DPP. *See* Differential pulse polarography (DPP)
DPV. *See* Differential pulse voltammetry (DPV)
Drug-eluting stent (DES), 25, 48, 62–68, 72, 75, 76
Dying cells, 407, 408

E
Electrochemical biosensors, 249
Electrochemical deposition, 38, 39, 47, 67, 71, 75, 94
Electrolyte, 2, 21, 45, 74, 102, 171, 172, 198, 205, 228, 238, 273, 285, 316, 403
Electronic noses, 40
Electrooxidation of glucose, 41, 125–130, 132, 134, 163, 166, 308
Electrophysiological, 40
Electropolymerization, 75, 98, 99
Electroreduction of glucose, 122–123
Endocytosis, 190
Endothelial, 10, 57, 63, 64
Endothelialization, 64, 69
Endothelium, 57, 69
Energy fluxes, 189
Ensemble, 91, 102, 103
Enzyme-FET (ENFET), 140
EPI. *See* Epinephrine (EPI)
Epidemiology, 35
Epinephrine (EPI), 308–311, 314, 325, 326
Everolimus, 63
Exactech, 42, 128, 129, 146
Excipients, 2
Exocytosis, 189, 190, 197
Extracellular fluid, 2, 4
Extravascular, 381–398

F
FADH$_2$, 123–129, 132, 135, 145
Ferrocene/ferrocenium, 146
FET. *See* Field effect transistor (FET)
FIA. *See* Flow injection analysis (FIA)
Fibrous, 4, 6–8, 11, 15, 59
Field effect transistor (FET), 140
Flexibility of a membrane, 191
Flip-flop rates, 191
Flow injection analysis (FIA), 96, 292, 301, 328
Fluorescence microscopy, 190, 194, 263

Foam cells, 57, 59
Forensics, 382
Frictional resistance, 191

G

Galvanic, 2, 11
Galvanic corrosion, 2, 45
GDH. *See* Glucose dehydrogenases (GDH)
Genotoxic, 7, 11, 12
Glia cells, 40
Gluconic acid, 107, 108, 122, 140, 291
Gluconolactone, 124, 125, 128, 130, 132, 140, 174
Glucose dehydrogenases (GDH), 124
Glucose oxidase(s) (GOx), 41, 87, 88, 92, 97–99, 107, 123, 124, 127, 132, 135–137, 165, 170, 173, 304, 308
Glutathione peroxidase (GPx), 257, 258, 273
Gold, 2, 24, 38, 39, 46, 92, 94–96, 100, 108, 110, 126, 137, 141, 195, 197, 254, 255, 259, 264, 303, 320, 323, 343, 344, 346–349, 354, 357, 360, 363–364
Gompertzian model, 405, 406
Gompertz's function, 404, 405, 414
GOx. *See* Glucose oxidase(s) (GOx)
GPx. *See* Glutathione peroxidase (GPx)
Graphite, 84, 88, 89, 92, 95, 98, 100, 127, 129, 139, 162, 264
Growth curves, 260, 406, 407, 414, 415
Growth rate, 41, 258, 259, 405–409, 411, 412, 414, 436

H

Hamaker constant, 191
Hemostasis, 3
Heparin, 66
Histopathological, 70
Home blood glucose monitors, 143–159
Hormone(s), 2, 41, 144, 310, 311

HSB. *See* Hue saturation brightness (HSB)
Hue saturation brightness (HSB), 386
Human HeLa cells, 415, 430, 431
Human systems, 1
Hydrogel, 43, 130–137, 162, 163, 165, 167, 173, 174
Hydrogen peroxide, 10, 23, 87, 96–98, 101, 108, 109, 168, 252, 253, 264, 304–307, 346
Hydrolysis, 16–18, 24, 140
Hydrophilic, 25, 66, 68, 132, 153, 167, 193–198, 219, 226, 236, 289, 347
Hydrophobic, 18, 25, 66, 68, 70, 74, 75, 153, 167, 190, 191, 193–198, 208, 210, 211, 216, 218, 219, 226, 230, 233, 236–238, 289, 312, 319
Hydroxyapetite (HA), 47
Hyper-coagulability, 43
Hyperglycemia, 41
Hyperplasia, 23, 24, 413
Hypochlorous acid, 253
Hypoxic cells, 407

I

IDES. *See* Integrated electrode structures (IDES)
Immobilized structures, 196–198
Immunosensors, 107–111
Impedance, 3, 38, 39, 43, 45, 68, 69, 156, 197, 259–262, 264
Implants, 2–12, 14–23, 26–28, 44–48, 50, 66, 76, 161, 172
Incubation, 69
Indium tin oxide (ITO), 94, 138
Inflammation, 9, 23, 24, 27, 69, 162, 228, 273
Inflammatory cells, 10, 57
Infrared, 345, 392
Insulin, 41, 50, 144, 159, 169, 170
Integrated electrode structures (IDES), 261
Interaction docking, 206, 208, 210–212, 229, 230, 236
Intercellular, 173, 190

Interfacial potential difference, 191
Interfacial tension, 191
Interstitial fluid, 4, 136, 160, 163, 169–171, 173, 403, 407–408
Interstitial fluid pressure, 408
Interventional cardiology, 60–62, 76
Intimal thickening, 59
In vivo, 4, 18, 21, 22, 25, 27, 28, 65, 70, 164, 167, 168, 197, 252, 297, 310, 312, 389, 391, 395, 414, 433, 434
Ion-beam-assisted deposition, 28
Ion channel, 190, 192
Ionizing radiation, 414, 415, 430–431
Ions, 2, 4, 6, 10, 20, 28, 96–97, 102, 130, 134, 137, 140, 151, 173, 174, 197, 255, 268, 285, 289–291, 295, 298, 306, 319, 329, 345, 437
Ion-sensitive, 140, 261
Ion-sensitive FET (ISFET), 140, 261
Iron, 2, 19, 23, 42, 43, 128, 140
Ischaemic, 55–57
Isosbestic, 390, 391

K
Kleiber's law, 405, 409

L
L*a* and b*, 387
Langerhans, 41
Lateral diffusion, 191
Late stent thrombosis (LST), 64
Layer-by-layer (LbL), 302, 303, 317, 321
LDL, 57
LET. *See* Linear energy transfer (LET)
Ligand, 128, 133, 134, 193, 303
Ligand exchange, 133, 134
Limit of detection (LOD), 91, 94, 97, 99, 292, 301, 302, 306, 313, 314, 316, 317, 326, 328
Linear energy transfer (LET), 414
Linear quadratic (LQ), 415, 420, 437
Lipid(s)
 bilayer, 194, 195, 197, 221, 228
 mono layers, 193, 195, 198, 201, 208
 vesicles, 189–238
Liquid crystal, 167, 173, 203, 221, 223, 226, 228, 391–392
LOD. *See* Limit of detection (LOD)
LST. *See* Late stent thrombosis (LST)

M
Macroelectrode, 91, 96, 97
Macrophages, 57, 59, 63, 162, 253, 265, 383
Magnesium, 19, 23
Magnetic resonance (MR), 1, 23
Malignant tumors, 407
Martensitic transformation, 10
Mass balance, 404, 406, 409
Mechanical cardiopulmonary, 60
Medical devices, 35–37, 44, 46, 48–50, 52–76
Medical diagnostics, 285, 310, 327
Medical studies, 285
Medical treatment, 1, 36, 37, 46, 49, 285, 310
Medicine, 1–28, 35, 36, 40, 42, 44, 46, 50, 55, 251, 311, 327, 343–376, 386, 389, 392
Melanin, 384, 393, 394, 398
Membrane fluidity, 191, 192, 223, 227
Mercury sulfate electrode (MSE), 283
Metabolic processes, 189, 259, 409
Metal cladding, 5
Metallothionein (MT), 254, 255, 268
Metastasize, 407, 414, 436
Methacrylate, 71–73
Michaelis constant, 123, 166
Microband electrodes, 104–106
Microcrystallites, 98
Microdomain, 91, 227
Micro kinetics, 204–216, 228
Micropipette aspiration, 190
Microwires, 39
Minimal invasive, 44
MIP. *See* Molecularly imprinted polymer (MIP)
Mitochondria, 191–193, 226, 228, 230–233, 253, 264

Molecularly imprinted polymer (MIP), 328
Monocytes, 57, 253
Monte Carlo method, 393, 394
Motility, 259, 403
MSE. *See* Mercury sulfate electrode (MSE)
Multiwalled carbon nanotubes (MWCNT), 98–100, 139, 325, 328
Mutagenic, 12, 96
MWNT. *See* Multiwalled carbon nanotubes (MWCNT)
Myocardial infarction, 41, 57

N
Nanoparticles (NPs), 26, 91, 94–96, 102, 103, 319–321, 329, 344, 346, 356–358, 361–363
Nano-technology, 46
Near-infrared (NIR), 335, 385, 391
Necrosis, 8
Necrotic cells, 407
Necrotic core, 59, 404, 408
Neointimal, 10, 23–25, 62, 63, 76
Nephropathy, 41
Neurotransmitter, 37, 39, 40, 285, 290, 308–326, 329
Nickel, 5–7, 11, 12, 23, 39, 46, 48, 92, 98, 101, 139
Nicotinamide adenine dinucleotide (NAD+), 124, 127, 128, 146, 147, 299–304, 389
Nitinol, 5, 7, 10, 11, 23, 38, 46, 47, 65
Nitrogen monoxide, 253
Nonenzymatic, 122–123, 298, 305, 307
Norepinephrine (NOREPI), 290, 308, 310, 311, 314, 325, 326

O
Occluded arteries, 55
Ontogenetic, 405, 409, 412
OPS. *See* Orthogonal polarization spectral imaging (OPS)
Optoacoustic spectroscopy, 383, 391

Orthogonal polarization spectral imaging (OPS), 385
Osseointegration, 15, 26–28
Osteoclastic, 48
Osteoconduction, 11
Osteogenesis, 11, 16, 47
Oxidation, 2, 42, 67, 91, 123, 192, 253, 286, 347
Oxidative, 41, 57, 191, 271, 273, 274, 285, 305, 322
Oxygenated cells, 307
Ozone, 253

P
PAA. *See* Polyacrilic acid (PAA)
Pacemaker, 4, 24, 36, 38, 50, 122
Paclitaxel (Taxol), 62, 63, 66, 68–73
Pad printing, 85, 86
Palladium, 91, 139, 303
Pancreas, 40, 50
Parkinson's disease (PD), 36, 37, 40, 290, 310
Passivation, 5, 21, 26, 46
Patency, 55, 56
Pathological, 15, 50, 55, 237
Pathophysiological, 191, 192
PD. *See* Parkinson's disease (PD)
PDLA-PLLA, 18
Pediatricians, 381
PEDOT. *See* Poly(3,4-ethylenedioxythiophene) (PEDOT)
Peptide, 41, 74, 190, 197, 216, 218, 219, 221, 222, 268
Peritoneal dialysis, 158, 159
Peroxynitrite (ONOO−), 252–254, 265
PEVA:PMBA. *See* Poly(ethylene-vinyl acetate):polybutyl methacrylate (PEVA:PMBA)
PGA. *See* Polyglycolic acid (PGA)
Phenomenological, 404
Phenomenological models, 404
Phenyltrimethoxysilane (PhMOS), 75
PhMOS. *See* Phenyltrimethoxysilane (PhMOS)
Phosphorylcholine, 66
PLA-PGA, 18

Index

Plasma, 154, 156, 157, 191, 227, 229, 297, 309–311, 346, 347
Plasma-spray, 15, 23, 26, 27, 47
Plasmonic, 345, 346, 352
Platinum, 3, 10, 13, 21, 23, 26, 38, 39, 42, 46, 93, 95–97, 100, 124, 138, 139, 141–143, 162, 165, 173, 174, 258, 272, 273, 305, 306, 308, 312
P3MT. *See* Poly(3-methylthiophene) (P3MT)
Poly(2-acrylamido-2-methyl-1-propane-sulfonic acid) (PAMPSA), 295, 296
Poly(*p*-aminobenzenesulfonic acid), 314, 319
Poly(3,4-ethylenedioxythiophene) (PEDOT), 39, 284, 293, 297, 302–306, 308, 314–317, 319, 320, 322–324, 327–329
Poly(3-methylthiophene) (P3MT), 300, 303, 311–313, 315–317, 319, 320, 325, 326, 328, 329
Poly(*o*-methoxyaniline) (POMA), 292, 294, 296, 297
Polyacrilic acid (PAA), 300, 301
Polyaniline (PANI), 285–287, 289, 292–297, 300–303, 305–308, 312, 315–317, 319–321, 327–329
Polyanion, 134, 135, 172, 289, 294, 295, 300, 329
Polycation, 135
Polydioxanone, 17
Polyethylene, 7, 23, 67, 133, 136, 153
Polyethylene-vinyl acetate:polybutyl methacrylate (PEVA: PMBA), 69
Polyglycolic acid (PGA), 17–19
Polymer, 7, 39, 62, 88, 130, 265, 283, 344
Poly(ethylene-vinyl acetate):polybutyl methacrylate (PEVA:PMBA), 69
Polypyrrole (PPY), 39, 70, 74, 75, 139, 285, 287, 289, 291, 292, 294, 300–302, 305, 306, 308, 311, 312, 314–320, 325–328
Polysterenesulfonate (PSS), 293, 295, 300–303, 326

Polythiophene (PTHI), 285, 287, 289, 297, 313, 318
Polyurethane, 23, 24, 161, 167
Polyvinylsulfonate (PVS), 292, 293, 295, 297, 300, 301
POMA. *See* Poly(*o*-methoxyaniline) (POMA)
Potentiometer, 3
Potentiometry, 124, 140
PQQ, 123, 124
PQQ-GDH. *See* PQQ-glucose dehydrogenases (PQQ-GDH)
PQQ-glucose dehydrogenases (PQQ-GDH), 41, 42, 123, 124, 126, 128, 130, 146, 147, 159
Preexponential factor, 206, 219, 220, 226
Propyl-trimthoxysilane (PrTMOS), 75
Proteins, 3, 4, 7, 21, 25, 48, 65, 69, 74, 108, 110, 123, 127, 130, 132, 134, 141, 146, 162, 192, 249–251, 254, 255, 257, 264, 267–269, 274, 275, 299, 354–356, 376, 383
Proton-sensitive field-effect transistor, 140
PrTMOS. *See* Propyl-trimthoxysilane (PrTMOS)
PSS. *See* Polysterenesulfonate (PSS)
PTHI. *See* Polythiophene (PTHI)
PVS. *See* Polyvinylsulfonate (PVS)

Q

Quiescent, 104, 408, 409
Quinoid, 42, 126, 127
Quinone, 42, 99, 127, 318

R

Radiographic visibility, 38
Radio resistance, 409, 425, 426, 430
Rapamycin. *See* Antiproliferative agents sirolimus (Rapamycin)
RDE. *See* Rotating disc electrode (RDE)
Reactive nitrogen species (RNS), 251–254, 263–267, 275

Reactive oxygen species (ROS), 251–254, 263–267, 275
Reductase, 259
Reduction, 2, 37, 40, 42, 61, 69, 92–97, 108, 109, 124, 134, 142, 143, 171, 199, 221, 253–255, 257, 268, 269, 272, 286, 291, 299, 303–305, 390
Reflectance spectroscopy, 383, 386–387
Reocclusion, 55
Retinopathy, 41
RNS. *See* Reactive nitrogen species (RNS)
ROS. *See* Reactive oxygen species (ROS)
Rotating disc electrode (RDE), 292, 294, 295, 301, 306
Rupture spreading, 206, 208, 210–213, 217, 218, 221, 223, 224, 230, 237
Rupture spreading macro kinetics, 206, 207, 212, 213

S

Saliva, 2, 108
SAM. *See* Spectral angle mapping (SAM)
Saphenous vein, 60
Scanning force microscopy (SFM), 194
SCE. *See* Standard calomel electrode (SCE)
Screen printed electrodes, 83–111, 265
Screen printing, 83–85, 88, 89, 101, 102, 104, 107, 108, 111, 149, 151
SDS. *See* Sodium dodecylsulfate (SDS)
SeCys. *See* Selenocysteine (SeCys)
Selenium, 249, 251, 252, 255, 257, 258, 268, 273, 275, 276
Selenocysteine (SeCys), 255, 257, 267, 268, 275
Selenoprotein, 257, 258
Selenoprotein P (SelP), 257
Selenoprotein W (SelW), 257
Self-assembled monolayers, 72, 346–347, 349
Semi-conductor, 38
SFM. *See* Scanning force microscopy (SFM)
SHE. *See* Standard hydrogen electrode (SHE)
Shear elastic modulus, 191
Shear force, 44, 191
Shear strain, 191
Shear stress, 133, 191
Shear viscosity, 191
Silicon, 7, 16, 20, 108, 140, 261, 345
Silver, 12, 23, 27, 28, 137, 151, 165
Single walled carbon nanotubes (SWCNT), 100, 139, 325
Sirolimus, 62, 63, 65
Skeletal reconstruction, 15
Small unilamellar vesicles (SUV), 194
SMC. *See* Smooth muscle cells (SMC)
SMDE. *See* Static mercury drop electrode (SMDE)
Smooth muscle cells (SMC), 46, 57
Sodium, 2, 22, 27, 102
Sodium chloride, 2
Sodium dodecylsulfate (SDS), 306, 319, 326
Spectral angle mapping (SAM), 72–75, 390
Spectrophotometer, 387
Spraying, 27, 66
Square wave voltammetry (SWV), 302, 306, 313–315, 317, 326, 328
Staining method, 260, 263
Standard calomel electrode (SCE), 90, 139, 297, 304
Standard hydrogen electrode (SHE), 123, 124, 128, 129, 287
Static mercury drop electrode (SMDE), 198, 199
Stem cells, 16, 49, 50, 411, 412, 430, 431

Stent, 5, 8, 10, 20, 24, 25, 44, 46–48, 55–75
Stenting, 24, 62, 76
Stimulating, 36–39, 59, 311
Stimulating electrodes, 36–40
STN. *See* Subthalamic nucleus (STN)
Sub-curative, 407
Subcutaneous, 43, 161–165, 167, 169–171, 173, 174, 383, 393, 394, 396, 397, 418
Subthalamic nucleus (STN), 37
Sugar levels, 41
Sugars, 2, 41, 123, 124, 158
Sulfur, 249, 251, 252, 254–258, 263, 267–272, 275, 276
Superoxide, 252, 253, 264
Surface grafting, 67
SUV. *See* Small unilamellar vesicles (SUV)
SWCNT. *See* Single walled carbon nanotubes (SWCNT)
SWNT, 316, 320
SWV. *See* Square wave voltammetry (SWV)

T

Tantalum (Ta), 5, 6, 13, 15, 23–25
Taxol. *See* Paclitaxel (Taxol)
T(4)-5'-deiodinase, 257
Thickness, 6, 21, 23, 25–27, 66, 67, 70, 75, 85, 138, 152, 191, 289, 294, 300, 319, 322, 324, 329, 354, 367, 393, 394, 404, 411
Thionine, 124, 126, 127, 301
Thioredoxin (Trx), 257, 268, 369
Thioredoxin reductase (TrxR), 257
Thiosulfinate, 256, 257, 271
Thiosulfonate, 256, 257, 271
Thixotropic, 8
Thrombosis, 3, 10, 23, 25, 58, 62, 64, 67, 76
Thrombus, 3, 7, 10, 46, 57, 228
Thyroxine deiodinase, 257

Titanium, 2, 5, 6, 10, 23, 46
Transcutaneous, 163, 164, 166–168
Transient receptor potential, 192
Transient receptor potential channels (TRPC), 192
Transmembrane potential, 191
Transmission spectroscopy, 383
Trauma, 16, 168, 382, 383, 388, 389, 397
Tremor, 36, 37
Tristimulus method, 387
Trisulfide, 256, 270
TRPC. *See* Transient receptor potential channels (TRPC)
TrxR. *See* Thioredoxin reductase (TrxR)
Tumor, 16, 43, 403–437
Turbulent flow, 56, 57
Type I, 144, 159, 170
Type II, 144, 159

U

UA. *See* Uric acid (UA)
Ultra sound, 44, 383, 398
Unencapsulated field effect transistor, 140
Unilamellar vesicles, 194, 201, 202
Unmineralized, 11, 14
Uric acid (UA), 92, 99, 101, 292, 293, 297–299, 301–304, 306, 312, 314–317, 319–322, 325, 326
Urine, 2, 20, 108, 250, 287, 297–299, 309, 310, 329

V

Vaccine, 2
Vanadium, 5, 15
Vapor-phase, 28, 349
Viscosity, 43, 84, 133, 152, 156, 166, 191
Voltametric, 69, 70, 251

W
West's model, 405–407, 412, 436
Wound healing, 7, 8, 10, 228, 290

X
X-ray photoelectron spectroscopy (XPS), 68, 92
X-rays, 1, 60, 68, 415–418, 421, 425, 431, 437

Y
Yttrium, 20

Z
Zirconium, 15, 20
Zotarolimus, 63

Printed by Books on Demand, Germany